Normality

Normality:
A Critical Genealogy

PETER CRYLE
AND ELIZABETH STEPHENS

The University of Chicago Press

CHICAGO AND LONDON

The University of Chicago Press, Chicago 60637
The University of Chicago Press, Ltd., London
© 2017 by The University of Chicago
All rights reserved. No part of this book may be used or reproduced in any manner whatsoever without written permission, except in the case of brief quotations in critical articles and reviews. For more information, contact the University of Chicago Press, 1427 E. 60th St., Chicago, IL 60637.
Published 2017
Printed in the United States of America

26 25 24 23 22 21 20 19 18 17   1 2 3 4 5

ISBN-13: 978-0-226-48386-3 (cloth)
ISBN-13: 978-0-226-48405-1 (paper)
ISBN-13: 978-0-226-48419-8 (e-book)
DOI: 10.7208/chicago/9780226484198.001.0001

Library of Congress Cataloging-in-Publication Data

Names: Cryle, P. M. (Peter Maxwell), 1946– author. | Stephens, Elizabeth, 1969– author.
Title: Normality : a critical genealogy / Peter Cryle and Elizabeth Stephens.
Description: Chicago ; London : The University of Chicago Press, 2017. | Includes bibliographical references and index.
Identifiers: LCCN 2016051200 | ISBN 9780226483863 (cloth : alk. paper) | ISBN 9780226484051 (pbk. : alk. paper) | ISBN 9780226484198 (e-book)
Subjects: LCSH: Norm (Philosophy)—History. | Social sciences—Europe, Western—History—20th century. | Social sciences—Europe, Western—History—19th century. | Social sciences—United States—History—20th century. | Social sciences—United States—History—19th century. | Medicine—France—History—19th century.
Classification: LCC B105.N65 C79 2017 | DDC 302/.1—dc23
LC record available at https://lccn.loc.gov/2016051200

♾ This paper meets the requirements of ANSI/NISO Z39.48-1992 (Permanence of Paper).

Contents

Introduction • 1

I THE NORMAL IN NINETEENTH-CENTURY
SCIENTIFIC THOUGHT

1   The "Normal State" in French Anatomical and Physiological
Discourse of the 1820s and 1830s   23
2   "Counting" in the French Medical Academy during the 1830s   63
3   Rethinking Medical Statistics: Distribution, Deviation, and
Type, 1840–1880   100
4   Measuring Bodies and Identifying Racial Types: Physical
Anthropology, c. 1860–1880   142
5   The Dangerous Person as a Type: Criminal Anthropology,
c. 1880–1900   180
6   Anthropometrics and the Normal in Francis Galton's
Anthropological, Statistical, and Eugenic Research, c. 1870–1910   212

II THE DISSEMINATION OF THE NORMAL IN
TWENTIETH-CENTURY CULTURE

7   Sex and the Normal Person: Sexology, Psychoanalysis,
and Sexual Hygiene Literature, 1870–1930   261
8   The Object of Normality: Composite Statues of the Statistically
Average American Man and Woman, 1890–1945   294
9   Sex and Statistics: The End of Normality   333

Conclusion • 353

Acknowledgments • 361
Notes • 363
Bibliography • 407
Index • 433

Introduction

The idea of the "normal" is now so familiar that it is hard to imagine contemporary life without it, and at the same time it is so thoroughly ingrained in our thinking that we have difficulty paying analytical attention to it. In everyday speech, it is the most commonly used term for the common, the ordinary, the usual, the standard, the conventional, the regular. It occupies a broad expanse of the cultural middle as a set of functionally equivalent ways of everyday thinking. Although the idea of the normal is now so culturally pervasive as to seem ubiquitous, its meaning is nonetheless vague, and difficult to define: "If it is hard to deny something called normality exists," Robert McRuer aptly observes, "it is even harder to pinpoint what that is."[1] In part, this is because there is a slippage in the idea itself. The word "normal" often suggests something more than simply conformity to a standard or type: it also implies what is correct or good, something so perfect in its exemplarity that it constitutes an ideal. The meaning of the normal encompasses both the norm, understood as a descriptive (or positive) fact, and normativity, understood as the affirmation of cultural values: "the normal is the effect obtained by the execution of the normative project, it is the norm exhibited in the fact," asserts George Canguilhem.[2] As a result, "The concept of the normal is itself normative."[3] One of the remarkable features of contemporary views of the normal is that the middle point in a range of qualities and characters should ever have come to be invested with such great value. After all, middling qualities might on the face of things have been considered mediocre or nondescript, as they indeed have been and continue to be in certain cultural contexts. However, this privileging of middleness might also be seen as an important source of the power of the normal, facilitating its conceptual slide from the descriptive to the prescriptive, from norms to normativity. As Ian Hacking perceptively observes in his influential essay on the normal state: the normal "uses a power as old as Aristotle to bridge the fact/value distinction, whispering in your ear that what is normal is also right."[4] This is the dull charm of the word and the key source of its unspectacular strength.

While the word "normal" may sound innocuous, even bland—serving as it does as a synonym for the ordinary and unremarkable—this unassuming word can have a significant effect on the lives of those defined in contrast to it as abnormal, pathological, or deviant. To identify a norm is to impose a rule, writes Canguilhem. To normalize is "to impose a requirement on an existence."[5] The effects of such impositions and normalizing practices on those identified as outside the normal have been the subject of much recent scholarship, especially in queer and gender studies, as well as critical disability and race studies. Much of this work has cast new light on specific instances in which, following Hacking, "the benign and sterile-sounding word 'normal' has become one of the most powerful ideological tools of the twentieth century."[6] The majority of such work hence focuses on practices of normativity and normalization, on the compulsory imposition of the normal, drawing attention to the way these serve to reinforce existing cultural systems of privilege and power. This book seeks both to contribute to such work and to depart from it. It focuses specifically on the normal itself, aiming thereby to bring greater historical precision to our understanding of the term and its cognates. In so doing, it attempts in the first instance to understand better the persistent but elusive influence of the normal by recovering the conditions of its historical emergence. By bringing its nondescript qualities into a sharper focus, we expect to be able to show clearly the specific conditions in which it emerged, and the cultural and intellectual practices that have allowed it to flourish.

As the title of our book indicates, this study was born out of impatience and yet produced by dint of persistent labor. We were worn out by the ubiquity of the normal, alienated by its casual self-assurance, sidelined by the manner in which it took its comfortable place in the middle of our lives. But—that last sentence apart—this book is not a diatribe against normality. It is rather a careful genealogy of the concept, a critical history that begins by examining just how, in the early nineteenth century, people first came to use the word "normal," and how it came over the course of the twentieth century to acquire its current cultural authority. We have found no golden age of normality, not even a steady rise to prominence, but rather a series of hard-fought intellectual contests with equivocal outcomes. In this book we tell the nineteenth-century scientific contests from all sides, and telling it that way has itself been an exercise in patience. Yet it is precisely because this is a history of contest and contradiction that we have been able to give space to critical thinking while continuing to build our genealogy. For all its invasiveness, normality has never been simply triumphant. The first part of our book is in fact an analysis of the controversies that attended and constrained the concept of the normal throughout the nineteenth century. In the second part, which focuses on the twentieth century, we show how the notion of normality entered into general usage, eventually becoming so widespread that it could function as a working equivalent of "widespread," a commonsense term for commonsensical. Our central thesis, which we offer as a genealogical summary with critical import, is that the concept of the normal as we know it today dates from no earlier than the mid- twentieth century. It is in no sense historically ubiquitous, and does not deserve to be considered a timeless idea. The intellectual and cultural history that made the normal what it is today is in fact encumbered with enduring questions and unresolved issues. Our wager is that those questions, if brought to light in the way we are seeking to do here, will help to understand the perpetuation of its cultural power in ways that will allow us to undermine its authority.

Although the current ubiquity of the normal can make its significance seem assured and timeless, in fact the history of the notion is surprisingly short. For most of the nineteenth century, the word "normal" was used almost exclusively in scientific contexts. Its etymological antecedents can be traced back to the mid-eighteenth century, when it first appeared in geometry, being used as a less common synonym for a perpendicular line. However, we will show that the term was first used in its modern sense in France in comparative anatomy, around 1820. A literally vital matter was the question of whether quantitative thinking, particularly statistical calculations, could be applied to the field of medicine. French medical thinkers continued to affirm in their great majority that statistics involving the calculation of averages and probabilities did not deserve a place in clinical practice. There were no average patients, no reliable means of calculating the probability of recovery from sickness, and no standard illnesses. The threat posed by statistical method, as they saw it, was a double one: quantification left no room for the doctor's finesse, and the standardization it entailed denied patients their distinctiveness. While it might sometimes be affirmed, after Foucault, that the nineteenth century witnessed the inexorable rise of statistics as a method of governmental normalization, our history shows at this point that numerically based medicine was often strenuously resisted by professionals. Much has been written about the "medical gaze" as a privileged locus of power in clinical situations, but our history will show that the all-knowing glance was involved in a sustained contest with the quite different knowledge practice of counting. Individuating knowledge remained in competition with standardizing knowledge, and their coexistence sometimes had the effect of dividing the profession against itself.

The word "normal" did not appear in the *Oxford English Dictionary* until 1848, and at the end of that century its meaning was still so unfamiliar that the French *Grand Larousse* described it as "new in the language," requiring "from the person who hears it for the first time a certain effort of attention."[7] In statistics, use of the word "normal" is even more recent. Despite the increasing importance of probability calculations and other quantified forms of knowledge over the course of the nineteenth century, it was only at the end of the century that the word "normal" acquired a statistical meaning. Formulated in the context of the "normal curve" and "normal distribution," the normal here referred to a range or variation of data determined by a mathematical law according to which the majority of data of a certain kind will cluster in the middle of their range. For Hacking, statistics understands the normal as a limited numerical range, a numerical average or standard, whereas medicine understood it as a dynamic biological state. Until the end of the nineteenth century, Hacking argues, the medical and mathematical concepts of the normal remained quite distinct. For Hacking, it was only when these two concepts of the normal converged at the start of the twentieth century that the word came to acquire its current cultural authority, moving from its specialist use in scientific discourses "into the sphere of—almost everything."[8] Our research supports Hacking's first claim: that the medical and mathematical ideas of the normal constitute two distinct genealogies, with the latter emerging much more recently than the former. However, we have found that the meanings of average and healthy remained separate until well into the twentieth century, and that the term "normal" moved into widespread popular use before these different meanings came together. The authority of the normal, as the term moved from professional discourses into the popular sphere during the first half of the twentieth century, did not derive from its newer quantitative meaning.

Much remains to be said about the precise history by which the word "normal" was invested with such cultural power and prestige during the first half of the twentieth century, and our book describes some of the defining trends and decisive moments of that history. We will examine here the specific historical circumstances and cultural networks in which the idea of the normal first appeared, and set out the forms of knowledge and the cultural practices to which it gave rise. What we have found is not the uninterrupted, inevitable cultural ascension of a unified concept, but rather a much messier and more contested history. By finding division where others have tended to see unquestioned authority, we anticipate that our work will bring historical grist to the mill of scholars who have made the normal an object of resistance. The centrality of norms, normality, and the normative to the conceptualization and practice of particular disciplines—including philosophy, medicine, linguistics, and political theory, as well as queer theory, disability studies, and critical race studies—has been the subject of considerable recent scholarly attention.[9] Where much of this work takes the relationship between norms, normality, and normativity for granted, our critical genealogy is concerned to interrogate this relationship by focusing specifically on the development of the normal itself. We note at certain points that being taken for granted came precisely to be one of the key features of the normal. Ours is a discursive history that shows how its effects were materialized and institutionalized, how they were productive of new subjectivities and bodies. By doing this work, we are seeking to contribute to, but also to reframe, the insightful political and cultural histories of the normal on which it builds, such as Lennard Davis's *Enforcing Normalcy: Disability, Deafness and the Body*, which shows how the medical treatment of deaf subjects throughout the twentieth century constitutes the enforcement of normative assumptions about hearing, and Rosemarie Garland-Thomson's *Extraordinary Bodies: Figuring Physical Disability in American Culture and Literature*, which introduced and denaturalized the concept of a "normate" body.[10] Cultural historians and scholars of sexuality such as Elizabeth Reiss, Dean Spade, and Alice Domurat Dreger have similarly demonstrated how concepts of the normal have informed the medical and legal treatment of intersex, trans, and conjoined subjects.[11] As Julian Carter observes in *The Heart of Whiteness: Normal Sexuality and Race in America, 1880–1940*, the normal derives its disciplinary force from the way it obscures its political partiality behind an appearance of scientific impartiality. In the late nineteenth century, Carter shows, medical and scientific writing began to adopt the word "normal" as a way to describe "whiteness" and "heterosexuality" in apparently neutral terms, scrubbed free of their political history of violence. Although the normal is veiled by its "appearance of blank emptiness and innocence,"[12] Carter notes quite accurately, closer analysis reveals that it "has a long history as a covertly political phenomenon."[13] What this history makes clear, Carter shows, is the extent to which "from its inception, modern 'normality' involved both a positivistic claim about the pure neutrality of facts, and a distinctly eugenic element of judgment about which human bodies and behaviors were best."[14] The norm is a value disguised as a fact, Carter argues—although, as Mary Poovey has shown, the fact itself cannot be understood as neutral or impartial but is rather inextricable from the political or cultural context in which it came to be recognized as a "fact."[15] The normal, used in its contemporary senses, carries great cultural force. Jeanette Winterson's memoir *Why Be Happy When You Can Be Normal?* exemplifies this common understanding of the normal as a system of dreary conventionality, showing how the normal functions as a hidden system of compulsory conformity.[16] Despite all this, the normal retains its privileged cultural position and its discursive poise. Michael Warner has famously contended that normal is what most people aspire to be: "Nearly everyone, it seems, wants to be normal," he writes with brio, "and who can blame them, if the alternative is being abnormal, or deviant, or not being one of the rest of us? Put in those terms, there doesn't seem to be a choice at all. Especially not in America, where normal probably outranks all other social aspirations."[17]

Our critical genealogy seeks to rethink and reframe political critiques of the word "normal" by placing them within their longer historical context, in order to understand better how the normal has come to play such a widely influential role. Often its ubiquity and stringency are taken for granted even as the normal is made an object of critique and identified as a cultural force to be resisted. This is especially evident in queer theory, in which some of the most severe critiques of the normal have been articulated. Michael Warner's often-cited definition of queer theory as that which opposes "not just the normal behavior of the social but the *idea* of normal behavior" is an example of the declarative antinormativity that can be said to characterize queer theory.[18] As Warner makes clear here and elsewhere, it is not a matter for queer of siding with the abnormal against the normal: that would amount to taking up a position already inscribed within the logic of normality. To make the distinction in the terms of Aristotelian logic, queer seeks to oppose the normal, not as its contrary, but as its contradictory—that is, not as its routine, complementary opposite, but as its opponent. That appears to be very much what David Halperin means by his own well-known definition: "As the very word implies, 'queer' does not name some natural kind or refer to some determinate object; it acquires its meaning from its oppositional relation to the norm. Queer is by definition *whatever* is at odds with the normal, the legitimate, the dominant."[19] Oppositionality has thus become so central to queer theory that Annamarie Jagose is able to say: "these days it almost goes without saying that queer is understood to mean 'antinormative.'"[20] It does indeed seem, as Jagose observes, that antinormativity now "stands, mostly unchallenged, as queer theory's privileged figure for the political."[21]

Yet despite being resolutely "opposed" in this way, the normal can often remain untroubled by detailed analysis, serving by dint of its supposed stability as a point of reference. Our book is intended, not to denounce or renounce the normal, nor even to envisage concepts of sexuality, gender, race, or embodiment outside the confines of normativity, but rather to put the normal under pressure by turning attention to the cultural construction and historical specificity of the notion itself. In Michael Warner's definition of queer, the normal is synonymous with the respectable and the conventional; in Halperin's, with the legitimate and the dominant. In many other recent accounts it is taken as interchangeable with the typical, the average, the ordinary, or the healthy, although it is not always clear to what extent that broad range of terms can be understood as a set of even approximate synonyms. This book is intended to bring precision and specificity to our understanding of the meaning of the term "normal" across cultural time and space, in order to understand better its function and its durability.

As the work of the queer and disability studies scholars cited above demonstrates, the focus of much recent research has been the operation of normalization and normativity as powerful dynamics of enforced conformity. This emphasis is also found in Foucault's influential account of the history of normalization, to which much work in the history of sexuality is indebted. Foucault contextualizes norms and normalization as constituent elements of the new regimes of power that emerged at the end of the eighteenth century. These new forms of power were characterized by their focus on life, and institutional practices in which "the law operates more and more as a norm."[22] Such a power, Foucault famously states: "has to qualify, measure, appraise, and hierarchize, rather than display itself in its murderous splendor; it does not have to draw the line that separates the enemies of the sovereign from his obedient subjects; it effects distributions around the norm. That is not to say that the law effaces itself or that the institutions of justice are prone to disappear, but that the law operates more and more as a norm, that the juridical institution is increasingly incorporated into a continuum of apparatuses (medical, administrative, etc.) whose function is primarily regulatory."[23]

Normalization, in Foucault's broadly persuasive account, became one of the key disciplinary technologies of modern biopolitical institutions, integral to their administration and operation: "Like surveillance and with it, normalization becomes one of the great instruments of power at the end of the classical age," he argues. "For the marks that once indicated status, privilege and affiliation were increasingly replaced—or at least supplemented—by a whole range of degrees of normality indicating membership of a homogeneous social body but also playing a part in classification, hierarchization and the distribution of rank."[24] In this way, normalization described a double movement: on the one hand homogenizing and aggregating individuals into a population, on the other distributing and identifying differences between them. Normalization thus worked on two levels at once: a biopolitics of the population, which operated at the level of large groups, and an anatomo-politics of the human body, which operated upon the individual body.[25] So although normalization was, as Foucault observed, "primarily regulatory" in its effects,[26] it did not function simply to standardize or to enforce a conformity, but rather to individualize and differentiate at the same time: "In a sense, the power of normalization imposes homogeneity; but it individualizes by making it possible to measure gaps, to determine levels, to fix specialties and to render the differences useful by fitting them one to another. It is easy to understand how the power of the norm functions within a system of formal equality, since within a homogeneity that is the rule, the norm introduces, as a useful imperative and as a result of measurement, all the shading of individual differences."[27]

The function and the purpose of normalization, in Foucault's account, are not merely to move subjects toward a norm—thus making them more conventional—but also to measure the gaps and differences by which they deviate from that norm. The norm marks an ideal that can never be embodied, but around which minutely differentiated distances can be charted: "when one wishes to individualize the healthy, normal and law-abiding adult, it is always by asking him how much of the child he has in him, what secret madness lies within him, what fundamental crime he has dreamt of committing."[28] The purpose of normalization, understood in this way, is not to make subjects more normal but to establish distance from the norm as a distinguishing characteristic of individuals. Measuring the distance between an individual and the norm becomes a key figure of knowledge.

Despite Foucault's emphasis on the way normalization individuates as well as homogenizes, much of the subsequent scholarship on normalization has focused on its repressive and standardizing function, rather than its productive or individuating ones. One consequence of this is that the cultural function of the normal can be misapprehended, in a way that limits the possibilities of effective resistance. We see this when critically minded thinkers suppose that problematizing the coherence of the normal will undermine its cultural function. For instance, J. Jack Halberstam argues in *Gaga Feminism: Sex, Gender, and the End of the Normal* that the statistical frequency of nonnormative sex acts and gender performance has the capacity to undermine the cultural power of the normal, exposing its internal contradictions and thereby "bring[ing] the whole crumbling edifice of normative sex and gender tumbling down."[29] We acknowledge that polemical statements like this can serve as rallying cries, but we are quite unpersuaded by the underlying assumption that the concept of the normal relies on logical coherence, and that exposing its contradictions will fatally undermine its functionality. As Laura Doan has perceptively noted, queer theory appears to have "devoted its considerable intellectual energies to opposing and destabilizing a category already intrinsically riddled with inconsistency."[30] We will show in our history that at the point of its popular emergence the normal was already understood as paradoxical and polemical: as an average that was also an ideal, a typicality so perfect that no individual could ever embody it. Pointing out that normality is elusive and precarious does not undermine its operation. On the contrary, its conceptual incoherence is the source of enormous cultural productivity and often reinforces its power. Our research has shown us how heterogeneous and variable the meaning of the normal has been over time, and how consistently concepts of the normal gain cultural authority and strength precisely through being challenged. To put it baldly, the normal is not undone by criticism: it positively thrives on contestation. Recognizing this allows us to cast new light on the contemporary concept of the normal, and to suggest new lines of approach by which to reevaluate and contain its cultural position.

Our approach is a genealogical one. We are well aware that for some queer scholars historical studies have little to contribute to the political and theoretical project of queer critique, whose goal is to envisage different forms of knowledge production. Those who hold that view may wish to put an end to historical studies as such and, if so, our book will be of no interest to them.[31] They may see chronological accounts as "chrononormative," to use Elizabeth Freeman's neologism, or may prefer to engage in philosophical investigations of temporality that problematize the sequentiality of historical projects.³² In spite of that, we persist in the belief that a genealogical approach has something to offer a critical understanding of normality. A critical genealogy such as that undertaken here allows us to attend to the historical specificity of the term "normal" and its transformations across time in a way that enriches our understanding of the subject, while avoiding totalizing or teleological narratives. We can thus support the interrogation of the normal that underpins so much contemporary cultural and critical analysis by marking the fissures and fault lines in the concept itself. Our work is genealogical in the Foucauldian sense. Its aim is not to look for a single, unified history of the normal that produced the particular types of subjects understood as normal today, but rather to recover the complex processes, the professional accidents or misunderstandings, and the unpredictable developments that must be understood in their own time and place.³³ To those focused on polemics against the normal we caution, in the light of our history, that critique does not undermine the function of the normal, but on the contrary often contributes to its power and ubiquity. Nearly all of the episodes studied in this book involve determined contestation, and nearly all contributed over time to the strengthening and broadening of normality.

To those who practice historical inquiry we offer here a sustained contribution to the history of medicine and anthropometry in the belief that our history will find a place alongside their work.³⁴ We are going to show that the normal was an unstable set of concerns and practices subject to questioning for a century and a half, and we will lay out the uncertainties and worries that continually beset it. We recognize from the outset that there will be no decisive moment in which the discursive prestige of normality finds itself brought to ruin by our efforts. Yet while we cannot hold out the hope of putting normality to flight, we will be making a sustained effort to put it in its place and keep it there. Our aim is to circumscribe the normal, and our expectation is that the details we bring to light will have the cumulative effect of hedging it about, narrowing its epistemic scope, and inhibiting its invasiveness. We have put together this long genealogy in the belief that attention to particular historical circumstances can serve to define and confine normality as a phenomenon. In showing, as we will come to do, that the historical becoming of the normal was entirely contingent, we believe we have produced an artifact of knowledge that deserves to stand. May it stand in the way of any tendency to think of the normal as either natural or universal.

In thinking about the interaction between history and queer, we have been much helped by Laura Doan's *Disturbing Practices*. The academic practicalities of Doan's engagement are rather different from ours. She spans a narrower historical and geographical range, but with a method that is in many ways more complex. We are not seeking, as she bravely is, to bring the analytical practices of queer to the very business of writing history.[35] Our approach simply deploys recognizable elements of intellectual and cultural history as they have been practiced in recent decades, but it does so in a way that ought to offer something of thematic substance to queer studies. Our aim is to open a conversation about normality between history and queer, and Doan marks with clarity and tact some differences in perspective. She makes a general point with which we heartily concur: "Arguably (and perversely), queer studies—a field that often defines 'queer' in relation to the normal—has devoted its considerable intellectual energies to opposing and destabilizing a category already intrinsically riddled with inconsistency."[36] "Historians express amazement," she adds, "that queer studies has too often sidestepped the historicity of the normal."[37] Doan recognizes that "a major challenge for the practice of queer critical history is to articulate fully the particularized relational values of the norm, normal and heteronormativity," and to "demonstrate how the normal shapes, consolidates and reifies the concepts of the average and the healthy in highly specific ways that differ from a host of other discursive realms."[38] It is precisely such a history that our book is intended to map out. Our historical examination will show that the normal and whatever counted as its opposite have changed remarkably since the beginning of the nineteenth century, and that examining those changes at close hand can serve the purpose of resistant thinking. However stable normality might now seem, or be made to seem, it has a highly contingent cultural and intellectual history.

Doan actually takes Halperin's influential definition of queer as an indication of the need for further historical work on the normal. Having drawn attention to the fact that Halperin aligns three adjectival substantives as if they were synonyms—the normal, the legitimate, the dominant—she makes the historian's point that those three terms may have only recently become approximate synonyms. What a careful history requires, she says in effect, is to understand how synonyms might have shifted over time, and our own study can be read as a sustained attempt to respond to that challenge.[39] Over the period from about 1820 to about 1950, we will in fact tell a story of changing synonymy and shifting antonyms. There were during that time all sorts of articulations and disarticulations of the normal and its notional opposite. Around 1830, the term came to make sense in physiological discourse as the opposite of "ill," and for about fifty years the dominant usage was a medical one in which the "normal state," sometimes called the "physiological state," was opposed to pathological conditions. At around the same time, comparative anatomists had begun to speak of bodily "compositions" in which normal organs functioned in law-governed, systematic ways. What they called a normal organ was by definition neither dominant nor subordinate: it was characterized by thorough structural integration. Comparative anatomy thus differed from physiology in that it did not deploy a normal-abnormal binary. The scientific language of teratology developed by Isidore Geoffroy Saint-Hilaire had little use for the term "abnormal," preferring to speak only of specific anomalies. For teratology, nothing guaranteed that anomalies in their diversity could ever be gathered into a singular condition of abnormality, and there was in any case no taxonomic value in such an aggregation. Here is one clear indication that the intellectual history of the normal cannot be reduced to that of the normal-abnormal binary.

It would be entirely reasonable to suppose that the term "normal" must have come quickly to the fore after about 1830, when statistics was claiming a significant place in scientific and governmental practice, but that is not in fact what happened. Leading statisticians at the time did not speak of a "normal curve," although they did refer to other, pertinent kinds of curve that we will identify and discuss in chapter 3. Statisticians were interested in predicting the patterned outcomes of long series of chance events, and their primary focus was on average values. In their work, "normal" appeared rather belatedly, sometimes as an approximate synonym for "probable." And in physical anthropology, which came to prominence in France and elsewhere after about 1860, the normal meant something different again. By "normal characters," physical anthropologists understood unvarying physical features passed on through the process of heredity. Skeletal normality of this kind was taken by them as evidence of a stability in human bodies enduring from one generation to another. These anthropologists were equally preoccupied with typicality, and they gathered what they had to say about stability and typicality under the notion of race, even though they could not strictly agree about how race was to be defined. In their discussions, "normal" and "racial" became near synonyms. A different group based in Italy devoted itself after about 1880 to building what they called criminal anthropology. Their research, preoccupied as it was with identifying the criminal type, turned almost all its attention to the physical traits of "born criminals." Studying people of that kind was such an urgent matter in view of the threat they posed to society that there hardly seemed to be time to consider normal individuals in any detail. Yet criminal anthropologists continually referred to normal people in passing, meaning by normal the unmarked, the unremarkable, and the largely unobserved. A further shift took place when the British eugenicist Francis Galton developed eugenics toward the end of the nineteenth century. The "normal curve" appeared then for the first time as a means of organizing observed patterns of heredity and identifying possibilities for racial improvement.

Galton's theorization of the "normal curve" and "normal distribution" marked a decisive moment in the genealogy of the normal, inscribing the term within statistics much more centrally than had previously been the case. It is in Galton's work, in fact, that we appear to find the convergence of the mathematical concept of the average and the medical concept of the healthy that Ian Hacking has so influentially identified as underpinning the cultural dominance of the normal in the twentieth century. The fact that this convergence seems to have taken place in the context of Galton's theorization of eugenics has been widely seen as evidence that the modern notion of the normal emerged in a political context where it served as an instrument of dominance and oppression, of enforced conventionality and standardization. While this association of the normal with eugenic theory does indeed identify a crucially important moment in the critical genealogy of the normal, it also focuses almost exclusively on its use as a tool of cultural authority and violence, in ways that overshadow other, less examined aspects of its genealogy.[40] This book seeks to recover those aspects, and to restore their importance to the significance and function of the term in all its messy conceptual complexity. Only in this way, we believe, can its cultural influence and effects be properly understood. The field of analytical activity has, in our view, been unduly narrowed by targeting standardizing concepts of the normal at the expense of individuating ones. In a similar way, existing studies have tended to focus on the large public institutions from which the genealogy of the normal developed, and to overlook the wide variety of institutional contexts and cultural conditions to which it gave rise. Where our own study is concerned, it might have been possible to engage in a long and potentially fruitful study of the rise of the *école normale* or normal school in Napoleonic France. That was, after all, a new institution whose specific purpose was to serve as a place of training—as both a model and a locus of disciplinary constraint—for school education. But, instead of governmental systems, we have chosen to focus on knowledge practices that have to do with individual bodies and the ordering of differences between them—not just medicine, but anthropometry.

The second half of our book focuses on these, and challenges the widespread assumption that the popular concept of the normal emerged as a result of its uptake in statistics at the end of the nineteenth century. Margaret Lock, for instance, argues that an increasing number of medical interventions from the first decades of the twentieth century were undertaken to standardize bodily variations by making all bodies conform to the same statistically calculated averages.[41] Julian Carter asserts categorically that "normality originates in, though it is not restricted to, its mathematical meaning,"[42] while Annamarie Jagose argues that medical writing on the orgasm produced new taxonomies of heterosexuality and homosexuality at the end of the nineteenth century, which were underwritten by "the rise of statistical methodologies and analysis in the social sciences" that were "designed to produce erotic behavior as quantifiable, and hence more easily administrable."[43] Although it was through scientific writing on sex produced at the turn of the twentieth century that the word "normal" first began to move into the public sphere, as we will show in chapter 7, it was not statistical, but medically individuated concepts of the normal that dominated in this literature, continuing to do so well into the twentieth century. Moreover, we will also argue that quantitative concepts of the normal cannot be assumed to have a standardizing effect. The cultural networks to which the large-scale anthropometric projects of the twentieth century gave rise were not simply those of institutional administration or population management but rather a new and complex commercial culture of mass production, efficiency, and time management, and of industrious self-improvement. This sphere is not inhabited primarily by discipline but by commerce. It is focused not on standardization, but on the individual consumer. Large-scale anthropometric studies, like those undertaken by the Grant Study of Normal Young Men examined in chapter 8, or the quantitative sex research developed by Kinsey that is the focus of chapter 9, played a pivotal role not simply in producing standardized "normate bodies," although they certainly did that. These projects also contributed to the emergence of consumer culture and to the very conception of democratic capitalism, as Sarah Igo shows admirably in her study of surveys and participatory citizenship *The Averaged American: Surveys, Citizens, and the Making of a Mass Public*.[44] We will demonstrate that contemporary popular ideas of the normal emerged not in the context of the large disciplinary institutions and biopolitical systems that came into being at the start of the nineteenth century, where Foucault locates the emergence of normalization, but much later, in the emergent self-improvement and consumer cultures of the mid-twentieth century. This shift is a significant one. By relocating the moment at which the idea of the normal appeared in the public sphere, we help shift the focus away from the disciplinary regimes on which scholarship in this field has primarily focused and toward new practices of self-management and self-improvement. These practices are certainly disciplinary, and constitutive of contemporary forms of subjectivity as a site of self-care and cultivation, but they work in very different ways from the large disciplinary institutions that were the focus of Foucault's studies of normalization. The normal, as it emerged in popular culture in the middle of the previous century, was part of very different cultural networks from those of the nineteenth century. What a twentieth-century consumer culture required was not the "docile bodies" produced by the nineteenth-century prison and other disciplinary institutions, but the highly adaptable and flexible bodies of twentieth-century capitalism. Its location was not that of the nineteenth-century prison, but the twentieth-century office. While it grew out of the nineteenth-century disciplinary society, it gave rise to something much newer: the twentieth-century data society. The disciplinary society and the data society are in many ways mutually reinforcing regimes, of course, but their differences are also significant for understanding the specific genealogy of normality this book seeks to recover.

Against an emerging critical tendency to read the history of the normal as one in which quantified forms of thinking gradually became more culturally dominant and practices of standardization more ubiquitous, our long genealogy allows us to recognize how the normal was conceived in medical terms as a state of the individual body subject to continual change. That is, the normal describes not one part of a binary condition in which the other term is the deviant, the pathological, the abnormal, but a state of balance or a range of variations deemed appropriate. Accommodating this view of the normal alongside a history of statistical thinking will allow us to reconsider the cultural networks and "particularized relationalities" in which the terms "norm," "normal," and "normalization" came to have popular significance in the middle of the last century.

The chapters in this book follow the reframing and terminological adaptation of the normal over the period from 1820 to 1950, but while our analyses are ranged in chronological order, it cannot be said that our historical narrative advances in linear fashion. To tell a story of straightforward progress would be quite misleading, in fact. It is rather a matter of accumulation and accretion, as questions and issues linger from one chapter to the next. At the end of our story, in the midtwentieth century, there remained a collection of discursive and conceptual leftovers from nineteenth-century scientific thought: notions of the average, the healthy, and the typical. All of these were now being served up as approximate synonyms for the normal. Having been contested during the nineteenth century in quite specific ways, the normal had come to function as a conceptual aggregate, gaining power by its versatility, and avoiding contestation by its very imprecision.

Chapter 1 finds a historical beginning for the modern conception of the normal around 1820. The key concept in French medical writing was that of the normal state, which emerged in comparative anatomy before being taken up and disseminated in physiology. We show that in anatomy the normal state of an organ was defined by its functionality and its degree of structural integration, whereas in physiology the normal state was thought of in binary terms, being opposed to pathological disorders. Georges Canguilhem has given a strong lead in the history of nineteenth-century French medical thinking, and we engage with his work in this chapter while parting company with him on some key points. He speaks of a general failure of nineteenth-century French medical thinkers to develop what he considers a properly qualitative understanding of the normal. Medical thinkers continually lapse, he says, into quantitative mode, losing sight of the vital as they are drawn into the broadly statistical. Our contention is that irregular oscillation between these two kinds of thinking served in fact to make of the normal a quite particular figure of knowledge, so that equivocation between the qualitative and the quantitative recurs throughout our history. The unresolved coexistence of the two appears from the first as a constitutive complication, an enabling contradiction at the heart of our theme.

In chapter 2, we examine a debate in 1837 in the Académie de Médecine during which France's leading doctors attempted, with no great success, to rid themselves of this complication as they engaged in a public contest between qualitative and quantitative views of clinical medicine. The majority of academicians refused to allow in principle that their concepts and practices might be translated into averages and probabilistic calculations. They were comfortable enough with the notion of the normal state, which had by then become a familiar element in their professional discourse, but they did not accept that a properly medical concept of the normal might be grounded in quantitative thinking. It was possible, as a minority of so-called numerists pointed out, to pay more attention in clinical practice to quantitative data by keeping patient records and ordering them into tables, but that emerging practice was confronted by a well- organized intellectual and professional discourse extolling the "art" of the doctor, which consisted in perceiving at a glance the unique qualities of each patient and the specificities of their condition. The idea of identifying the average patient was held up to ridicule by the opponents of number. From 1837 onward, statistics offered new means of talking probabilistically about long series of events and large classes of people, but ranged against statistics throughout that time was a strong body of professional opinion that considered such calculations improper to clinical medicine. The emergence of a numerically based discourse of normality in the medical field was cognate with the emergence of a counter-discourse that refused to allow that mathematical forms of knowledge might properly be applied to living beings.

We show in chapter 3 that during the following decades, even as statistics became an intellectual and administrative routine in some fields, attempts to practice statistical analysis in medicine were fraught with difficulty and attended by controversy. The primary requirement of statistical analysis was for data in sufficient quantities collected according to unvarying procedures, and clinical "counting" was almost never adequate from that point of view. During the middle decades of the nineteenth century, statisticians with some interest in medical matters focused their attention on population health rather than on clinical treatment. They found their preferred terrain in the study of such things as epidemic mortality rates and the effects of vaccination on whole populations. A key figure in this regard was the Belgian Adolphe Quetelet, in whose thinking the average and the normal came effectively to coincide, although the proximity of the two was never fully owned by him as a matter of statistical method. There was in Quetelet's work a theoretically unresolved cohabitation of the average and the typical that further complicated, and in that sense further strengthened, the theme of normality.

Chapters 4 and 5 shift the focus from medicine to anthropology, although it should be noted that nearly all the leading anthropologists we consider in these chapters were trained as doctors. The French physical anthropologists discussed in chapter 4 made hereditary normality their principal object of study and established the notion of race, at least in principle, as the main gathering place for their knowledge. But they did not do that job well, since there was considerable uncertainty among them about how race was to be defined. They became experts in the measurement of bodies, especially in craniometry. Their observations were most often organized in series and regular distributions produced by the use of methodical protocols that excluded irregular individuals. But they were unable to find or confirm the existence of races simply by averaging the data for each physical "character" measured, then summing up the averages into a type of person. Whatever "race" was, it was not able to be known in that quantitative manner. Anthropological science was committed from the outset to speaking of racial types, but found itself unable to arrive at the type by computation. It did retrieve a concept of the type from natural history, but the type in that sense was not describable in quantitative terms. So anthropologists found themselves speaking of a hybrid entity that was to be understood as a typical norm. Once again, a mixture of styles of knowing was productive, if clumsily so, for thinking and talking about normality.

In chapter 5 we consider some ways of thinking that partook of both the numerical and the proverbial. The Italian "positive school" was preoccupied with the description of socially dangerous individuals as a class of person, and indeed as a "race." In this milieu the normal was not made an object of scientific study: normal subjects were taken as known and supposed data about them referred to in passing without actually being gathered or tabulated. Taking the normal for granted became itself a figure of knowledge, allowing it to serve as a stable reference without being subject to interrogation. In that way, the normal was able to find a loosely drawn but conceptually fundamental place, thereby absolving criminal anthropologists of paying close scientific attention to it. After all, they noted, ordinary people had been perfectly able to develop contrastive knowledge of dangerous and normal individuals, as shown by the collections of everyday sayings quoted in the writings of Cesare Lombroso. Taking all these kinds of evidence together, the positive school developed a genre of writing in which the authority of science was mobilized in support of a range of claims about humans that were often not grounded in experimental or statistical evidence. Numbers were used in opportunistic ways to confirm what was already believed; full credence was given to established verities of common sense; the natural and the organic were invoked in order to privilege certain physical and psychological characteristics over others; and strongly constraining generalizations were produced about "natural" human sexuality. That generic and epistemic mix quite often reappeared in twentieth-century talk about normality.

In chapter 6, we see how anthropometric studies of race and the type led to the development of the statistical concept of the normal, which found its fullest form in the work of Francis Galton. Galton himself intended this research to be applied through the implementation of eugenic policies designed to regulate sexual and social life for the betterment of the race, but we show how his work marked an epochal shift from the collection of anthropometric data to biometrics and psychometrics. At the turn of the twentieth century, the focus of measurement was extended from physical or racial features, understood as fixed external "characters," to mental faculties and capabilities, or internal states. This extension had two key consequences. The first is that Galton's aspiration "to grade physically and mentally [all] mankind,"[45] required the development of new technologies for the recording, storage, and retrieval of data, including an early punch-card machine.[46] The second is that, in order to acquire the vast databases on which his research depended, Galton relied heavily on voluntary participation by members of the general public, thereby playing an instrumental role in establishing a culture of participatory anthropometrics in which willingness to contribute to scientific and social studies was increasingly understood as a measure of responsible citizenship.

In chapter 7, we examine writing about sexuality produced at the turn of the twentieth century, drawing attention to the fact that medical understandings of the normal, far from being overtaken by statistical ones, often served as a first reference for such work. Focusing on sexology, psychoanalysis, marital advice, and sexual hygiene literature, this chapter problematizes the assumption that the medical and scientific writing on sex at this time either established or reproduced a binary understanding of normality, defined against abnormality or pathology. Rather, we show how writing on sexuality at the turn of the century came to focus increasingly on the idea of the normal itself, which it theorized relatively as a dynamic state. The normal and the abnormal were not opposed, but were part of a spectrum of possibilities. Moreover, a subject's position on the scale of normality and abnormality was often subject to change. As the normal came under greater scrutiny, it seemed that no one could be perfectly or consistently normal: increasingly, the normal described a state that was assumed to be common while proving highly elusive in actuality.

Whereas chapter 7 focuses on the normal understood as an interior condition or psychological state, chapter 8 returns to anthropometrics. This chapter considers the representation of statistical data in the United States during the 1920s and 1930s in the form of composite statues. It shows how closely these representations helped to translate scientific conceptions of the normal into terms and practices that became available for popular consumption. It was in the context of public health discourses emerging during the first half of the twentieth century that the idea of the normal first moved into popular culture. We bring this to light through the historical concurrence of two key moments in our genealogy, both of which took place as the Second World War came to an end in 1945: the publication of the reports on the Grant Study of Normal Young Men—the first full-length, large-scale study of the normal person—and the exhibition of the composite statues Norma and Normman in the first permanent health museum in the United States, in Cleveland. Over the course of the twentieth century, anthropometrics was increasingly promoted as a tool for self-improvement and used in the production and marketing of consumer products. At the same time, normality itself was seen as a highly complex and provisional state: an average that was also an ideal, a typicality so perfect that no one could actually embody it.

Chapter 9 examines the impact of Kinsey's work and the ways in which it was methodologically framed and popularly received. Kinsey's first report, *Sexual Behavior in the Human Male*, caused a sensation when published in 1948. Unlike the writings of sexologists, psychoanalysts, and sexual advice authors examined in chapter 7, Kinsey's reports relied heavily on statistics and numerical data. This first large-scale study of sexual behavior among ordinary Americans showed how high was the incidence of sexual practices considered abnormal or perverse, and how statistically rare was apparently normal sex. Just a few years after it had first emerged in the public health discourses of the postwar period, the idea of normal sexuality was denounced in Kinsey's reports as a "moral" judgment that had no place in scientific thinking.[47] The idea of the normal moved into the public sphere as though it had always been there, emerging as an influential theme even as it was being made the subject of resistance and critique.

PART I

The Normal in Nineteenth-Century Scientific Thought

CHAPTER ONE

The "Normal State" in French Anatomical and Physiological Discourse of the 1820s and 1830s

At the outset, we can see two ways of engaging with the history of normality. The high road is mapped out in the work of Georges Canguilhem, whose *Le Normal et le pathologique* offers a powerful philosophical history of our general topic. Canguilhem's essay fully deserves to be called historical, since it examines the work of the most influential writers on biological questions in nineteenth-century France. Yet it remains unswervingly philosophical in that it puts forward normativity—the requirement that the normal serve as a rule—as its guiding concept, arguing that the capacity to impose order on surrounding organisms is a constitutive property of living creatures. Even an amoeba, says Canguilhem, "prefers and excludes." All organisms are shaped and oriented by "the propulsion and repulsion of the objects of satisfaction proposed."[1] This is what he means by "vital normativity," and he uses that as a conceptual yardstick by which to measure the value of nineteenth-century biological writing.[2] It follows that he is not particularly concerned to describe historical patterns of thinking in their own terms. His interest is in the underdeveloped presence of a vitalist understanding of normativity in the nineteenth century, and he regularly points to its weakness or its absence in places where in his view it ought to have played a role. So Canguilhem does not simply offer an account of what took place in the past but frames his analysis with considerations about what people ought to have thought if they had given full value to the philosophical principle that grounds his own position. His stated concern is less with the historical accidents that may have befallen thinking about normality than with "the notion of norm in the normative sense of the word."[3] Normativity, for Canguilhem, is thus more than a topical matter: it is a proper constraint on thinking. To write a philosophical history of the norm is in his view to respect and rehearse the self-confirming circularity of the central concept. That makes his own history exemplary in a specific philosophical sense. As he talks about the norm, he everywhere affirms and applies its significance as a literally vital constraint on biological thinking. From Canguilhem's point of view, the only proper way to write about the normal is to respect and reflect this fundamental truth. To put it in an appropriately circular fashion, the history of normativity he produces is itself a normative one, imposing a philosophical order on some rather disparate historical productions.

Our book is shaped by other ambitions, and that is why we will not engage directly with Canguilhem's philosophical view of the norm. We will continue to refer to his work throughout our study, but we will not take the path he surveyed. We will not put forward a philosophical theory of normality but will attend instead to whatever concepts or notions of the normal present themselves along the way, seeking as far as possible to understand them in their own terms. That choice of method, it should be said, does not reflect any lack of ambition on our part. We will begin roughly where Canguilhem began, focusing on some French medical thinkers he discusses and others he does not consider, before moving beyond the geographical and disciplinary regions he mapped. Our search for emergent ways of talking and thinking about normality will lead us to analyze developments in medical thinking, in anthropometrics, and sexology. We aim to produce a quite broad history of intellectual and cultural change over the nineteenth and twentieth centuries, focused initially on France, then expanding to include Belgium, Italy, Britain, and the United States. But at every point the steps we take toward our goal will be measured ones. We will simply attempt to understand what the normal meant in context, how it functioned then and there, and we will make our history out of the incidents of its becoming. We will describe the intellectual formations and cultural practices that shaped normality through time.

To mark this difference of method and to indicate what might be at stake in our choice, consider, for example, Canguilhem's discussion of an article in Diderot and d'Alembert's *Encyclopédie* that deals with the standardization of artillery equipment in eighteenth-century France. The article, says Canguilhem, contains "almost all the concepts utilized in a modern treatise on normalization, with the exception of the term 'norm.'" So, he concludes, the article can be said to contain "the thing without the word."[4] He thus accepts with no apparent difficulty that there might indeed be abstract "things"—here, figures of knowledge or knowledge practices—in the absence of words that refer to and implement them.⁵ Our own position is much closer to that of Michel Foucault, who argued at length in *Les Mots et les choses* that words and knowledge formations are caught up by their very nature in a sustaining circular relationship.⁶ In keeping with Foucault's view, we will be engaging in a history that allows lexical developments to count as events. Certain terms, we will suppose, become by their emergence markers of conceptual change. As might be expected then, our history of normality will pay close attention to shifting uses of the word "normal," making the assumption that the thing we now know as normality has been molded by discursive usage.

We can build here on the work of Caroline Warman, who has studied uses of the word "normal" in French and English in the late eighteenth and early nineteenth centuries, commenting as she did so that there was no need to make any great difference between the two languages because "the countries were so intellectually entwined."⁷ Following a method consonant with our own, Warman is able to confirm Canguilhem's broad historical account while providing fine lexical detail that significantly inflects it. She tracks the word through dictionaries, observing that "the term 'normal' entered [them] relatively late: having been a synonymous but less common alternative for 'perpendicular' in geometry."⁸ The geometrical usage to which she refers did not simply disappear at the end of the eighteenth century, and we have found "une normale" still used in that way in a physics treatise published in 1822.⁹ But the import of "normal" was much extended when it began to be used as an adjective in conjunction with some key nouns. What Warman calls the first "figurative" use of the term occurred in the expression *école normale*, normal school, as a place for teacher training that served to standardize educational practice.¹⁰ Whether such usage should indeed be understood as figurative is a moot point. Was it thought that the perpendicular or the upright signified discipline? Whatever the semantic basis for the term's extension from geometry to education, there emerged in the normal school a particular kind of thing that has remained relatively stable in French and English until the present day.

It was not, as it happened, the notion of the normal school that served during the early decades of the nineteenth century as a locus of complicated and contested thinking about normality, but rather that of the "normal state" (or condition). That term emerged in medical discourse around 1820. And while Canguilhem has nothing to say about adjectival as opposed to nominal uses, he does identify the field of medicine as the critical one for his history. Philosophical considerations about norms and the normal, he says, "ought to benefit, for the purposes of question-asking and clarification, from direct medical culture."[11] His reasons for focusing on this field are manifest. Seeking to understand how the normal and the pathological functioned in the nineteenth century as a binary pair, he advances the general thesis that medical thinkers failed to produce a concept of the norm grounded in biology, and were thus unable to define the pathological as its proper theoretical contrary. Instead, he argues, physiologists typically conceived of the relation between the normal and the pathological as a matter of degree. Pathology was understood as excess or inadequacy, as quantitative divergence from the "normal state."[12] Our research has persuaded us that the field of nineteenth-century medicine is indeed the domain that will most reward close examination—for more reasons, indeed, than Canguilhem himself acknowledges. So that is where we will begin our analysis. In this chapter, we will examine how the term "the normal state" came to be used, and we will consider the theoretical import of its emergence.

We can attest that "the normal state" had no place in French scientific discourse before about 1820. Yet by the 1830s it was in quite widespread use. Indeed, the term was routinely used in the medical academy during the mid-1830s. So we have found evidence of a discursive event, although of course the significance of that event is yet to be made clear. Even as we ask how and why the change came about, we will not be preoccupied with matters of agency, since we do not consider that widespread discursive change is brought about by singular intervention. Our primary interest here is in the shift itself, rather than in identifying an originary moment or an agent of change. Yet precisely because our focus at this point is on discursive fields as such, we must do all we can to identify the intellectual places in which the shift occurred. We have spoken so far of medical discourse as the locus of contest and complexity, and we will follow through on that engagement, but we need to mark from the outset the areas of scientific activity that had a bearing on the field of medicine. The most prominent, very much to the fore in Canguilhem's history, was physiology. But leading writers in that field did not come to speak of the normal state before about 1830. François-Joseph-Victor Broussais may have been the first to use the term freely, doing so in an essay published in 1828, but it was entirely absent from the essays Broussais himself published during the preceding twenty-five years, including one published as late as 1822. Could it be that such a theoretically explicit and highly polemical author simply slipped unaware into a new discursive pattern? Or did Broussais, as Canguilhem might suggest, simply find a new word to name a topical thing that was already present in his work? Those are questions that will help to shape a later section in this chapter devoted to the normal state as it was conceived and spoken of by physiologists.

But there is another field pertinent to medical thought that will be given priority in our history, precisely because the term "the normal state" appeared there in 1818 as a self-conscious terminological innovation, ten years before it was used in Broussais's essay in physiology. The field in question is comparative anatomy. Canguilhem's history has relatively little to say about anatomy, although he does make some passing observations about it to which we will return. That does not constitute reprehensible omission on his part, since the stated topic of his history is the binary opposition between the normal and the pathological, which affords him a direct entrée into medical thinking. But we are going to argue here that the normal state as conceived in comparative anatomy was defined in a rather different way that owed less to binary logic. Canguilhem might well have contended that developments in anatomy mattered less in the long run for the history of normativity, and we cannot claim for our part that anatomy was more important for medical thinking than physiology. The dominant figures in physiology were all practicing physicians, whereas that was not generally the case in comparative anatomy, where the leading exponents were natural historians and zoologists.[13] But the fact remains that the concept of the normal state first took shape in anatomical writing, and that this change occurred about a decade before physiologists began to use the term. It does not follow from our analysis that physiologists "took" the concept of the normal from anatomy, or indeed that the term signified exactly the same thing within the two fields. But there is every reason to believe that physiologists, wherever they may have happened on the term, thrived on its usage, finding it remarkably convenient for their own purposes. So when the term came into prominent theoretical use in both fields during the early 1830s before being taken up by doctors as part of their professional routine, it seems reasonable to suppose that discursive authority for the medical usage was derived from both fields—if not from their narrow convergence, then at least from some conceptual overlap. We cannot be sure whether physiologists coined the term more or less independently of anatomists, but we can clearly observe congruent changes in both fields accompanied by a shift in everyday medical terminology. The appearance of a widely successful new expression, we will argue, marked the emergence of a rather new conceptual thing.

THE NORMAL STATE IN COMPARATIVE ANATOMY

A prime example of the emerging usage of the "normal state" can be found in a treatise by Étienne Geoffroy Saint-Hilaire, *Philosophie anatomique*, the first volume of which was published in 1818. Geoffroy's scientific work located itself somewhat restively within a long tradition of "natural history" that looked back through the prestigious eighteenth-century productions of Linnaeus and Buffon to the classical writings of Aristotle. The central purpose of this learned tradition was to study living things in a manner that was both descriptive and comparative, putting like with like while noting shades of difference so that each creature observed could be placed as nearly as possible to those that most resembled it. Reasoned proximity of this kind was itself an artifact of scientific knowledge. Furthermore, observations so arranged could be readily grouped into classes. An array of living beings could be put on display, and the regularities of nature manifested as an orderly disposition of nuanced similarity and difference. But when Geoffroy offered an overview of natural history in his own time, he made a critical distinction that threatened to disturb the stability of this whole field of inquiry. Or rather, he asserted that natural history was already being divided against itself by the coexistence of two knowledge practices that could only be expected to produce divergent effects. The first of these approaches—a classical one that could be traced back to Aristotle, although Geoffroy did not need to name the great predecessor—always began with mankind, taking humans as the point of reference for its account of natural regularities and making them the standard for every measurement of difference. Human organs were declared to be "the most perfect" by definition. The "organs" in question, it should be noted in passing, were simply the constitutive elements of bodily "organ-ization" as understood by all natural historians of the time: they included not only such things as the heart and the spleen, but also bones and muscles. The critical point made by Geoffroy was that when humans were taken as the point of departure for scientific description, organs as they were found in lesser animals were routinely described according to the degree to which they fell short of the human standard. Organs were thus diversified and graded precisely insofar as they could be considered "deformed."[14]

To this classical anthropocentric approach Geoffroy opposed a form of analysis that did not take for granted the generic perfection of human anatomical forms. He thus gave "comparative anatomy" a rather different orientation. It was now a matter, he said, of "considering organs wherever they are at the *maximum* of their development, and following them from degree to degree as far as zero existence."[15] The concept of (human) perfection was displaced by that of maximal organic development, which might happen to occur in any species. It was patent that the "maximum" of which Geoffroy spoke had nothing to do with size. Development had to do with what he called the "compositional" (or structural) importance held by the organ in its anatomical environment.[16] Comparative analysis would then proceed by considering what an organ "became" as one moved from one class of animal to another. At one point, Geoffroy took the example of the furcula (forked) clavicle. In mammals, this bone took an uncertain or inconsistent (*équivoque*) form, occurring in a "more or less rudimentary state." But as one moved from mammals to birds, the furcula clavicle became "something more precise and more persistent." It played a greater role and took on a less variable form. In describing it, Geoffroy deployed some of his key terms: "we see it take on a more classical character, justifying this high degree of composition by its great usefulness."[17] To enjoy a "high degree of composition" was to play a well-integrated set of anatomical roles, contributing significantly to the activity and stability of organs around it. But the furcula clavicle actually found its most complete development in fish: "it is only in fish that these clavicles reach their full possible growth and attain the fullness of their functions."[18] The maximum development of this organ, the full realization of its structural anatomical possibilities, was not to be found in humans but in a supposedly lesser class of vertebrates. In humans, as it happened, the organ was nondescript and rather rudimentary.

For Geoffroy, to study organs as discrete forms was necessarily to misunderstand the importance of "composition," for what really mattered was anatomical organization as such: "the organization has rights that cannot be overridden by anything; if they are impinged upon, the principle of unity comes under attack and the machine ultimately becomes totally disorganized."[19] Without due recognition of this principle and careful study of its workings, anatomical observation was bound to be unproductive in the long run. As Geoffroy saw it, the whole discipline of comparative anatomy had now reached a crisis of method. Buoyed by a diffuse sense of scientific progress, researchers were engaging in busy observational work that could not be expected to lead to worthwhile discoveries: "A new age is beginning. People are setting out to lay a foundation for the edifice of science, and since they believe that one can lay down a solid foundation only by abstaining from all abstract propositions, they spend all their time doing observational work."[20] Enthusiasm for ever more observation was leading people to "misunderstand one of the fundamental principles of natural philosophy" as they allowed themselves to be carried along by "a kind of infatuation with details."[21] Observation, Geoffroy insisted, would never suffice of itself, and more needed to be done in the name of science than the mere gathering of details, no matter how finely examined. There were great natural laws to be understood, and that could only be accomplished by the study of "relations."[22]

To show what was at stake in relational or analogical thinking, Geoffroy offered the example of the forelegs of ruminant animals. Proceeding in the routine, classical manner, a veterinary anatomist would note how appropriate their design was to purpose: "He perceives a finished design, a work in which all parts are admirably suited." But if he saw no more than that, he would be failing to perceive a more general analogy: "Would he think of the arm of man, and of what might be learnt by that comparison?" No he would not, because he would be preoccupied with the "newness" of these forms with respect to other kinds of animals. He might actually grasp the relations in ruminants between the forelegs and other organs. He might even be led to "ideas of harmony." But he would not in effect have moved beyond his first impressions. He would still "believe in the existence of new organs" and find himself creating a new language–new anatomical terms–to describe what he felt.[23] That would amount to an undue fixation on detail and would result in a failure to perceive compositional analogies. In ruminants such anatomists would identify what they saw as a foot, in humans and apes they would think of it as a hand, but the change of terms would hide the "almost complete" resemblance in the composition of these limbs. The better, indeed the true, method of natural history was to look for analogical similarity, and use that as a base for measuring differences: "That is the idea that frames the methods of natural history: the beings in a group are connected by the closest of relations, and are composed of organs that are entirely analogous. Having admitted the almost complete similarity of a great number of species, the ingenious art [of the true natural historian] is then simply to differentiate them according to slight characteristic marks."[24]

It might well have been thought by naturalists wedded to the classical tradition that finding analogies between organs was a quasi-metaphorical distraction from the business of close anatomical observation, but for Geoffroy the search for analogies actually claimed a place at the heart of the discipline. Only when analogies were properly discerned could rational differentiation occur, and naturalists who did not focus on analogies tended to lose the thread of anatomical intelligibility. They were then likely to find themselves without a guiding theory, at sea in an infinite variety of animal forms.[25] The truth of natural history as Geoffroy saw it was that "nature tends to make the same organs reappear in the same number and the same relations; it simply varies their forms ad infinitum."[26] The unvarying regularity of nature was not to be found in detailed forms but in compositional relations, and the purpose of a "philosophical anatomy" was precisely "the comparative study of details."[27]

Studying anatomical compositions analogically was not simply a matter of remarking on general similarities that might happen to appear when kinds of animal were compared. It was for Geoffroy a quite specific form of analysis that brought to light the changing roles of particular organs by comparing organizations as such. That is why, for all the natural historian's focus on a stable world, Geoffroy's anatomical texts were filled with narrative. In moving from one organization to another, even as he worked within the classificatory space laid out by the discipline, he continually spoke of the development or regression of an organ. This was not development within the life of an animal, or indeed evolution over time within a species: it was compositional variation, the becoming of an organ that could be seen when one followed it analogically from one organization to another. The task of a revised natural history was to provide a scientific account of the compositional stages through which organs might progress. That was where the concept of the normal state came in: it was for Geoffroy a typical stage of organic development.

The first occasion on which Geoffroy used the term "normal state" was in a footnote to his discussion of the hyoid (upsilon-shaped) bones as they developed in different classes of vertebrates. The hyoid bones are located in the neck, and the upright stance of humans, he said, led them to take on a certain form. That might have caused humans to be "different from the normal state of mammals," he said, had it not been for certain compositional arrangements that occurred around the bones.[28] The normal state was by implication something uniformly found in mammals, and its presence was being referred to here as a property of that whole class of vertebrates. The composition involving hyoid bones was typical in the narrow, scientific sense: it was characteristic of a type. Furthermore, Geoffroy was suggesting that the normal state had a capacity to reestablish itself and maintain typicality, even though the particular variation found in humans meant that they tended to occupy a rather eccentric place within the class.

Geoffroy's second use of the term, in the body of the text about seventy pages later, was entirely consonant with the first. He used it when comparing birds' lungs with those of mammals. Lungs in birds, he explained, could not pass in front of the trachea, so that the trachea needed extended double branches in order to reach the chest. "This extension," he added, "relates to a slight modification of the tube that introduces air; it is a very slight modification of what exists in mammals, if it is in the latter class that the normal state of this system of organization is to be found."[29] Geoffroy did not take it for granted that the normal state of pulmonary composition in vertebrates should be the one found in mammals. A certain state of pulmonary composition was indeed generally found and maintained in that class and was thus "normal" within it, but there was no theoretical compulsion to take that as the state of reference when discussing birds. Birds were in principle no less worthy of attention than mammals, but one could not provide a description of pulmonary development across all vertebrates without choosing a particular class of animals as a reference. In the absence of a concept of human "perfection," the normal state of a particular class—in this instance of mammals—could be called into service as the standard.

The normal state provided a standard for characterizing a class of animals and for calibrating variations with respect to it, but the term had a second usage in Geoffroy's work that proved to be conceptually bound up with the first. It was used to refer to organs that had themselves reached a certain stage of development. Having previously used the term almost in passing, Geoffroy came to put it forward with some discursive ceremony: "The class of fully developed [*accomplis*] or *normal* organs: that is the term that I shall use from this point onwards in order to call to mind the high degree of composition of any organ which, in the principal subdivisions of the vertebrates, maintains a fixed, invariable character. This is the kind of organ that can be regarded as *totally accomplished* [*totalement achevé*] insofar as, without its useful intervention, one could not conceive of a classical, fundamental organization."[30] Here it was the organs themselves that were qualified as normal, and they were so named because they were usefully engaged in a high degree of composition. Their organizational roles were so well defined, so steadily maintained that the organs had become fixed and invariable. They were certainly not "perpendicular" in the eighteenth-century sense of the word "normal," but there were echoes here of a notion of rectitude. These were organs that were fully in their place. They were able to be so because their development was "accomplished," and their degree of development was demonstrated precisely by the capacity to occupy their place to the full.

So there were two dimensions to the "normal state." Because of its very stability within a particular anatomical composition, it made an organization distinctively and invariably recognizable; and by the same token, because of its tendency to be widespread, it could serve across a whole group of animals to characterize a type. Dynamic normality of organization offered support for classificatory normality, so that normal organs functioned in constraining and constrained, defining and defined ways. For every group of animals, there was a type corresponding to the normal state, and for every individual within that group there was a regulating force that functioned as if it were a type in and for itself. To put it another way, there was no effective difference for a fully developed organ between being normalizing and being normalized. That was exactly why and how organs came to be "fixed."

In the second volume of his treatise, published four years later in 1822, Geoffroy used the word "normal" more frequently, as might have been expected after he had reintroduced it with such care. He now allowed it to have a place in his terminological routine. Many bones in a "monstrous" human described by a colleague were thus said to have "remained in the forms proper to the normal state."[31] Their forms could be spoken of as if they had been distributed and maintained by the organization in question. And Geoffroy was able to follow through his theory by transferring the epithet "normal" from the state to the organs themselves. He referred to "the normal state of those parts" and later, more succinctly still, to "normal foetuses" and to "a normal brain."[32]

It is important to note that even as Geoffroy made "normal" a familiar term, he did not ever disturb the theoretical foundation he had laid for its usage. A certain discipline is called for here on the part of modern readers in order to keep in mind that the term had not yet taken on the range of meanings that has become familiar in our own time. At no point, for example, did Geoffroy use "normal" as a synonym for "habitual." Normality as he understood it had nothing to do with frequency of occurrence. It was defined as a natural propensity to organizational stability. Remembering that, one is better able to make sense of a claim made by Geoffroy that might otherwise sit oddly with modern readers. He was given to describing his chosen discipline as "transcendental anatomy."[33] By that expression he meant to signal his determination to make anatomical organization the main object of study, setting aside a preoccupation with zoological details as such. The "normal state" was one of his conceptual tools for carrying out the task of transcendental anatomy. As he defined the term, it could never have been known by the mere aggregation of details or by some calculation of their recurrence. The concept of the normal state helped to identify and name the things that were, by their uniformity and their generality, transcendental in the field of anatomy. The normal state was abstract insofar as it constituted and exemplified "regular" anatomy. It was rule-governed and in an important sense self-regulating. Clearly, "transcendental" rules of that kind had nothing to do in principle with calculable frequency.

There is another stumbling block here for modern readers. It is that Geoffroy did not simply define the normal state in binary terms. He spoke of stages in the development of organs as they moved from the rudimentary to the normal, and sometimes beyond that to a "dominant" form.[34] He spoke too of "anomalies" that appeared when organs and organizations were compared across species and classes of animal. Reptiles were a particularly unsteady class, often presenting "those anomalies that one must expect to find at every step when examining the organs of reptiles."[35] But anomalies of this sort were, so to speak, a by-product of Geoffroy's comparative anatomy: they were surprises, apparent irregularities that called for his analytical attention. There was no "abnormal state" defined in principle as the opposite of the normal one. Normality was not a singular condition to be identified as the contrary of another condition or set of conditions thereby defined as non-normal. Anatomical composition was everywhere subject to laws, and the normal state, wherever it occurred, showed those laws at work. For an organ to be normal was to serve as an agent of law for organs around it. That anomalies occurred at all might well be considered remarkable when one remembered how the normal state tended to impose a pattern of regularity. But despite this tendency, the irregularity of certain forms was undeniable: "We must certainly consider these to be irregular, for it would be a mistake not to recognise that we are dealing here with the strangest of anomalies. One can be all the more certain that this is so if one has become familiar with the characters of the normal state, and if one has learnt to perceive a general tendency to reproduce a better state of things and retrieve the dominant forms."[36] In other words, it is precisely because one had learnt to identify the normal state that one could perceive anomalies as such. But there was more to it than that. Anomalies themselves were able to be accounted for in the very terms of the laws that governed anatomical regularities.

This manner of addressing and accommodating anomalies was perhaps the most striking aspect of Geoffroy's contribution to comparative anatomy and his most telling use of the concept of the normal state. In the second volume of his *Philosophie anatomique*, published four years after the first in 1822, he paid particular attention to anatomical monstrosity, repeatedly making the point that even in these circumstances "a certain order continues to reign in disorder."[37] Defects and irregularities could occur, but they did not constitute a "departure from the ordinary laws."[38] The laws of normality were "ordinary" in both relevant senses of the word: they were productive of order, and they applied in every circumstance. When regularity, the rule-governed nature of anatomy, was understood thus, it was equally manifest in apparently irregular compositions. Étienne's son Isidore took up this aspect of his father's work and built on it in his own four-volume treatise on teratology, the scientific study of monsters.[39] And in a further essay celebrating his father's life and work, Isidore showed how this reconception of monstrosity had served as a point of departure for configuring the whole field of teratology. It was now clear, said Isidore, that "monsters are not whims of nature. Their organization is subject to rigorously determined rules, and those rules are identical to the ones that govern the animal series. In a word, monsters are other normal beings; or rather, there are no monsters, and nature is unified."[40] Isidore gave a fuller account than his father had of the biological processes thought to be responsible for this pattern. Observations made by the emerging science of embryogeny revealed that radical "acephalic" monsters were simply displaying "normal conditions for the first ages of the life of the embryo."[41] Order was always perceptible in monstrosity: "monstrosity is not blind disorder; it is another order that is just as regular. Or, to put it another way, it is the mix of an old order and a new one, the simultaneous presence of two states that, ordinarily, would occur in succession."[42]

The main point was clear. It was that the principles governing the composition of human monsters were exactly the same as those governing the composition of normal humans. People had often made the mistake, said Étienne Geoffroy himself, of identifying the normal human state as a genus in its own right. This had led them to see monsters in an equivalently generic way, as standing outside the human. But it was wrong to think of the normal state in this global manner, "as if the only possibility were to have a single subdivision between regular man and irregular man!"[43] He mocked the binary view, rejecting it out of hand. Instead, he said, anatomists needed to take on the task of describing "specific modifications" as they occurred in each case of monstrosity.[44] To make a broad generic distinction between monsters and true humans was to fail to apply a dynamic concept of normality. This objection, it should be noted, was not grounded in a humanist ethics: the problem was a classificatory one.[45] If the principles of zoology were properly applied, the differences between certain human monsters and the normal state of humans as a species might appear to be as great as, say, between a reptile and a mammal. But those differences could only be properly known by detailed reference—in this example, to the normal state of reptiles and mammals. It was far better to think of "the normal state of man" as "the abstract, generic being, and of its different pathological deviations as the species making up this ideal genus."[46] In that perspective, the deviations would have their own form of belonging. "Monstrous" organs could always be understood in principle as adopting the normal state of organs found in other species: "It never happens that man leaves the line assigning him determinate forms without taking on others that correspond more or less to the forms of certain animals. That is because, when the natural progression of developments and formations is disturbed, if the first disturbance does not give rise to a second, and thence to others that are successively more severe, everything comes back to the accustomed order."[47] In other words, "disturbances" were continually subject to normalizing influence, and whenever they did occur, the normal state of other species remained available to serve as a matter of rule even in the most strikingly anomalous cases.

When, for example, a human fetus displayed a monstrous condition in which the genital, digestive, and excretory functions were combined, that same anatomical composition could also be found in the normal state of monotremes such as echidnas, who had "remained true [*fidèles*] to their classical organization."[48] And when a particular disorder in humans took the form of an anomalous shape of the posterior spheroid bone at the base of the skull, that could be seen as "presenting in the pathological state the normal conditions of birds."[49] Anomalies were to be found on all sides, but so were normal states, and a vast but orderly range of normal states was being made available to serve as a referential matrix for the scientific description of anomalies, even as the laws of organization "ensured the periodic return of regular productions."[50] Geoffroy spoke of the "resources of general anatomy" from which "this science draws its elements of comparison."[51] What made anatomy "general" in this way was precisely its capacity to bring rules to light by the practice of compositional analysis. Scientific "transcendence" was to be achieved by the eventual reduction of irregularity to a dynamic regularity. The normal state of a given organ or a given species was characterized by a tendency to fixity but was, according to compositional circumstance, subject to variations in intensity and function that would on occasion allow organic roles to be thoroughly transformed.

We have so far attended exclusively to scientific writing in French, and we will continue to do so for most of this chapter. There are actually a number of reasons for favoring French texts at this point in our history. The essays and treatises we are discussing here contain the earliest scientific understandings of the normal that we have found, and are elaborate enough to bear detailed conceptual analysis. They also allow us to sustain a conversation with Canguilhem, who takes almost all of his examples from French. We trust that our analysis will gain in scholarly resonance as a result of that choice, but we do not wish precision of focus to be mistaken for narrowness of geographical outlook. Throughout our book we will have occasion to note that the normal—the word and the thing—appeared at approximately the same time in a number of places in Europe. And we can note here in passing that during the 1830s the normal found a place in a major British scientific text. That this should have happened only a decade or so after Étienne Geoffroy's first writing on the subject might be thought of as a chance event, but it is just the kind of apparent coincidence that occurs when a significant new way of thinking is appearing across a broad front. In 1835, John Stevens Henslow published an essay in which he appeared to reinvent for the purposes of botany a key concept introduced by Étienne Geoffroy in comparative anatomy. Speaking in particular of the structure of flowers, he spoke of their "normal condition," which could be modified, he said by three kinds of development: abortion, degeneration, and adhesion. Each of these might produce forms that "disagreed with the normal type."[52] Identifying a normal condition and its accompanying "normal characters" allowed Henslow to speak of "deviations from the ordinary form," and indeed of "monstrosities."[53] He referred with some regularity to the work of leading French and German scholars, but those of whom he spoke were botanists, not anatomists, and he did not identify in their work any discussion of the normal.[54] Henslow was in effect introducing into botany some of the analytical apparatus made available in anatomy by the concept of the normal, but he was doing so without reference to Étienne Geoffroy. This can be taken as exemplary of Caroline Warman's point, quoted above, about "entwined" intellectual developments in France and Britain. The normal was emerging in different fields and in different countries in ways that were discursively cognate without being agentially coordinated.

To return to Geoffroy's comparative anatomy, it has to be said that Geoffroy's concept of the normal state has more in common with Canguilhem's concept of normativity than Canguilhem himself appears to recognize. At the beginning of *Le Normal et le pathologique,* Canguilhem indicates that it would have been possible for him to set about his analysis rather differently, taking up "the teratological problem" rather than "the nosological problem"—the classification of diseases—which was in fact his chosen focus.[55] That the two "problems" might be in some sense equivalent for his purposes can be seen when he refers to principles of order brought to light in the separate fields: "There is an order in diseases, according to Sydenham, as there is a regularity in anomalies according to I. Geoffroy Saint-Hilaire."[56] Canguilhem's reference here is not to Étienne Geoffroy, but to his son. It is noteworthy that, when Canguilhem refers to comparative anatomy, he speaks only of Isidore's work. So he appears to have missed the self-conscious early use of the term "normal state" in Étienne's writing. That cannot be considered philosophically momentous if, as Canguilhem affirms, one problem is as good a point of entry as another for the study of normativity in medicine. But it does matter from a historical point of view that these developments in comparative anatomy should have taken place unrecognized by Canguilhem a decade before the physiological writing he analyzes so carefully. And it ought to matter to Canguilhem that Étienne used the concept of the normal state to construct a theoretical account of anatomical regularities with no particular reference to a contrary state. Canguilhem does acknowledge at one point that, having subsequently read Étienne Wolff's *La Science des monstres*, published in 1948, he would now insist, when speaking of teratogenesis, on "the possibility and even the obligation of casting light on the understanding of monstrous formations by studying normal formations."[57] But he need not have waited until 1948 to find that line of argument: it was present in Étienne Geoffroy's treatise of 1818-22.

Had Canguilhem taken the teratological path, he would have been engaged in a rather different enterprise, since the pathological would no longer have been central to his inquiry. Étienne Geoffroy, when speaking of a particular form of monstrosity, questioned the use of "pathological" to describe it. A pathological state, he said, could only occur in "an accomplished [*achevé*] organ that had already existed in an earlier form and had undergone alterations, moving from the healthy state, its ordinary condition, to an unusual organization."[58] When his son Isidore set about building teratology as a discipline, he systematized this position. "Teratology," he said, "could no longer be considered a division of pathological anatomy."[59] There was indeed a whole class of anomalies, called by Isidore *hétérotaxies*, that was characterized by a "regular disposition" allowing life to continue while being "different from the disposition that constitutes the normal state."[60] Such was the case, for example, when the position of organs within the body was completely inverted without any visible deformity or loss of functionality.[61] Comparative anatomy, when it took the monstrous as its object, had no compelling reason to be interested in pathology, even as it remained centrally preoccupied with the normal. Canguilhem was fully aware of the difference: "the anomalous," he said, "is not the pathological. The pathological implies pathos, a direct, concrete feeling of suffering and powerlessness, the feeling that life is being thwarted."[62] The divide between teratology and pathology thus became an intellectual watershed: to follow the normal state as it developed in relation to teratology, as the Geoffroys sought to do, was not ipso facto to advance the understanding of the normal understood in opposition to the pathological. For the moment, we will continue to follow the course of teratology, although we will return to pathology in the final section of this chapter when we ourselves come to deal with physiology.

RETHINKING THE DISCIPLINE OF ANATOMY

While our interest here is not in the long history of anatomy as a discipline, we will briefly discuss a high profile public debate in which the issues raised by Étienne Geoffroy's proposed revision of the discipline brought him into direct conflict with the leading figure in French anatomy at the time, Georges Cuvier. Throughout our study we will continue to draw attention to debates of this kind, since they serve the purpose for intellectual historians of revealing the questions that mattered most at the time, framed in the ways that were most telling in context. The formal debate between Geoffroy and Cuvier was itself a major scientific event. It took place in 1830 in the Académie des Sciences, and polemical exchange continued in print for some months after that. This particular debate, by contrast with some that followed, will not occupy a great deal of space in our genealogy, largely because Cuvier himself did not use the term "normal" and did not directly express a view about it. Our interest is simply in the manner in which the exchange served to delimit Geoffroy's position. Geoffroy, as we have already seen in our examination of his major treatise, was advocating a shift to analogical examination and transcendental anatomy. Cuvier, on the other hand, represented and defended the established method, invoking Aristotle as he did so.[63] It was clear at this point that Cuvier's method had been serving anonymously as the primary target of Geoffroy's criticism.

We can encapsulate the difference between their positions using expressions that have already claimed a place in our discussion. This was a contest between "perfection" and "the normal state." Cuvier did not reject analogical thinking out of hand. There were, he conceded, identifiable similarities between certain animals and humans that might on occasion be considered analogous, but those could not form the basis of a method. Organic similarities could only be studied in an orderly manner as "the descending degrees [*dégradations*] of a common form."[64] These were the downward steps that led eventually from the highest (human) forms to those of the lowest animals: "to discuss these animals in an orderly way," said Cuvier, "we will examine those that resemble us by their overall organization, then we will move on to the others, according to how far distant they are from this first type."[65] This method provided a stable base for the making of fine distinctions, and Cuvier's distinctions continued to make sense by the extent to which they departed from the "perfection" of human organs.[66] There can be little doubt, however, that Cuvier's continued use of this term in a variety of senses placed his thinking under strain. He regularly drew on the concept of perfection but did so in ways that were not articulated theoretically. In his view, man exemplified "the most perfect and most complicated organization," thereby supposing that the height of perfection was accompanied by maximal complication.[67] By that same logic, animals had different faculties "according to their different degrees of perfection" and all other animals were simpler than humans, presenting decreasing degrees of perfection as one moved down the scale toward the lowest.[68] But Cuvier also spoke of perfection within a genre: "perfect insects," for example, had "three parts separated by constrictions."[69] Many insects did not in fact have this "perfect" organic formation, so that they had to be considered significantly less than perfect in two different, unrationalized ways. That they were less perfect than humans according to one measure had nothing to do in theory with the fact that they were less perfect than other insects according to a lower-level generic rule. Furthermore, it happened within the life cycle of a given less-than-perfect insect that it attained its "perfect state," where the term of reference for perfection was drawn from within its own developmental process. That is how it was with dragonflies, which remained in water in the larval state before emerging from it to take on the "perfect state."[70] The earlier state was presumably not perfect, even in itself, for reasons that certainly owed something to Aristotle, but that Cuvier left unexplained. By using the term in these different and potentially dissonant ways, he was stretching its significance to breaking point. Or at least, he was requiring perfection to be continually redefined, thereby unsystematically readjusting the base of his descriptive method. Geoffroy's concept of the normal state, on the other hand, was not subject to the same conceptual stress. The normal state was not perfection. It named a dynamic tendency to analogous similarity across species and, so to speak, a dynamic tendency to fixity within organic compositions.[71]

Geoffroy also criticized the talk of perfection on what he called philosophical grounds: "people will no doubt get rid of old phrases like 'the most perfect beings' as they come to be persuaded that the best and surest way for philosophy is not always to put oneself forward as a term of comparison."[72] Naturalists, he said, had to give up this habit. By taking account of the full set of organizations that were so close to that of mammals, they ought surely to come to the realization that "man was only one of the species in this great herd."[73] For Cuvier, the type was given generically by human anatomy, while for Geoffroy types of organs were everywhere perceptible through analogy and could be identified even in monsters as normal states belonging to species other than their own. Cuvier's anatomy located organs in a taxonomy shaped by a reasoned hierarchy, whereas Geoffroy's built his classification through analogies taken by him as evidence of dynamic laws governing the composition of all organs and all animals. By thinking and talking in terms of the normal state and related concepts, Geoffroy was bidding to put an end to all talk of "perfection" within the field of anatomy.

Whether or not Étienne Geoffroy Saint-Hilaire was absolutely the first to speak of the normal state, there can be no doubt that his work served as a point of reference for its dissemination. In March 1830, the French daily *Le National* contained an article on the great debate in which Geoffroy's position on monsters was summarized accurately enough to be reported later by Geoffroy himself: "Monsters, who have for so long been regarded as strange whims of nature, are simply beings whose regular development has been halted in certain of their parts. It is admirable that an organ in an individual never loses the normal characters of the species to which it belongs unless this deformation imprints on the organ the normal characters of an inferior species."[74] Supported by this specific anatomical usage, the normal was now claiming a place in the everyday, cultivated interpretation of scientific thought.

THE NORMAL AND THE ANOMALOUS

Pursuing a line of inquiry opened up by his father Étienne in a volume that had appeared in 1822,[75] Isidore Geoffroy published a four-volume treatise over the period 1832–37 entitled *Histoire générale et particulière des anomalies de l'organisation chez l'homme et les animaux* (*A General and Particular History of Anomalies of Organization in Man and Animals*). It is of particular interest to our genealogy that Isidore Geoffroy should have made "anomalies of organization" the topic of his research. He announced from the outset his intention to show how "several principles, established on a somewhat unstable base by normal facts, can be fully demonstrated by the study of anomalies."[76] There were, he said, two sets of facts to be examined that were "intimately bound up with each other," the normal and the teratological.[77] This was not the binary that serves in our own time to shape common sense talk about normality: it was not the normal versus the abnormal. Rather, Isidore set up his problematic in terms of the normal and the anomalous, so that "normal organised beings" were thought of in opposition to "anomalous beings." Those were his binary terms: "every determination of an anomaly, every examination of its essential characters comes down in the final analysis to a comparison with the normal order, and therefore to the expression of a relation."[78] Since any individual creature who presented anomalies would, when studied closely, be describable according to "the normal forms of another age or another species," it was never a matter for Isidore of identifying a singular condition that could be characterized as the "abnormal state."[79] By definition, anomalies occurred within species, and what was anomalous in one species was likely to be normal in another. So it followed from the same definition that anomalous qualities could not be aggregated into a singular condition. There was in principle no general abnormal type that could sum them up conceptually. Anomalies were multiple and diverse, and Isidore Geoffroy's self-appointed task as a teratologist was to give a full descriptive account of their range. To oppose the normal to the abnormal would have been to create a false theoretical symmetry, as if the "abnormal state" had somehow been in its own right a countervailing principle of anatomical order. There were at bottom just two things to be understood by anatomical science: the regulatory force inscribed in the normal state and those apparent departures from it that presented as anomalies.

But if anomalies were to be the focus of scientific attention, there needed to be agreement about just how they could be identified. Isidore addressed this difficulty in two quite different ways. The first was an appeal to etymology in the classic manner of learned inquiry. He discussed the Greek-derived *anomalie*, which he preferred to a competitor term used by a German colleague writing in Latin: *abnormitas*.[80] It was important, however, Isidore observed, that the preferred Greek term not be taken in its literal sense, which was "disorderly, not subject to law." He explained that "there are no organic formations that are not subject to laws; the word *disorder*, taken in its true meaning, could not be applied to any of nature's productions."[81] So there was a paradox inscribed in the etymology of his key term, and a practical need to keep the word at some remove from French or Latin equivalents that might prove misleading by their very transparency. Moreover, this effort had to be sustained despite Isidore's general inclination to adopt more straightforward scientific terminology in preference to obscure, Greek-derived neologisms. To put it simply, the "failures of law" named somewhat obliquely as his object of analysis could not prove as a matter of principle to be actual failures. So the appeal to etymology produced a rather complicated outcome.

The complication becomes all the greater if one takes into account Canguilhem's claim that Isidore's understanding of the Greek derivation was inaccurate. Lalande's prestigious *Vocabulaire philosophique*, quoted by Canguilhem, affirms that *anomalia* is derived from *an-omalos*, meaning uneven, not smooth, rather than from the negation of *nomos*, law, as Isidore and many others had supposed. Understood in this light, *anomalia* does not signify the absence of law, merely the absence of regularity. And because the Greek *nomos* and the Latin *norma* are usually taken to be approximate equivalents, the standard mistake about Greek, says Lalande, further blurs the distinction between the concept of law and that of rule, as terms derived from Latin and Greek are found in play together. Given Canguilhem's insistence that the norm should be understood as a form of law rather than just a frequent recurrence, it is not hard to see why he turned to Lalande's etymology for support. Lalande actually states that the Greek-derived *anomal* (anomalous) is a properly descriptive term signifying irregular, whereas the Latin-derived *anormal* (abnormal) is "evaluative or normative." In practice, Isidore did not collapse all difference between law and regularity despite his mistaken belief about the Greek etymology, and Canguilhem gives him credit for maintaining some difference between the two.[82] Nonetheless, it must have seemed clear to Isidore that attending to etymology offered no particular theoretical insight.

Needing a working definition of anomaly for observational purposes, Isidore deployed a form of quantitative thinking. He did not launch wholeheartedly into the numerical and statistical methods that were coming to prominence in other fields during the 1830s, which we will discuss below. But he did attempt to define anomalies by the mere fact of their infrequency or rarity: "*anomalous* means nothing other than *unusual, unaccustomed*. In general, an anomalous being should be thought of as one whose organization stands at a distance from that of the great majority of beings with whom it is to be compared."[83] In speaking thus of frequency and rarity, Isidore was finding an intellectually convenient way through his definitional difficulty, but he was arguably turning away as he did so from the logic that had guided his father's method. Étienne Geoffroy had undoubtedly practiced anatomical mensuration. Having measured encephalic mass in a set of individuals, he was able to compare his measurements with those of certain cranial bones, concluding that "according to whether the encephalic masses stand closer to or further away from the conditions of their normal state, the bones that cover them are affected in a directly proportionate ratio."[84] But while quantitative, these were not assessments of frequency: they were measures of proportion allowing every individual under observation to be referred back to the normal state. Étienne had not attempted to define anomaly as such in quantitative terms.

By resorting to the notion of frequency, Isidore was leaving himself open to a fundamental criticism levelled by Canguilhem at almost all nineteenth-century French writers in the fields of biology and medicine. All those considered by Canguilhem appear to him to have made the same set of conceptual mistakes. They proved unable to respond to biological questions without reducing them to matters of quantity, thereby failing to give an account of life itself. A pathological condition was typically conceived at that time, says Canguilhem persuasively, as mere excess or inadequacy of those very qualities taken to characterize the normal state. There was no qualitative difference between the pathological and the normal, no developed understanding of biological normativity. So was it not a mistake of just the same kind—this time within the field of teratology—to define the anomalous by the mere fact of its standing outside the range of the majority? Canguilhem had no doubt that it was, and that was the main point he had to make about Isidore Geoffroy's work. He quoted a passage from Isidore's *Histoire des anomalies* in which an anomaly was defined as "any organic particularity" in an individual with respect to "the great majority of individuals of its species, its age, and its sex."[85] On this Canguilhem commented: "it is clear that when defined in this way, anomaly considered in general is a purely empirical or descriptive concept; it is a statistical deviation."[86] Canguilhem's own commitment to "the logical independence of the concepts of norm and average" led him to identify this as a form of confusion.[87] Statistics were for Canguilhem a distraction from the central theoretical issue: the task as he saw it was to identify the anomalous in itself without, so to speak, measuring it into existence as a concept by calculating the frequency of its occurrence.

If Canguilhem's normative position is accepted, his critique of Isidore Geoffroy's work cannot be considered unfair, although it does not give a full account of the complexity—or at least the complications—of Isidore's thinking. At times, Isidore simply took statistics as a source of facts. Speaking of the length of fetuses, for example, he drew on statistics of children born at the Maternité hospital in Paris, using the average as a basis for generalization. For his purposes, that served perfectly well to determine what was normal. One and a half feet was the length "that must be considered normal according to the results that have been deduced at the Maternité from the statistical comparison of several thousand cases."[88] He went on to speak in more general terms about the average height of a given population. It was always possible to identify three groups, one "in very large numbers, with the same height, or differing slightly in their dimensions," and two others, "in very small numbers, by contrast, who are either much shorter or much taller. The first are of ordinary height [*ordinaire*], the second dwarves and the last giants."[89] This was no more than intellectual routine for demographers of the time. Isidore was simply using data of this kind as it came to hand, but he was doing so—Canguilhem is certainly right about that—in a manner that was not theorized as it was later to be by some statisticians. Where height was concerned, said Isidore, three types could be identified: "all individuals of the same race, when compared with regard to height, can be reduced to three types: the normal type, the anomalous type through defective or generally arrested development, and the anomalous type by excess."[90] But this was not the concept of anatomical type defined in Étienne's work according to the normal state of organization in a species: it was simply a way of ordering data that identified the middle range of a population, declaring that range to be the "normal type" without ever saying what actually constituted it anatomically as a type. It was certainly not the case that Isidore had no conception of an organizational type. He was given to speaking of *le type spécifique*: the type of the species, the type that, when described, served to characterize a species as such. "It becomes difficult and almost absurd," he said as if to confirm his commitment to building anatomical knowledge using the type, "to admit the existence of several types within a single species."[91] The "specific type" was singular in principle. It was not the product of a calculation. It was not an average.

The unresolved tension between these two different ways of framing the normal was characteristic of—and in a sense constitutive of—Isidore's thinking. While he turned to statistics at various points in order to describe normal properties as the ones that occurred most frequently, he also maintained a commitment to comparative anatomy grounded in the strong theory of the normal state developed initially by his father. A proper method of natural history, he reminded his readers, had to begin with the normal state: "the natural order dictates that we take the normal state as the point of departure and move step by step from the least severe deviations to those that imprint on the organization the most remarkable and extensive modifications. That will mean proceeding from the known to the unknown, and also from the simple to the composite."[92] According to this declaration of methodical principle, the normal state needed to be analyzed at the outset. Only then could anomalies be properly identified as departures from it: "Should not the explanation of the normal state precede that of the anomaly? Should not the rule be in place before the exception is considered?"[93] When one set about it in this manner, there might not in fact prove to be any reliable correlation between frequency and monstrosity. Isidore observed of a particular kind of monster he was describing that, "while obviously more anomalous than the preceding ones," it was "less rare."[94] Monstrosity did not reliably correspond to rarity, and some of the most extreme modifications might occur relatively often. Indeed, when one considered those teratological presentations in which organs were symmetrically transposed across the midline, it might sometimes happen in other species that the inverted, anomalous order became the more frequent one: "The only difference between the normal disposition and the inversion is that one of them presents in the immense majority of individuals of a species and the other in a very small number. In fact, since there is no reason why, of two equivalent states of the organization, one should always be the more common in all animals and the other always rare, there are species in which the inverse disposition is the one that is more ordinary [*ordinaire*]. It is the one that is generally present, and becomes the normal state."[95] In cases like these, frequency could hardly be taken to define the normal state: it was no more than a shifting indicator of dynamic normality.

Isidore Geoffroy deployed statistical thinking of a kind to deal with the question of perfection that had been at issue between his father Étienne and Cuvier, using it to dismiss a theory of monstrosity that dated from the eighteenth century. Charles Bonnet, in his *Considérations sur les corps organisés* (*Considerations on Organized Bodies*) of 1762 had defined monsters as "all organized productions that do not follow the ordinary rules."[96] That definition made no sense to Isidore: "I need make only one remark to show how thoroughly unsound this definition is. If a man were to present the forms, the facial angle and the features that the sculptors of antiquity attributed to the statues of the gods, or if he even came close to those types of physical perfection and human beauty, Bonnet's definition would place him in the category of monsters."[97] Isidore was effectively demonstrating the impact of statistical thinking on more classical views of human anatomy. Bonnet, when he had used the word *ordinaire*, may well have invested that word, as indeed Étienne Geoffroy had later done, with a sense of productive regulation. But Isidore was taking it polemically as a mere reference to frequency. Considered in statistical terms, monstrosity was rare, just as perfection was. For that reason alone, frequency of occurrence could not serve to define either of them in theory. The ideal type needed to be defined in its own right, and monstrosity identified as anomalous with respect to it. It was certainly true that the type of physical perfection was rare, but that was for Isidore a secondary consideration. The type belonged always to a "race," as an ideal inscribed in and fashioned from its best qualities: "For each people, the type of beauty is the type of the race. Each time that men have created the ideal type of superhuman perfection, they have found it in a slight exaggeration of some of their characteristic features."[98] This is where the complication of Isidore's thought allowed him to find a way through. Even as he was committed to the concept of the ideal type as a particular set of anatomical qualities in a given species, he was using statistical reasoning against those he saw as its opponents.

An eighteenth-century version of quantitative discourse may have appeared to be in play when Isidore considered the work of another naturalist of that time, Georges-Louis Buffon. Buffon, when writing around 1760, was widely known to have identified three classes of monsters: "monsters by excess, monsters by defect, and monsters by the wrong position of parts." The number of actual parts in the third class of monsters was "normal," said Isidore in his summary of Buffon's text: it was merely their position that was monstrous.[99] This was an accurate enough account of Buffon's view on the matter, though a somewhat anachronistic one in its terminology. Buffon had indeed produced a tripartite division, but without using the word "normal," which in his time was confined to the geometrical usage discussed at the beginning of this chapter.[100] He had referred to an autopsy performed on a man with "all his internal parts situated the wrong way round [*à contre-sens*]. Those which, in the common order of Nature, are found on the right side were located on the left, and those from the left side on the right."[101] Isidore did not attempt to dismantle Buffon's position. It was, after all, broadly compatible with his own: the "wrong way" identified by Buffon might well have been renamed anomalous, and the naturally occurring "common order" the normal state. But the classificatory method was fundamentally different, and Geoffroy marked that difference in a historically condescending manner. It was all very well in the early stages of a natural science to do as Buffon had done: "That is the path that has been followed initially in all the natural sciences because it is the simplest, and perhaps the only possible one as long as only a small number of facts are known."[102] But a fully developed science of teratology could no longer be content to begin with such a broad classification. What was required now was "a slow and difficult, but assured way of proceeding, moving from genera founded on observation to orders and classes, instead of descending from classes established a priori to orders and genera." The business of classification was now required to proceed inductively, rather than be framed by a set of possibilities based on a priori logic: "the newer method, although quite unable to be applied when the natural sciences are in their infancy, is in fact the only one that can be used when they are progressing as they are today."[103] In other words, the very fact that Buffon had spoken from the outset in terms of defect and excess was taken by Isidore as evidence of the immaturity of the descriptive method he had used. The science of teratology, Isidore was saying in effect, had now moved beyond the stage where broadly quantitative talk of that kind ought to frame its inquiry. Its task was now to produce a scientific description in which differences of type were established on a qualitative basis.

Whereas Canguilhem's normative history leads him to critique Isidore Geoffroy for failing to maintain a properly biological view of the normal, our own genealogy leads us to value the complications and contradictions of Isidore's thinking as an artifact of intellectual history. Canguilhem points to "confusion" in nineteenth-century medical discourse between two notions of the normal state: that of "the habitual state of the organs" and that of their "ideal state."[104] But in Isidore's work the coexistence of the two offered its own form of convenience, allowing him to slip back and forth between two theoretical perspectives and two kinds of scientific rhetoric. Isidore's writing could be called unfussy in this regard. Far from introducing the term "normal state" in the self-conscious manner of Étienne, he seemed prepared to take it as known, using the expression quite unceremoniously. A further mark of conceptual familiarity in his writing is the easy rotation of quasi-synonyms: not just *normal*, but *ordinaire* and *régulier* were rostered for duty, even though the equivalence of those expressions had not been properly established in theory. Isidore would speak, for example, of organs that were located higher or lower in the body than they were *à l'ordinaire*, or he would refer to the "regular state" rather than the normal one.[105] At times indeed, any or all of these adjectives might find themselves being treated as equivalents of the term *moyen* (average). We sometimes find a whole bouquet of terms in one sentence: "the *ordinary* or *normal* height is necessarily the *average* height of this race, or at least it differs so little that there is no drawback here in taking the one for the other."[106] The tendency in Isidore's writing was not in fact to produce a closely argued convergence of explicit anatomical theories drawn from his father's work and implicit theories of frequency drawn from statistics, but to treat those two kinds of thinking discursively as if they were approximate equivalents. Ease of movement between the two—the normative and the frequentist—is not, as we will come to show, a quality unique to Isidore Geoffroy. The normal came to be thinkable in the nineteenth century and into the twentieth through the loosely theorized cohabitation of these two modes of thought. To know the normal was often to know it as somehow both "ideal" and "habitual," or alternately one and the other. The loose coexistence of the two appears to have been constitutive of the normal as a figure of knowledge.

THE NORMAL STATE IN PHYSIOLOGY

The term "normal state" appeared in the field of French physiology around 1828, about ten years after its first appearance in comparative anatomy. There was no high-profile debate, no discussion of the term advocating or justifying its adoption. Rather, it was used freely from the time it first appeared, to the extent that freedom of usage quickly became one of its discursive characteristics. It is clear, looking back, that for physiologists the normal state was both topically compelling and lexically accessible. In this section, we will attempt to understand how that came to be so. There is some evidence that the new term was able to be taken up so readily because it drew on other expressions that had been used in physiology over the preceding decades and was able to include their signification within its own. To assess that evidence, and to determine just what kind of intellectual history it calls for, we will first go back to the beginning of the nineteenth century, to the work of Xavier Bichat, a pioneer physiologist whose great reputation was sustained through the decades that followed.

In his *Recherches physiologiques sur la vie et la mort* (*Physiological Investigations into Life and Death*), first published in 1799, Bichat distinguished systematically between two kinds of life. The Geoffroys made no primary distinction between organs according to function, considering that any concern with function was a distraction from the proper tasks of comparative anatomy, but the study of functions was precisely the domain of physiology as framed by Bichat. His scientific interest was in life as such, and his key theoretical contribution was to distinguish between "organic" and "animal" life. Both kinds were at work within the human body, he said, the former in organs with a continuous function such as the heart and the liver, and the latter in such organs as limbs and the eyes. By contrast with organic life, animal life acted in discontinuous ways, under the influence of the will, and needed rest. Organs operated by animal life were also able to function independently of each other, whereas those with organic life could not.[107] This led Bichat to have a particular view of the concept of perfection, a term and a notion that served us earlier in this chapter as a kind of litmus test for historical theories of the normal. We will apply it again here.

Bichat drew attention to "the perfection of the animal forms," which was, he said, visible in "the symmetry generally observed in the various organs." Symmetry was a property of animal life, such that "whatever disturbs this symmetry will cause an alteration in those functions."[108] Bichat was allowing in principle for occasional disturbances of the symmetrical pattern, although he had no scientific reason to be interested in monsters. His physiological point was that symmetry was required if the functions of animal life were to be properly carried out. The "perfection" to which he referred was not the finely proportioned geometry invoked by Isidore Geoffroy and others as the classical Greek model of sculptured beauty. It was a living, functional perfection, but a thoroughly precise one for all that: "everything is exact, precise, rigorously determined in its form, size, and position."[109] This set of properties belonged, however, to only one class of organs. It was not characteristic of "the organs of inner life." Anyone who practiced dissection knew, to quote just one of Bichat's examples, that "the organs of absorption, the lymphatic glands in particular, rarely display in two individuals the same proportions of number, of volume, etc."[110] So there was for him a marked contrast between the geometrical regularity of animal life and the "frequent aberrations," indeed the "irregularity," found in organic life.[111] He observed that "nature is far less given to deviations of form [*écarts de conformation*] in animal life than in organic life."[112] So there was not just one kind of order in the body, but two: "harmony is the character of external functions ... discord, on the other hand, is the attribute of organic functions."[113] Whereas for Cuvier's comparative anatomy the perfection of human organization had functioned as a definitional given, Bichat found physiological perfection in only one of the two classes of human organs. The "natural state" of humans thus involved a quite fundamental complication of symmetry and asymmetry, order and discord.[114]

When Canguilhem comes to examine Bichat's influential essay, he points to the term "natural state," observing that Bichat was using it where later writers would use the term "normal state."[115] We are going to argue here, as might be expected following our earlier discussions of Canguilhem's work, that this comment is not so much wrong as misleading. Bichat did say, for example, when speaking of a change whereby inflammation subsided: "it becomes gradually milder and returns to its natural degree."[116] Where Bichat says "natural degree" or "natural state," later writers might well have said "normal state," but we are wary of collapsing the historical distance between the terms. We will show in this section that the term "normal state" did in fact come to stand somewhere near the place of some earlier expressions like "natural state," but we will not suppose that it did so because the thing was already there waiting for the word to join it. Our present example can serve to make that point. The word "normal," as Caroline Warman has shown, was available in Bichat's time only in its geometrical sense, and could hardly be used to name the natural state as he described it. Its geometrical sense could have applied well enough to animal organs, which were indeed characterized by their regularity, but no use of the geometrical term, no matter how liberal or how figurative, could make it serve to characterize "irregular" internal organs. Bichat could only have spoken of the normal state if that term had seemed to him capacious enough to include both the organically regular and the organically irregular. At the time it did not have that capacity.

When speaking of the unaltered state, Bichat used a number of other adjectives that also had a place in comparative anatomy. The most prominent of these was *ordinaire*. We can circumscribe his usage of that word by noting a nonmedical example in his writing. At one point, he spoke of breathing "l'air ordinaire," literally "ordinary air," as opposed to breathing oxygen, so that *ordinaire* signified "occurring standardly in nature."[117] When speaking directly of physiological matters, he was able to use the adjective in the same way, particularly by contrast with exceptional phenomena. A certain fluid was said to be "foreign to the arteries in the ordinary state" and "white organs" were ones into which "blood does not penetrate in the ordinary state."[118] Furthermore, the quality of ordinariness could be conjoined with that of regularity. Together they were characteristic of life: "life connects up its phenomena with their ordinary regularity."[119] Bichat also spoke of the "habitual." The flow of arterial blood, for example, produced "habitual arousal" in the organs that depended on it.[120] This talk of the ordinary and the habitual ought not to be conflated with the frequentist thinking that informed the practice of statistics. Bichat, as Canguilhem points out, mistrusted statistics, and when he spoke of the habitual, there was no assessment of a majority of cases and no calculation of averages.[121] Whatever the habitual was exactly, it was a quality of life, a quality that allowed life to be identified and described over time. And the regularity to which Bichat referred was not necessarily symmetry, although it sometimes took that form. It included both the harmony of animal life and the productive discord of organic life. The "natural state," along with ordinary forms of perfection, included things that were ordinarily irregular.

The discursive cohabitation of these key adjectives ought to be read as an artifact of history. In supporting and constraining each other, they constituted a kind of professional repertoire for physiology, and our genealogy requires us to understand how they functioned as a set. *Naturel*, *ordinaire*, and *régulier* all dwelt in the conceptual neighbourhood where *normal* later came to hold a place, and all helped to map the signifying space in which it emerged. So how was the adjective *normal* able, at a particular point in time, to sum up their various qualities to the extent that it did, becoming the most prominent, the most widely used of all? Caroline Warman, in an article on the history of the word referred to earlier, asks a question that is of close interest to us here. Focusing on physiological usage, Warman puts forward the challenging—and in some ways awkward—notion of the "pre-normal." The term is theoretically unwieldy for the reasons we stated earlier when objecting to Canguilhem's talk about "the thing without the word." What can it mean to speak of the concept of the normal before the word comes into use, given that, as Warman herself observes, "we cannot inspect a concept when it has no terms to support it"?[122] We will not use Warman's term, but we will indicate why it might have seemed appropriate for her purposes, and why we are allowing it to linger here. She seeks to understand "the conceptual and linguistic conditions that prevailed just before 'normality' became a concept that was used to indicate standardisation and regulation," and it is the "just before" that seems to us most worthy of attention.[123] In what follows, we will examine a series of texts published over a period of twenty-five years by the high-profile physiologist François-Joseph-Victor Broussais, with a view to tracing the emergence of the adjective "normal" in this proximate manner. We will examine four essays on physiology published by Broussais over the period 1803–1828. Only the last of these contains the term "normal state."

The first essay ever published by Broussais was entitled *Recherches sur la fièvre hectique* (*Investigations into Hectic Fever*). It appeared in 1803, only a few years after Bichat's. Hectic fever was akin to tuberculosis, and Broussais described cases of the illness that he had encountered while serving as a doctor in the French army. He did not in fact dwell on or otherwise exploit the distinction between organic and animal life, although he did speak deferentially of Bichat.[124] But he did continue to build the set of approximately equivalent adjectives that, we argue, served to support and constrain the word *normal* when it eventually emerged in his writing. Near the beginning of the essay there was a tribute to a medical figure whose reputation was not sustained in later decades. Jean-Noël Hallé, who had taught Broussais at the Paris École de Médecine, was acknowledged here for the conception of his course on "hygiene" in which a distinction was made that is not developed in Hallé's published work, but was highlighted by Broussais and deserves to be highlighted in our history. Hallé distinguished between "the physiology of the healthy man and the physiology of the sick man." His general point as understood by Broussais was that "the order that nature shows in resisting the agents tending to destroy it appears to him no less admirable than the order it uses to maintain the felicitous harmony that constitutes health: *order can be observed in disorder itself.* That is a valuable idea that can be applied to all the phenomena of nature!"[125] In other words, the physiology of the healthy person and that of the sick person resisting destruction could be understood in the same systematic terms. The order of nature was present across the full range. Just how "valuable" this idea proved to be historically can be seen when we consider that much the same point was made in the field we have already considered. That was in essence what the Geoffroys were arguing in comparative anatomy when they sought to make scientific sense out of disorder and put forward the concept of the normal state to ground their general theory. For the Geoffroys, an individual being that could not be aligned with the characteristic type of the species could nonetheless be described analogously in terms of a normal state of some kind. So it does appear that something of the same pattern was now appearing in physiology at a high level of generality. Insofar as physiology sought to accommodate the pathological and the healthy within the same theoretical framework, it may have eventually found in the normal state the conceptual means to do so.

We can begin to gain a clearer idea of the concepts available to physiological thinking at the beginning of the nineteenth century by looking at the shifting terminology used by Broussais to characterize pathology and its opposite. Just what kind of opposites were these two things, in fact? If the physiological order manifested in illness was the same in principle as the order at work in the felicitous condition of health, then health could hardly be thought of as the mere absence of all pathology. Rather than perfection negatively defined, health needed to be thought of as a set of functions at work in healthy bodies, functions that were transformed during illness without being simply abolished or "destroyed." To affirm the orderly nature of pathological phenomena was to give up a static notion of health. Broussais did not directly confront this set of difficulties until his later work but, from the time of his first essay, he used expressions that brought him into contact with them. In his essay on hectic fever, he did not speak of the natural state, as Bichat had, and he certainly did not speak of the normal state. But he did refer often to health, attaching adjectives to the noun *santé* that effectively constrained what it could mean in his work. He spoke, for example, of one patient's "première santé," his initial health, as the starting point from which an illness could be identified as such, and of another's "confirmed state of health" as the conclusion of his illness.[126] In the simplest terms, "health" was being used to mark the temporal boundaries of sickness. Broussais also used the word *parfait* (perfect) in this context, but he used it in a manner that rather changed its meaning, adapting it to the requirements of a theory that understood health and pathology in the same terms. He spoke of "l'état de parfaite santé" the state of thorough, unmistakable health, rather than *l'état de santé parfaite*, the state where health was perfect. What mattered to him was to identify the condition in which health was "perfectly" visible to clinical observation.

More than a decade later, in 1816, Broussais published a much more polemical essay, adopting the argumentative manner that was to characterize all of his subsequent writing. The work in question was entitled *Examen de la doctrine médicale généralement adoptée, et des systèmes modernes de nosologie* (*An Examination of the Medical Doctrine That Is Generally Adopted, and of the Modern Systems of Nosology*). A noteworthy change with respect to the 1803 essay was not so much the emergence of new adjectives as the fact that the key noun to which they were attached became itself an object of close attention. Broussais insisted that the very concept of the "state" (or condition) was fundamental to clinical method. Other doctors, he claimed, were given to making a quite surprising mistake. They were failing to carry out a task that ought to have been central to their clinical practice: "How can it be that, in cases where the morbid state is acknowledged by all doctors, they do not even have the idea of comparing the influences of these organs in the state of health with those that are before them in the cases under observation?"[127] Logic dictated a comparison of states: the "state of (perfectly visible) health" needed to be compared analytically with the "state of illness" or the "morbid state."[128] The "state" was for Broussais a fundamental unit, the observational gathering place of physiological knowledge. But there was one striking new adjective added to the set of words used to qualify it: *l'état physiologique*, the physiological state. This is the first of many uses of that term in Broussais's published writing: "I am prepared to prove that life can be excessive [*en plus*] in one organ at the same time as it is deficient in several others. I shall no longer do this by resorting to violent phlegmasias, which so often produce a very real weakness in the muscles and the pulse. I shall do so by turning to the physiological state."[129] Here, as elsewhere in the language of Broussais and his colleagues, the "physiological state" did not mean a patient's current state assessed in physiological terms. It was a general concept opposed to that of the "pathological state," one that proved to be an increasingly successful lexical rival for *l'état de (parfaite) santé*. While the term may appear oddly and unhelpfully reflexive to modern readers, it served around this time to name, not a state defined by the absence of any pathology, but one defined by the undisturbed functions of life.

In 1822, six years later, Broussais published the third edition of an essay entitled *Histoire des phlegmasies ou inflammations chroniques* (*A Classification of Phlegmasias, or Chronic Inflammations*). The second edition of this text was contemporaneous with *Examen de la doctrine*. It is interesting to note when one is tracking changes over time in Broussais's terminology that the third edition actually contained footnotes commenting critically on expressions used in the second edition. He remarked for example that the word "invasion" was misused in the description of a change in one patient's condition: "'exaspération' [violent irritation] is the word that should have been used."[130] But at no point did the "normal state" appear either in the body of the text or in the footnotes. That adds weight to our hypothesis that the term "normal state" did not appear as a correction of earlier terms, but rather served to confirm and sum up their proximate equivalence. "Normal" appears to have emerged eventually as a *primus inter pares*, as if to stand for a whole paradigm of adjectives. The terms that had appeared in the earlier essays simply held their place in the essay on phlegmasias: he referred to a patient's "habitual physiognomy," to another's "ordinary" degree of plumpness and, quite straightforwardly to "the state of good health."[131]

The final essay we will consider here is in fact Broussais's magnum opus, *De l'irritation et de la folie* (*On Irritation and Madness*), published in 1828. Reference to the normal state appeared in this work at every turn although there was no initial moment of solemn definition. It was as if the normal had become available to Broussais in the intervening six years since his last publication and was now regarded by him as a widely shared concept—so much so that, far from reserving the term "normal" for his own physiology, Broussais even imputed it to his adversaries when they did not in fact use the word themselves. The adversaries in question were a putative school of philosophers, lumped together under the label "Kanto-Platonicians," about whom Broussais declared that they were preoccupied with the normal, although not in the proper manner. These people, including France's Victor Cousin, were said to be mistakenly attempting to build knowledge of humanity on the basis of "normal physiology." They were contemplating humans in the healthy state without realizing that "anyone who has studied only normal physiology is not in possession of enough facts to solve these problems. Man is only half-known if he is observed in the healthy state alone. The state of illness is as much a part of his moral existence as it is of his physical existence."[132] This was not so much a dispute within philosophy as an aggressive refusal of contemplative or critical philosophy in the name of physiological medicine. As it happened, Cousin himself had shown no particular interest in physiology as such, and did not in fact use the word "normal" at all except to refer occasionally to a normal school. Nonetheless, he and all others of his ilk were said by Broussais, even as they were supposedly preoccupied with the normal, to understand it as mere "ontologists." As such, they were *étrangers*—foreign to and ignorant of—"normal and abnormal knowledge."[133]

What was needed as an antidote to the semi-ignorance of such philosophers was a particular form of knowledge that coupled the normal and the abnormal, understanding one in relation to the other. That happened to be the business of physiology, and of doctors, who were required to understand illness as such: "Establishing a precise idea of *illness* is the primary aim of the doctor."[134] So at or near the point where the normal emerged in physiological discourse, it did so by asserting the normal and the abnormal as a pair of concepts. That was rather different from what had happened in comparative anatomy, where the Geoffroys fought shy of the notion of the abnormal: they were suspicious of what appeared to them a false symmetry between the normal, which they had carefully defined, and the abnormal, which did not figure in their writings. Broussais, by contrast with the

Geoffroys, focused on physiological states and made the normal-abnormal binary central to his thinking. Even as he was imputing a concept of the healthy normal to his opponents, he was saying that they had not understood its significance because they had not grasped the importance—and the epistemic equivalence—of the abnormal. Whatever they might learn about the normal would count for barely half unless it was complemented by the study of the pathological state. There was in Broussais's view a battle to be fought against the Kanto-Platonicians by medical thinkers drawing on a native philosophical tradition that, while taking Locke as one of its references, had been developed in France by Cabanis and Destutt de Tracy.[135] That kind of philosophy, practiced by the exemplary doctor-philosopher Cabanis, was called to the support of physiology as it struggled against the foreigner Kant and his French agent Cousin.[136] Instead of "internal observation" and "metaphysics" as practiced by Kant and Cousin, what was called for was "external observation," the great art of the doctor described by Cabanis. Everything located by *psychologistes* like Cousin in the domain of the intellect could and should be redescribed within the French tradition of materialist philosophy. The intellect would then prove to be no more than "the excitation that is the normal state of the nervous-encephalic apparatus."[137] The material truth of humanity was to be known through the senses, and the fullness of knowledge could only be achieved by observing the normal and the abnormal states in their complementarity.[138]

Broussais's emphatic use of the normal-abnormal pair did not simply bring to an end his use of the other terms that we have been tracing, although they now appeared much less often. Certain of them were still occasionally present, notably derivates of the word *parfait* (perfect) in its lesser sense. At one point, nerves were considered "in their state of perfect development," where *parfait* was used to mean "complete," "accomplished," "clearly recognizable."[139] A little further on in the text an organ was characterized as "healthy and perfectly developed."[140] But the summary point in this part of his argument was made by referring to "the normal state," which now tended to displace the references to health, be it "perfect," "primary," or "confirmed," that had abounded in the earlier essays.[141] If frequency and prominence of usage are taken as a guide, Broussais was clearly persuaded that "the normal state" was the most economical and accurate expression for his purposes, and that it had rendered most of the others redundant.

Broussais was able to think of the normal and abnormal states in binary terms and to insist on their epistemic coupling because he supposed that both sets of phenomena were of the same order. Throughout his militant advocacy of binary thinking and his insistence on the importance of contrastive clinical observation, Broussais was thoroughly committed to understanding the relation between sickness and health, not in terms of fundamental otherness but as a matter of degree. That was the whole point of his theory of irritation. It allowed him to describe healthy and pathological states in the same terms, using the same framework of reference from start to finish. By focusing on irritation, he claimed to have established "a regular system applicable to health as well as to illness" in which "irritants" were defined as "all the modifiers of our [physiological] economy that increase the irritability or sensitivity of the living tissues, and raise these phenomena above the normal degree."[142] *Excitation*, said Broussais, was a general phenomenon present in living bodies. From one body to another and over time within a given body, it differed only by degrees. Illness and health could thus be located on the same scale, and the supposed irregularity of illness described in measurable terms. A pathological state did not involve any profound alterity: it could simply be understood as a variation, and its intensity assessed by distance from the normal state. That was the order underlying apparent disorder, and that was how the description of pathological conditions could be theoretically integrated into physiology.

When illness and health were understood thus, the normal state was not to be thought of as the acme of perfection, but simply as a certain *degree* of activity: "whenever arousal [*excitation*] or stimulation move *beyond the limits of the normal state*, they can be qualified with our term 'irritation.'"[143] In Broussais's writing, the normal state was identified in principle, but it was not closely bounded by definitional work. That kind of attention was not called for in principle, we must suppose, since the normal state was a range rather than the singular point of primary health or the zero point of pathology. So it was entirely appropriate that the concept itself should be relatively capacious. That was how Broussais could fail to define the normal as a primary state—and indeed appear to reject the possibility of doing so from the outset as "Kanto-Platonician"—while continuing to use the term confidently and productively. *Excitation* was a property of living bodies, while excessive *excitation* was of itself sufficient to produce a pathological state: "After setting out the phenomena of arousal [*excitation*], we are led to ask how this arousal can deviate from the normal state and constitute an abnormal or sickly state."[144] A change in degree produced a change in state, directly and unproblematically. In strict logic, Broussais's definition of the normal state was not properly grounded. How could the abnormal be recognized if the normal range from which it departed was not defined in theory in the first place? And even if the normal was understood in this convenient manner as a range, where did the normal finish and the abnormal begin? These are strong, perhaps intractable questions that continue to beset claims to know the normal in our own time. We will return to them throughout our history, but our interest here is first of all in understanding how it was possible in practice for Broussais and other medical thinkers of his time to take up the notion of the normal state without perceiving any need to define it from first principles. Étienne Geoffroy had provided such a definition, but his use of the term appears to have been less influential in the long run than Broussais's enthusiastic binarism. That the question of definition was not asked by Broussais and many who followed him should not be considered a simple matter of neglect. Taking the answers for granted and folding them into the ongoing use of the normal-abnormal binary was precisely how the theme worked in this context. It can be said, in fact, that the nondefinition of the normal actually proved enabling when the normal state emerged as a figure of medical knowledge. The very ease with which it was referred to contributed to its signification: it became discursive routine—a habit that was taken to reflect what was "habitual" in life itself.

Canguilhem draws attention to Broussais's failure to define the normal, and attaches considerable significance to it. The issue as he sees it is essentially philosophical. Vitalist thinkers in the Montpellier tradition, said Canguilhem, had asserted that there was a qualitative difference between the "normal phenomenon" and the "pathological phenomenon."[145] They did not in fact use the word "normal," but we have already made our point about that. The further issue raised here by Canguilhem is that Broussais's physiology denied any qualitative difference between the normal and the pathological, attributing all illness to "simple changes in intensity in the action of stimulants that are necessary for maintaining health."[146] As a consequence of this move, said Canguilhem, the "ambition to make pathology and therapeutics equally scientific by having them follow simply from a science of physiology put in place beforehand" was doomed to failure. That enterprise could only succeed "if in the first place a purely objective definition of the normal could be offered as a fact, and if in addition all differences between the normal state and the pathological state could be translated into the language of quantity."[147] As long as Broussais failed to offer a qualitative definition of the normal, and accordingly a qualitative distinction between the normal and the abnormal, he could not succeed in providing a scientific account of life itself—that is, for Canguilhem, of life understood as productive of norms.

What Broussais shared with Isidore Geoffroy in particular, and what made them both representative of thinking about the normal state in the 1830s, was precisely an intellectual complication that Canguilhem finds philosophically regrettable. Both of them spoke of the normal in terms that Canguilhem could not consider theoretically sound. He is right to say that they put forward claims about the normal that suppose a qualitative difference between the normal and the anomalous, while actually relying on quantitative thinking in order to carry out the everyday business of teratological and physiological science. But even if this can be construed as a philosophical mistake, it was a pattern that recurred when French scientific writers first began to speak of the normal state. This is a key historical point, as Canguilhem recognizes in his own way.[148] The largely unresolved equivocation between qualitative and quantitative thinking was itself a characteristic figure of knowledge, and a challenge for our genealogy is to understand how it came to be productive, albeit in a philosophically maladroit way. It was usually a matter at this time of finding order in apparent disorder by positing the concept of a regular monstrosity or a regular pathology, and the concept of the normal state was put forward as a way of understanding such regularity. There were laws of nature, not yet fully known to science, that determined the composition and functions of living bodies. But laws of another kind were also in play, ones that took frequency or the habitual as givens for calculating the normal. Regularity of the latter kind might be subject to mathematical calculation via the practice of statistics, and it was by no means clear what the first kind of regularity had to do with the second. Uncertainty about the relative claims and the possible theoretical articulation of these two kinds of law was widespread in medical thinking. In the following decades, as our next two chapters will show, it seemed to be largely a matter of accommodating these difficulties and of dwelling within them, rather than overcoming them.

CONCLUSION

We have now found a historical beginning for the modern conception of the normal, but we have not found it in the form, or in quite the place, that might have been expected. That may be in part because we have chosen not to engage in a history of governmental normalization, even though that is a topic around which other studies might well develop. For our part, we want to examine the normal as a figure of knowledge applied to organs and states of the body, producing vital material effects. For such a history, medical writing can readily serve as a locus of inquiry, and we have followed the lead given by Canguilhem in his study of nineteenth-century French texts of that kind. But while sustaining a conversation with Canguilhem's work, we have found our history diverging from his in a number of ways. One of those has to do with the centrality of binary thinking. Canguilhem shows persuasively how the normal and the pathological functioned in French physiological writing of the nineteenth century as a mutually dependent, self-defining pair. This meant, as Canguilhem points out vigorously, that there was no first-principle, fundamental definition of the normal within physiology. But we have found a different, somewhat earlier understanding of the normal at work in the field of anatomy. Among comparative anatomists, there emerged a view of the normal organ according to which the normal was understood in its own right, defined by its functionality and its degree of structural integration. Once the normal state of an organ was thus identified, anomalies of various kinds could be observed, and classifying them became the central business of teratology. But there was no point for teratological science in pursuing binary thinking by gathering the various discrete anomalies into some general category of abnormality. "Abnormality" did not name anything that was of interest to comparative anatomy at that time.

Anatomy and physiology did, however, have something important in common. In both fields, the concept of the normal served largely to displace that of perfection. Instead of taking the supposed perfection of human forms as the key term of reference for comparative anatomy, it became possible to think of normal states and normal organs within each species. And instead of perfect health as a fixed condition of the body, it became possible to think of the normal state as one in which the physiological capacity of a body was at work in sickness as actively as it was in health. As both anatomists and physiologists came to speak of the normal state, the ideal in its perfection came to be displaced by something more mobile and more complex. So to the extent that the normal was marked or defined by stability, stability was in fact achieved by the developing organization of forms and the ongoing play of forces. The normal was conceived in the medical sciences as a dynamic tendency to fixity.

In this chapter, we have encountered for the first time an issue that will return throughout our study: the unresolved coexistence of quantitative and qualitative thinking. Canguilhem, as we saw, deplores the fact that the French medical writing typically failed to hold to a "properly" qualitative understanding of the normal, falling back as it did on quantitative notions like the usual, the habitual, and the most frequent. But our historical point is that this very mixture of kinds of thinking was, almost from the first, one of the most striking characteristics of talk about the normal. The equivocation between the qualitative and the quantitative, the oscillation between them functioned then, and doubtless continues to function, as the constitutive complication at the heart of our theme. The medical academy attempted, with no great success, to rid itself of that contradiction by engaging in a public contest between the two kinds of thinking and two corresponding views of clinical practice.

CHAPTER TWO

"Counting" in the French Medical Academy during the 1830s

The concept of the normal, particularly as it emerged in the writing of Étienne Geoffroy Saint-Hilaire, was a qualitative one, serving to describe a state of organic development in which anatomical, organizational fixity was achieved and maintained. But notably in the work of his son Isidore, this qualitative understanding of the normal dwelt alongside a quantitative one that tended to speak of the normal as that which occurred most frequently. Insofar as quantitative thinking played a role, it then became possible to turn to statistics as a source of scientific knowledge. But how such calculations were to be properly made and how they might be applied in the critical field of medicine were matters of contention and uncertainty during the 1830s. That is why our genealogy now finds itself drawn into questions that have to do with the history of statistics. It should be noted, however, that our historical observations about statistics will be given purpose by our history of the normal; we are not engaging in a long-term account of how medical statistics—and a fortiori statistics in general—became what they are today. We will in fact take particular care to avoid describing the period of the 1830s as if it deserved to be known only for some failure to anticipate modern thinking. It is patent that medical statistics in the 1830s was not the refined scientific discipline that it is today, but we see little point in describing the key debate of the time as a stumble along the road of progress. Our assumption is simply that close study of the issues as they presented then is the proper business of intellectual history. The aim of our study is not, of course, to find retrospective solutions to the problems raised at the time. Quite the contrary, in fact. We expect to show how and why those problems proved to be so thoroughly intractable that the positions adopted in the 1830s continue, mutatis mutandis, to be more or less available and adopted today as conflict continues to be sustained and renewed around them.[1] We will be arguing that a discourse about the value of statistics for medicine was cognate with a discourse of resistance to statistical thinking, so that a long-term intellectual history ought to give an account of the ongoing struggle between the two. Alain Desrosières comments in his own history of statistical reasoning that it is appropriate for historians of science to focus on "uncertain paths and moments of innovation."[2] We have chosen to focus on this debate because it seems to us that the path of statistical medicine in France was never more uncertain than at this point, and that innovation was everywhere accompanied by questioning resistance.[3]

No scientific debate around this time was grander or more resonant than the Geoffroy-Cuvier debate, which took place in and around the Académie des Sciences, but the Académie de Médecine was a decidedly more polemical place than the Académie des Sciences, if a somewhat less prestigious one. As George Weisz points out nicely, "the passionate debates for which [the academy of medicine] was best known reflected the strengths, weaknesses, and disarray of medicine in the mid-nineteenth century."[4] Sustained argument about therapy, nosology, surgical procedures, and the general conduct of the profession was the standard fare of this group, as the record of its proceedings shows. And while no one in the Académie de Médecine enjoyed the public standing of a Cuvier or a Geoffroy, there is clear evidence that these debates were not simply the bickering of a disengaged elite. News of their discussions provoked a flow of correspondence from the provinces and sometimes drew comment from abroad.[5] Academic debate of that specific kind was, as it happened, one of the main ways in which French medicine transacted its intellectual business.[6]

The debate that will concern us most here took place over several months in 1837, being carried over as the main item of business from one session of the Academy to the next. Participants and observers had no doubt that the issues at hand were of great moment, and that view was shared well beyond the borders of France.[7] The *London Medical Gazette* published one of the key speeches, and the *American Journal of the Medical Sciences* published several of them, commenting that the subject was important.[8] When the debate had finally come to a conclusion after several months, its significance was enthusiastically acclaimed in the *Gazette médicale de Paris*, the leading weekly circulating among the French medical profession, which had been regularly filled with reports and opinion pieces on the subject. "This discussion," said the editorialist, "will continue to be celebrated in the annals of the Academy."[9] It must be said, however, that historians have not borne out that prediction, at least insofar as they have shown no inclination to celebrate the debate for its own sake. Rather, they have assigned it a place within a narrative of progress. Some see it as a first, decisive step along the road to modern medical statistics, while others see it as something like the opposite: a formal expression of collective professional caution and a pretext for long delaying the systematic use of numbers in French medicine. Our own view is that the debate as such deserves the attention of historians precisely because it marks a strong moment of intellectual agitation and professional irresolution. That is just the kind of event that ought to claim the attention of a critical genealogy like ours. To put it as simply as possible, the issue of statistics was trouble for medical science, and we want to understand why.

We are conscious of being out of step with almost all the existing histories of this debate because we do not accept a key view that they appear to hold in common. They suppose, quite reasonably, that the history of science ought to be concerned with decisive steps taken toward the present state of knowledge, or indeed with opportunities for progress that might have been missed because conservative forces mobilized against change. It is noteworthy that, within this general framework, the 1837 debate has been interpreted by some as a story of progress gained and by others as a story of progress denied. The most conventional, and usually the least detailed accounts tend to describe the leading figure of the numerical school as a great innovator, while more careful histories sometimes hold up for examination the supposed conservatism of his leading opponents.[10] We will not contest those accounts in their own terms: we will simply refuse to engage with the linear conception of history that gives them shape.

That does not mean that progressive histories hold no interest for our genealogy. A study by J. Rosser Matthews entitled *Quantification and the Quest for Medical Certainty* is a case in point.[11] Matthews focuses, just as we will here, on "debates over the use of comparative statistics in a medical context," but his interest is in "certain epochal transformations within the history of Western medicine and science.[12] The purpose and the outcome of his history are found in the "triumph of the clinical trial as a standard procedure."[13] Our concern, by contrast, is not with the historical becoming of the modern randomized clinical trial, nor indeed with anything that might be called a triumph. We are interested in the particular circumstance of knowledge in which fellows of the French Académie des Sciences and Académie de Médecine found themselves in the 1830s as they engaged in strenuous argument about the relevance of statistics to medicine. What matters most for our genealogy is to discern the assumptions about statistics that were available then and there. The fraught discussion that went on around statistical medicine, we will eventually claim, displays the uncertainty and the difficulty of understanding the normal in quantitative terms at that time, within that disciplinary framework.

Matthews argues persuasively that speakers on both sides of the debate in the medical academy were ill informed. The "numerical method" attacked by some and defended by others, he says, did not properly reflect the most advanced thinking about statistical matters, which was proceeding in another place: "detailed knowledge of the subtleties of probability theory did not circulate among Parisian physicians. It was being developed at Polytechnique by engineers for the mathematical and physical sciences."[14] To put his view a little crudely, the doctors all got it wrong. They did not really know what they were talking about. But if indeed they all got it wrong—and we have no reason to disagree with Matthews if his terms of reference be accepted—the self-appointed task of a genealogical history is to examine their shared assumptions, mistaken as they may appear in the light of contemporary science, in order to understand the intellectual work that was going on at the time. Insofar as the ideas in circulation were shared by almost everyone in the medical field, they deserve our full attention. Erroneous they may have been, but they were certainly not without moment.

Having no wish to launch a polemic against progressive histories such as Matthews's, we simply reaffirm our allegiance to history of a genealogical kind. In developing our analysis, we have turned for guidance and for reference to a number of histories, not of medicine, but of statistical thinking. Intellectual historians Lorraine Daston, Alain Desrosières, and Ian Hacking all deserve in that regard to be taken as counterweight examples to Matthews's progressive history. It has to be said that none of them has actually paid sustained attention to medical statistics as Matthews has done, but all have aided our understanding by their work in an adjacent field.[15] Daston's work deserves attention first of all for the fact that it identifies and traces a historical formation, what she calls "the classical theory of probabilities," that was taken as known by most participants in the 1837 debate.[16] Daston shows that there was a shift during the first decades of the nineteenth century in the manner in which analyzing probabilities and making mathematical calculations about them were understood. We are persuaded by her argument, and will recast it here in terms that are applicable to our medical debate. Hacking and Desrosières take up the same historically located distinction, using it to do conceptual work to which we will return in due course.

Whereas Matthews's work seeks out adumbrations and antecedents of current medical practice, Daston attempts to trace an intellectual formation—what she calls "the classical theory of probabilities"—in the form in which it emerged in the eighteenth century and remained in play even as it was being reshaped during the nineteenth. She does discuss an academic debate that took place in the Académie des Sciences in 1835, although she refers only in passing to the academic debate on medical matters that will be the principal focus of our analysis in this chapter.[17] As it happens, her interest in the 1830s is not in the intersection and interference of statistical thinking with medicine. She focuses instead on the work of Siméon-Denis Poisson who, during the period 1820-40, further developed the calculus of probabilities at the École Polytechnique, applying it not just to physical measurements but to legal judgments.[18] Daston's most valuable contribution from the point of view of our study is to explain with care what "probability" meant at the time and how it stood in relation to certainty.

THE CALCULUS OF PROBABILITIES

A first lesson to be learned from Daston's work is the value for intellectual history of using the terms that were available at the time for thinking about what we now call statistics. The "calculus of probabilities" was in fact the most salient of those, and we will make it our first topic.[19] There is undoubtedly for many modern readers—at least for those unfamiliar with mathematics—an apparent contradiction inscribed in the expression itself. If one supposes, as nonmathematicians often do, that mathematics works with fully determinate entities and precisely defined quantities, it is not at all apparent that probabilities can be subject to any sort of calculus. But mathematicians in the seventeenth and eighteenth centuries undertook that task with confidence, and had the means to do so. The line of inquiry they followed found its most distinctive outcome in a theorem worked out by Jakob Bernoulli, published posthumously in his *Ars conjectandi* of 1713. Our interest here is not in the mathematical niceties of Bernoulli's work, but in its general significance for intellectual history. So we will proceed by translating the theorem into lay terms. The simplest way to do that is to consider a concrete problem that was the object of regular attention at the time. Let us imagine, Bernoulli and his colleagues were wont to say, an opaque urn containing a large number of balls, some black and some white. And let us say for the sake of argument that there are twice as many black balls in the urn as there are white. Bernoulli's theorem provided a way of calculating the chances of taking out a black ball rather than a white one. It was not a matter of removing the balls one by one, constituting two piles, and eventually establishing that the black balls outnumbered the white ones by a ratio of two to one. That would have been mathematically trivial. Nor was it a matter of predicting what would happen on any one occasion when a ball was withdrawn. That was deemed to be mathematically impossible. Rather, if a ball was taken out, identified, then mixed once again with the others in the urn, and that action repeated time after time, Bernoulli stated, the ratio of black ones to white ones in the cumulative count would itself be close to two to one. In mathematical terms, the ratio of black to white would approximate more and more closely to two-to-one as the number of trials approached infinity. That is how mathematicians were able to establish a predictable outcome for a series of experiments, on condition that the proportions underlying the data were stable. The theorem did not apply, for example, if there were any changes in the proportion of black balls to white during the period of observation. Its value was predictive, but it could also have an investigative function. If the ratio was in fact unknown at the outset, it could be inferred with an increasing degree of reliability as the number of experiments increased.

The calculus of probabilities, while largely elaborated in the first place as a mathematical account of games of chance, was applied during the seventeenth and eighteenth centuries to a range of problems. Among those, as Daston shows, were matters of legal judgment and actuarial assessment. The central assumption made by classical probabilists, she argues, was that "the original relationship between classical probability theory and the Enlightenment moral sciences was almost literally one of prearranged harmony."[20] That is to say, probability was not just philosophically compatible with an orderly world; the calculus was a demonstration of its rationality seen—eventually—through its ratios: "if nature is governed by determinate laws, patient observations will ultimately conform to those laws. The perturbations of chance will eventually give way to stable ratios."[21]

But there was an aspect of Bernoulli's work that came to take on greater importance toward the end of the eighteenth century, to the point where it eventually became the primary object of the calculus. Bernoulli's theorem, as Daston, shows, extended the mathematical study of probability well beyond games of chance: "Mathematical conjectures about far more complex and interesting situations like human diseases and the weather became possible."[22] The theorem took on more general value because it could be used, not just as a way of anticipating rational order in the world, but as a way of anticipating human errors in the measurement of that order. Given a certain determinate law, there must by definition be a stable underlying ratio, and through repeated trials in its measurement, error could now be predicted to decrease progressively. No matter how great the number of observations, they would never simply reveal the ratio directly—that was not the business of the calculus—but they would approximate to it more and more closely as their number increased. So there was no decisive moment of arrival at the truth: it was rather a matter of managing error in an educated way. That could never involve the exact measurement of actual error because the ratio that served as a reference for such measurement could not be used without taking as given a law yet to be established. Instead, errors could be managed mathematically by addressing and assessing their probability. Bernoulli's theorem made it possible to do that by deciding in advance just what degree of approximation was to be considered suitable for a given inquiry. To be precise, one could decide that the ratio or proportion being calculated must be, say, one hundred or one thousand times more likely to fall within a specified limit than to fall outside it. Once that was done, Bernoulli's law made it possible to compute the number of measurements required to ensure that outcome.[23] It was thus possible, when conducting a series of measurements, to fix the range of acceptable variance beforehand and calculate its effect, thereby managing error by deciding its agreed scope. There was, however, no general principle that could actually dictate the range to be chosen. In every case, the range was the outcome of a calculation of probabilities. There was no universal right answer, just answers that were more acceptable—sometimes far more acceptable—than others. How much more likely would it be, once a set of measurements had been made, that the underlying ratio lay somewhere within the range decided at the outset? That was a matter for the calculus of probabilities. The point was not just to exercise care in measurement. It was not even to ensure in general terms that the measurements were far more likely to be accurate than not: one could and should choose in advance the mathematical extent of that likelihood. It had to be decided beforehand that, after a given series of measurements had been conducted, the probability that the underlying ratio was within the chosen range was to be, say, one hundred or one thousand times greater than the probability of its being outside the measured range. According to the acceptable variance chosen, greater or lesser numbers of observations would be required. For the probability of error to be very low, a very large number of measurements would be necessary.[24]

What Daston calls classical probability theory can be said to have reached its high point in the work of Pierre-Simon Laplace, who began to expound his views on the question in 1795 and first published an influential essay on the topic in 1812.[25] Laplace provided his own description of the method to be followed in order to manage the probability of errors. The first task was to make sure that the elements under consideration were of the same kind. If they were not, there was presumably nothing to be done. These elements then had to be "reduced to a number of cases that were equally possible."[26] To return to our classic example, that would mean focusing only on the fact of withdrawing a black ball or a white one. Such events as the breaking of the urn or withdrawing two balls at once were not the object of this calculus. In more demanding applications such as astronomical observations — a particular interest of Laplace[27] — making sure that the events were equally possible would require some care. An observed eclipse of the moon and an observed eclipse of the sun could be considered equally possible in that sense, as long as there was no equivocation, for example, in the definition of an eclipse. Such a reduction having been done to ensure that comparability was indeed assured, the next step was to determine, in Laplace's words, "the number of cases favourable to the event whose probability one is seeking."[28] In other words, how often did the event actually take the form that was the object of the inquiry? How often, in the case of an eclipse, for example, was it a total eclipse? This was when a ratio became apparent: the relation between the number of particular events (in our example, total eclipses) and the number of all possible cases (all eclipses) was "the measure of that probability, which is thus simply a fraction."[29] In the fraction so produced, the number above the line was the number of favorable cases — those in which the particular event, say a total eclipse, occurred — and the number below the line was the total number of possible cases.[30] What had been found via the set of observations in our example would not be a strict law determining the frequency of total and partial eclipses. There might be such a law determining those effects, but what would be established by Laplace's method would simply be the overall probability that, among a set of observed events known to be eclipses, the eclipse would be total.

There was in Laplace a mixture of qualities that may well disconcert modern readers. In his calculus, a systematic methodological practice of uncertainty was underpinned by a principle of epistemic certainty. He took the greatest care to manage the consolidation and assessment of data without prejudging the outcome, while nonetheless displaying a principled confidence that the outcome would always be properly rational. Laplace was thus able to deal with irregularities in data by regarding them as anomalies to be envisaged within the framework of a more general probability. The probability was, in his classical view, that they would eventually be brought back to regularity and to reason. For example, it was known to mathematicians with an interest in population numbers that the birthrate of boys was somewhat greater on average than the birthrate of girls. Laplace noted, however, that Buffon had identified a discrepant fact: there were several communes in Burgundy that had shown a significantly higher birth rate for girls over a five-year period. Faced with this, Laplace had no doubt that the anomaly could be expected to disappear if the birthrate continued to be measured over a period of a century.[31] Such observations as the one made by Buffon, he said, were "anomalies due to chance,"[32] and the role of a calculus of probabilities was to regard them expectantly as short-term phenomena. If enough events occurred and enough observations were made, it was probable and anticipated that such anomalies would disappear. That is why Laplace was able to claim, in a statement quoted by almost every French proponent of statistics in the decades that followed: "The theory of probabilities is no more than good sense reduced to mathematical calculation."[33] The calculus represented, he said, "what right-minded people felt to be true by a kind of instinct."[34] This "classical" assertion came in the following decades to appear reassuring to some and provocative to others. Whether or not a calculus of probabilities could stand in for common sense was in the 1830s a highly contested matter.

The classical quality of expectation was not sustained through the nineteenth century, as the overall significance of the calculus of probabilities underwent a shift. Until early in the century, to quote Desrosières's neat formulation of Daston's account, uncertainty about the future and the incompleteness of human knowledge led to "wagers on the future of the universe." Probabilities thus provided "a reasonable person with rules of behavior when information [was] lacking."[35] When probability was understood in this manner, the calculus was a formal representation of "reason to believe."[36] But the French Revolution, as Daston suggests, may well have shaken people's confidence in a shared good sense common to all rational men. To the extent that rationality was no longer reliably shared, the classical theory of probability tended to lose its raison d'être, ceasing to be a theory of general expectation.[37] But probability found a new significance in the work of nineteenth-century mathematicians like Poisson, who were primarily concerned with ordering inquiry through the calculated anticipation of errors. The analysis of probabilities no longer served the purpose of patiently confirming an expected law. Instead, it had become a formal set of precautions against the effective certainty of error.

Poisson's error-focused version of the calculus came to be known as *la loi des grands nombres*, the law of large numbers. On 11 April 1836, he made a presentation on the subject to the Académie des Sciences, declaring at the outset that this law should not be confused with "the beautiful theorem due to Jakob Bernoulli."[38] Bernoulli's theorem, for all its beauty, had a significant limitation in Poisson's eyes. It showed that, in the course of a long series of trials, events would occur proportionately to their respective probabilities. That rule applied, for example, to the probability of taking black and white balls from an urn when the proportion was fixed in advance by the contents of the urn. But Bernoulli's theorem supposed that throughout such trials the chances for each event (such as the proportion of black balls to white) remained constant. Poisson's law, however, was applicable to more complicated, more worldly situations. It was known that both physical and moral phenomena[39] were almost always subject to irregular variation, often over a great range. A problem better representing such complication would involve a whole series of urns of varying compositions. In such a case, Poisson asserted, the law of large numbers still applied. Whatever the proportion of black balls to white in each urn, the mathematical probability is that, if many trials take place, the ratio of black to white withdrawn will approximate to the average proportion of all the urns taken together, within a small range of error. In other words, the law of large numbers is still valid as long as it is a matter of the proportion of black balls to white, no matter how much the actual proportion might vary from one urn to another. Given a particular kind of event, the relation between the number of times it occurs and the total number of trials is close to invariable when the numbers involved are very large. Even with seemingly more aleatory things such as national crime rates and judicial decisions, the ratio of one kind of event to the whole varied less and less, said Poisson, as the number of measurements increased. The proportions came ever closer to a calculable size that they would reach if the number of trials could be prolonged to infinity.[40] It was not just that events would recur, as Bernoulli had stated, in proportion with a fixed ratio. Poisson's calculation took into account the probable net variations of the probabilities themselves. This is what Hacking refers to succinctly as the probability of a probability.[41] In his own account, Poisson went on to give an example drawn from games of chance. If a perfectly regular coin is used in coin tossing, Bernoulli's theorem predicts that a large number of tosses will result eventually in equal numbers of heads and tails. But if after, say, 2,000 tosses the proportion of heads to tails were to be 11 to 9, Poisson's law of large numbers predicted that the proportion would continue to be the same after many more tosses.[42] His law would not actually provide a causal account of the coin's unevenness, but it would in the long run offer a measure of its effects as a matter of probability.[43] As he stated elsewhere, "we are not concerned with the nature of causes, but with the variation of their isolated effects; we want to find the number of cases necessary in order for irregularities in the observed cases to be balanced out in the average results."[44]

The line of thinking that goes from Laplace to Poisson is, so to speak, the high road of the history of statistics, and that line does in fact happen to pass through a debate that occurred in the Académie des Sciences in 1835–36. On 14 December 1835, Poisson presented a paper on the probability of judgments, especially criminal judgments. Instead of applying the calculus only to the physical sciences, he was self-consciously extending it to the "moral" domain. He claimed that one could calculate the probability of judicial error and indeed the likelihood that appeals would be upheld. Eight thousand, he had calculated, was a threshold number for the nation. At that point, the proportion of successful appeals became stable.[45] The move was not unprecedented, and Poisson found antecedents for it in Condorcet and Laplace,[46] but it was nonetheless received in the Academy as provocative. At a meeting held on 11 April 1836, Louis Poinsot led the attack. Poinsot offered no criticism of the mathematics as such, but expressed an ethical objection to the very fact of conducting mathematics in this domain. He found it "distasteful . . . to represent by a *number* the *veracity* of a witness, thereby assimilating men to so many dice, each of which has several faces, some for error and others for truth."[47] That the calculus of probabilities had served in the past to analyze dice-throwing was here being held against it, as Poinsot sought to confine it to that unworthy domain. How could a calculus that had been developed for the analysis of games of chance possibly be applied to such a high-minded business as judicial decision making? How could it be thought that men of conscience would behave in a manner calculable by such a law? Whether at the end of the debate Poinsot or Poisson might have been thought to carry the day, it can hardly be said that this extension of the law of large numbers was met with enthusiasm. This was no triumph, no decisive step in the onward march of statistics. Other mathematicians were later to carry on the enterprise, and it has been argued that Poisson's ambitions were fulfilled in subsequent decades.[48] Historians of statistics have regularly gone on to follow those further developments, but we will not hasten to follow them at this point in our history.[49] We wish to dwell further on the objections to extending probability methods that were so forcefully expressed in the 1830s. As we have indicated, our topical field will not in fact be the judicial domain but that of medicine, where the principled objections were more sustained and, at least in the short term, more telling. We have already quoted Matthews's opinion that doctors in the 1830s were ill informed about the methods of Poisson, but our claim is that opposition among them to statistical methods, such as they were understood, was well organized intellectually. The medical debate of which we will speak may not deserve to stand on the high road of a history of statistics, but the medical academy appears to have constituted the professional, disciplinary space in which resistance to statistical thinking took its most concerted form. In this milieu, statistical thinking was broadly taken as known, and most often firmly resisted.

"MEDICAL PROBABILITY"

In 1798, P. J. G. Cabanis published an essay that might easily have been misrecognized by its title as an application of classical probability theory to medicine. It was called *Du degré de certitude de la médecine*.[50] Was this to be a study in the manner of those classical jurists for whom the veracity of witnesses could be measured by degrees and represented by fractions, according to the standing of each and the extent of their mutual corroboration?[51] Would Cabanis's book establish a formal set of rules connecting medical evidence with rationalized belief? Nothing of the sort. He aimed rather to give the medical profession a "solid base" that would allow it to withstand the "reproaches of uncertainty" levelled at it by certain unnamed *philosophes*.[52] At the outset, he rehearsed a litany of criticisms that his readers would know by heart: doctors did not understand how life functioned; they did not understand the nature and causes of illness; they did not understand remedies and how they worked, etc.[53] The situation was grim: "except for doctors, everyone [that is, informed opinion] seems to have adopted an attitude of absolute skepticism where medicine is concerned."[54] "Absolute skepticism" of this kind was a mighty challenge to the standing of medicine, and it could hardly be answered in Cabanis's view by talk of graded probabilities. A strong counteraffirmation was called for.

Cabanis imagined for the sake of argument that medicine had begun in the distant past through blind experiment. Doctors must have offered sick people drinks of various kinds that were thought to have healing properties. They would then have made inferences about the treatment of other disorders on the basis of any successes that occurred.[55] In all this, he concedes, they must have followed probabilities: "It cannot be denied that they were guided as they did so by probabilities, having nothing better to put in their place. But soon experience transformed those probabilities into practical certainties."[56] Medicine, in the course of its progress, must have gone from utter empiricism to an assessment of probabilities, and thence to practical certainties. Not everything in present practice was certain, of course. Diagnosis was continually rendered difficult by the fact that "at every turn one is obliged to admit exceptions to the rules. . . . There is no fixity in their application, no constancy in the procedures to be followed."[57] Theoretical medicine typically failed at the bedside, but this made it all the more necessary for the doctor to be guided by "a kind of instinct perfected by habit."[58] Whatever certainty obtained in that situation depended on the personal skills and qualities of the doctor. French had long referred to medicine in generic terms as "the art," and the skills displayed by doctors resembled those found in artists. The doctor's instinct, developed by experience, allowed him to "see the illness at a single glance [*coup d'œil*], apprehending all its features at once."[59] Medical knowledge was quite certain in practice, but its very certainty was of an artistic kind: "In medicine, since everything, or almost everything, depends on the glance and a happy instinct, certainties are found in the very sensations of the artist, rather than in the principles of the art."[60]

Cabanis sustained this kind of instinctive, artistic certainty against the kind that was subject to calculation. He could not of course deny that the art of medicine was subject to error, but the nature of error, he said, was not the same across different fields of knowledge. In the exact sciences, error arose from mistaken calculations or the imprecise use of formulas. In medicine, everything depended on a "wisdom of the organs" in which the most telling perceptions were more like inspiration than logical reasoning.[61] If certainty were indeed to be measured by calculation, he could not claim that medicine would reach a high "degree" of it: "I am very far from thinking that the particular knowledge of illnesses or remedial effects can be carried to the degree of precision that characterizes the certainties of calculation, and I am even less inclined to claim that prognosis can lend itself to that precision, which is, one might say, purely intellectual."[62] The certainty offered by medicine had to be of a different kind: "Everything that has to do with the practice of medicine surely requires many operations of a kind quite different from those that can be dictated by a simple formula."[63] Calculated probabilities had their place in the physical sciences, but the quality of clinical certainty produced in the practice of medicine was not of the same order.

This defensive and celebratory discourse about medical knowledge remained in play for much of the nineteenth century, although it became progressively more difficult to argue that the art of medicine need have no truck with mathematical calculation. Historians of medicine with a particular interest in sexuality will recognize Cabanis's view as an affirmation of what is regularly called, after Foucault, the "medical gaze," and indeed Foucault himself based some of his analysis of the clinician's "glance" on the essay from which we have just quoted.[64] But what interests us here is the fact that this discourse, in its very eloquence, was effectively set against any systematic practice of numbers. The synthesizing interpretive power of the glance was literally incalculable. True mathematicians, said Cabanis in a postscript, know that "calculation is not applicable to everything,"[65] but that is precisely what Poisson and others seemed later to be contesting when they began to apply the calculus of probabilities to such matters as judicial decisions. It was not clear, in the decades that followed, that medicine could maintain its claim to practical artistic certainty in the face of encroaching probability theory.

In *The Birth of the Clinic*, Foucault tells of a historical shift whereby medicine gave itself over to probabilistic thinking, dating that shift from the very beginning of the nineteenth century: "In the period of Laplace, either under his influence or within a similar movement of thought, medicine discovered that uncertainty may be treated, analytically, as the sum of a certain number of isolatable degrees of certainty that were capable of rigorous calculation."[66] Our own view is that Foucault's historical narrative unduly compresses any shift that may have occurred, and is incorrect in making "medicine" the unified subject of the change. He argues that the "confused, negative concept of certainty, whose meaning derived from a traditional opposition to mathematical knowledge"[67]—meaning by that something resembling Cabanis's view—was displaced by probabilistic thinking, thereby giving a new shape to clinical knowledge. It should be noted that he does not claim that mathematical thinking put an end to the medical glance as defended by Cabanis. Rather, says Foucault, the very domain in which the gaze was practiced was itself given shape by probabilistic calculation, so that medical facts came to be perceived in series: "This conceptual transformation was decisive: it opened up to investigation a domain in which each fact, observed, isolated, then compared with a set of facts, could take its place in a whole series of events whose convergence or divergence were in principle measurable. . . . It gave to the clinical field a new structure in which the individual in question was not so much a sick person as the endlessly reproducible pathological fact to be found in all patients suffering in a similar way. . . . Through the introduction of probabilistic thought, medicine entirely renewed the *perceptual values* of its domain."[68] Insofar as probabilistic thinking applied in this field, a "series" of events could not merely have been an accidental succession. The series itself represented a mathematical order. By that logic, according to Foucault, "the hospital domain is that in which the pathological fact appears as an event and in the series surrounding it."[69]

Our disagreement with Foucault is not about whether this distinctive articulation of mathematical thinking and the medical gaze ever took place. We are quite persuaded that it did, and will later point to examples of it. But we do not accept that the view defended by Cabanis—what Foucault characterizes as "the traditional opposition between the art of medicine and the knowledge of inert things"[70]—was simply left behind when the calculus of probabilities began to be applied to clinical medicine. Foucault suggests that since the clinic was only in the process of being formed at the end of the eighteenth century, Cabanis could not "justify the instruments of the clinic" by appealing to this new mathematical understanding, and so found himself drawing on a much older understanding of medical certainty.[71] But our claim is that the view represented by Cabanis, according to which mathematical thinking had to be opposed to insightful certainty, cannot be so easily assigned to an "earlier" time. We will agree with Foucault that in some clinics things did appear to happen in the manner he describes, but we will insist that the probabilistic model was not accepted in principle by the majority of the French medical profession during the decades that followed, and that many clinics were organized according to principles closer to what Foucault calls "traditional" ones. In fact, forty years after the "decisive" and "entire" change claimed by Foucault, mathematical knowledge was still widely regarded by medical thinkers in France as improper, indeed inimical to the art of medicine.

During the period examined closely by Foucault, there were in fact clear differences of opinion about probabilistic reasoning among doctors themselves. He quotes Philippe Pinel soon after Cabanis to characterize a general view of the clinic shared by the two,[72] but on the point of mathematical calculation their views were entirely opposed. Pinel, in his *Traité médico-philosophique sur l'aliénation mentale*, first published two years after Cabanis's text, devoted a chapter to explaining how the calculus of probabilities could be applied to treatment of the patients under his responsibility at La Salpêtrière. The problem in his field, he noted, was that the workings of the brain were not well known and that the causes of alienation (loss of reason) could not be directly observed. As a consequence, no treatment could be determined by a process of deductive reasoning. It was therefore appropriate to think about treatment in terms of probability.[73] If that were not done, alienist medicine in particular might find itself practicing "blind empiricism." The only way for medicine in general to become a "true science" was to apply the calculus of probabilities.[74] The divide between Pinel's thinking and that of Cabanis can be seen in their differing uses of the word *expérience*. Whereas Cabanis had insisted on the manner in which the doctor built his knowledge through experience, Pinel kept his use of the word closer to the other main sense of the French, which is that of the English "experiment." *Expérience* in Pinel's sense had to correspond to "a regular succession of observations carried out with extreme care and repeated over a number of years with a kind of conformity."[75] As with Laplace and those who followed him, the "extreme care" did not simply reside in observational technique, but in seriality: Pinel drew up tables to organize the data, thereby making it possible to determine "the degree of probability of healing of the inmates."[76] He was seeking to determine a ratio, the proportion of cures to admissions, and he aimed to identify factors that would increase the probability of success.[77] The main value of his tables was to permit comparison between types of treatment, measuring the success rate of the method he was using at La Salpêtrière against that of the old method.[78] So while on the one hand Cabanis was affirming that the doctor's art was *stricto sensu* incommensurate with mathematics, Pinel was attempting to organize numerical data in tabular fashion so as to draw inferences in narrowly rational terms. Neither approach can be said to have won out over the other at the beginning of the century.

That this set of issues was still quite unresolved in 1836 was made clear when Joseph-Henri Réveillé-Parise published a lead article in the *Gazette médicale de Paris* devoted to "medical probability." "Medical probability" stood for a problem shared by readers of the journal, mostly doctors themselves: "So where medical probability is concerned, doctors sometimes find themselves limited to vague probabilities according to the signs offered by illnesses and the action of modifying agents, signs which are admittedly important, but impossible to evaluate in any rigorous manner. The alternative to that is to find oneself caught up in systematic ideas, with ready-made solutions for every case that presents itself."[79] On the one hand stood the pleonasm of "vague probability," on the other the false certainty provided by interpretive systems. "Medical probability" as Réveillé-Parise understood it was the name of a general problem, not the object of a calculus. That was why medical probability had nothing to do in principle with mathematics: "What gives mathematical probability an assured eventuality is the fact that the terms of the problem are always fixed and determinate."[80] Poisson would have demurred at this point, saying that the calculus of probabilities could be applied to quantities that were not fixed. But mathematics of that kind were not available to this cultivated and well-published medical writer. Réveillé-Parise simply saw no hope that medicine might be able to adopt the methods of the physical sciences. Doctors were condemned to less precise methods: "So we are forced in matters of medical probability to proceed by means of approximation, never losing sight, insofar as we are able, of the law of generation of phenomena. That law means that one is led by analogy and induction to whatever conclusions one can reach."[81] By the baldest of terminological ironies, the term "medical probability" was actually being used here to refer to the fact that medical knowledge had nothing in common with mathematics. Having none of Cabanis's enthusiasm for the everyday wonder of instinctive clinical perception, Réveillé-Parise was unable to articulate an affirmation of practical certainty. That hope now appeared to have gone, and yet mathematics offered no solace: "since we cannot determine these laws or their precise relation to observed effects, all we have left are mere conjectures or lowly [*infimes*] probabilities."[82] "Medical probability," it seemed, was no more than an unfortunate intellectual condition of medical practice.

AVERAGES

The calculus of probabilities was not the only mathematical notion referred to regularly during the 1830s debates. It was accompanied by, and sometimes articulated with, that of the average.[83] That term had taken on a particular mathematical signification through the work of the Belgian Adolphe Quetelet, who continued to make strong claims, not taken up wholeheartedly by the scientific community, for the social usefulness of statistical knowledge. Quetelet's rather patchy career might well serve to show how irregular the "rise" of statistics actually proved to be, but our interest at this point is in the resistant discursive work that went on within the field of medicine. Quetelet's name did not circulate freely in the Académie de Médecine any more than Poisson's, but some notion of the average was available to all, often as an object of considerable disquiet. In 1835, two years before the extended debate in the Académie de Médecine, Quetelet published *Sur l'homme et le développement de ses facultés ou Essai de physique sociale* (*On Man and the Development of his Faculties, or Essay in Social Physics*), which he presented as a summary of his work on statistics to date.[84] He began by stating a "fundamental principle" that revealed his debt to Bernoulli and Laplace: "The calculus of probabilities shows that, all things being equal, one comes all the closer to the truth of laws one wishes to grasp if the observations include the greatest number of individuals."[85] It followed from this principle that in the observation of societies "the greater the number of individuals one observes, the more individual particularities, be they physical or moral, fade from view and give way to the general facts by virtue of which society exists and maintains itself."[86] For a given population, these "general facts" could be quite literally summed up in the figure of *l'homme moyen*, the average man: "The man I have in view here is, within the society, the analogue of the centre of gravity in the body. He is the average (*moyenne*) around which the social elements oscillate."[87] This "ideal type"[88] could thus be identified by gathering statistical data about a given population and calculating the average of various sets of features. The consolidated average would then serve as an index of the population. There might be exceptions, "as with all laws of nature," but this calculation would "best express what takes place in the society, and that is the most important thing for us to know."[89] This was a significant step in the intellectual history of the normal. A brand of social knowledge grounded in mathematics was asserting that the average mattered more than the exceptions. Without constituting a great mathematical advance, Quetelet's preoccupation with the average was to have a strong impact on cultivated public discourse in the decades that followed. As Hacking says, "there is no doubt that the 'average man' stuck, even though almost no-one had favourable things to say about the concept when taken literally."[90]

As if to consecrate his epistemic valorization of the average, Quetelet took *l'homme moyen* to be a standard of beauty mathematically defined: "If the average man were perfectly established, one could consider him as the type of beauty."[91] By that logic, those most distant from the mean in a given population were the most deformed: "everything that was furthest from resembling his proportions or his manner of being would constitute deformities or illnesses; anything that was so different, not only in its proportions and its form, as to stand outside the limits observed, would constitute monstrosity."[92] Here, by contrast with the work of the Geoffroys, monstrosity was defined by its relation to the average. In this mathematically shaped aesthetics, closeness to and distance from the average were given inverse value. To be close to the calculated average was certainly not to be lost in indeterminate middleness: it was to stand near the perfection of the type. Quetelet was making an affirmation about the significance of statistics rather than identifying a conclusion that could be drawn from any set of actual calculations. But he was putting forward a prima facie reason why individuals closest to the average should routinely be considered superior. This was a move that served to constitute the normal and the abnormal, the normal and the monstrous, in the same mathematical terms and by the same set of complementary moves.

THE ATTACK ON MEDICAL STATISTICS IN THE ACADEMY OF MEDICINE

The great debate began quite ceremoniously in the Académie de Médecine on 25 April 1837, but the issues had already been put to the Academy two weeks earlier. Jean Cruveilhier, who held a chair in the Paris Faculty of Medicine, had called then for a discussion of the role of statistics in medicine. That the time was ripe and the intellectual circumstance propitious can be seen by the prompt response of the Academy in making room for a full-scale debate. What matters most from the point of view of our history is that the terms of the debate were already more or less agreed beforehand, and the key points of contention known to all. Some interventions in the debate proper would deploy greater erudition and have a sharper polemical edge than Cruveilhier's summary effort, but his account of the issues can be thought of as the unimproved position of French medical academicians in their majority.

Statistics applied to medicine might be excellent in theory, Cruveilhier conceded at the outset, but there was little point in his making such a concession since he did not ever consider the theoretical claims of statistics in their own terms. The bid by Quetelet to characterize clinical knowledge in terms of probabilities and averages had no place in his account. His only interest—and, as it subsequently became clear, the only interest of both opponents and advocates of statistical medicine in the Academy—was in generalizations about medical practice.[93] On that point, Cruveilhier's mind was made up even as he was calling for further discussion. The application of statistics to medicine, he said, was "impossible."[94] That was patently the case too for judicial decisions, where an ill-advised attempt had been made in the past—he was referring to Condorcet, rather than to Poisson—to use algebraic formulas as a substitute for written proofs or witness statements. As if one could decide the fortune and honor of citizens in such a fashion, he observed mockingly. Such calculations were "bizarre" in a judicial context, and would be equally so in medicine.[95] Medicine was not and could not be "une science à chiffre et à compas": it was not the kind of discipline that dealt in numbers and measuring instruments. It was, rather, a science of observation and *expérience*.[96] Numbers were inimical to medicine because they had nothing to do in principle with living bodies: "Imposing the arithmetical method on the study of life seems to me contrary to the eminently variable and elusive quality of the phenomena found in living bodies."[97] At the heart of this thinking lay an irreconcilable opposition between the numerical and the biological. Medicine, in its methods, in its very forms of knowing, had to correspond to the mobility of life. Medicine of that kind was simply incompatible with what Cruveilhier called "the inflexibility of numbers."[98] It seemed to follow that whatever medicine had to say, and to know, about the normal state could not be translated into statistical terms without being trapped in an irresolvable tension between the mobility of life and the fixity of numbers.

As it happened, the Academy had recently been approached by one of its correspondents who was asking to present an address on "statistics and its applications to medicine."[99] That fact barely deserves to be considered a coincidence in view of the currency of the topic, but the request did provide an opportunity for a more sustained attack on statistical method.[100] The attack was mounted by B. J. I. Risueño d'Amador, a Spaniard who held a chair in the Faculty of Medicine at Montpellier. The fact of being a foreigner was skillfully turned to advantage as Risueño displayed his great mastery of French scholarly rhetoric along with a refined knowledge of the history of medicine and mathematics. According to the *Gazette médicale de Paris*, his address was a triumph: "The reading of his text produced an unforeseen effect. It was listened to with marked sympathy by the immense majority of those present. Lively applause and strong, sincere expressions of admiration were lavished on the speaker."[101] But this was more than a bravura individual performance. Rhetorical brilliance found its place in intellectual history because the Academy was ripe for a staged confrontation. So while the history of medical statistics has paid scant attention to Risueño, who cannot be said in fact to have contributed positively to the discipline, he played a starring role at that time as the champion of antistatistical thinking. Indeed, he helped to make resistance to statistics into an identifiable cause, a cause that is entitled to its own history and is getting it here alongside the history of medical statistics. Historians, when they do mention Risueño, tend to call him a traditionalist, if not a conservative. Matthews refers to him briefly as "following in the Montpellier tradition of Cabanis."[102] Desrosières is no more explicit, saying that he was a traditional physician.[103] Those characterizations are not so much inaccurate as incomplete. Risueño did indeed appeal to a tradition of medical thinking, but our task here, rather than use the bland term "traditional," is to understand his position in its polemical articulation with medical statistics.[104] Far from simply expressing reluctance in the face of scientific progress, he joined battle with statistical thinking on a number of specific points. His address marked a quite elaborate aggiornamento of the position taken decades earlier by Cabanis.

Risueño began by identifying his adversaries as a school that dealt in numbers and owed allegiance in its choice of method to the calculus of probabilities. In fact, as we shall see, his opponents in the debate hardly ever spoke of the calculus of probabilities, preferring to speak simply of "counting."[105] But Risueño attacked them simultaneously on both levels: "As you know, gentlemen, there exists at present a school that places numbers above all else, one that proclaims the calculus of probabilities to be the only rule of certainty possible in medicine. For this school, ideas appear only in the form of figures. Its members practice counting in the belief that counting is the true scientific activity, thereby reducing all knowledge about therapy to an addition or a subtraction correctly performed."[106] The theoretical position imputed to his adversaries was a commitment to the calculus of probabilities, and the routine evidence of that—its practical small change—was the currency of numbers in their work. This preoccupation with numbers, he went on to say, was a symptom of his colleagues' inability to manage the difficulties germane to medical practice. It all had to do with the "great question of certainty in medicine": "Struck by our uncertainties, discouraged by the lack of success of our rational methods, worn out by the trial and error of experience that always finds itself beginning again and by observation that is never complete, this school has found a new way to proceed. It counts facts and believes it can assess their value by their number."[107] In other words, numbers were being used by the school as a rather desperate remedy for the uncertainty inherent in medical practice.

The turn to mathematics, said Risueño, had necessarily meant a turn to the calculus of probabilities, but that was no remedy at all since "the probability considered by mathematicians is hardly anything other than the theory of chance." So doctors who played the numbers game were actually giving up "all medical certainty" of the kind that was based on induction, experience, and observation. They were putting in its place "mechanical, inflexible calculation."[108] Like Poinsot in the Académie des Sciences responding to Poisson's work on judicial decision making, Risueño sought to disqualify the calculus of probabilities by invoking a seedy past in which it had been preoccupied with games of chance. If medicine takes this path, he said, "it will no longer be an art, but a lottery. To adopt this method is to despair of the art and give up forever on knowing how it works. It means giving oneself up to chance on the strength of an illusory faith in arithmetic. This is skepticism embracing empiricism."[109] Skepticism was unworthy of medicine because it was contrary to the true medical certainty that was grounded in a cultivated method of observation and experience. And empiricism was doubly unworthy because it represented the complete absence of method. The most fundamental mistake made by the proponents of numerical method, said Risueño, was to choose an illusory form of certainty, claiming for medical science "a form of certainty that it does not and cannot have, while refusing on its behalf the only form of certainty that is proper to it."[110]

The numerical method was, in Risueño's view, insistently mechanical in its operation. It led physicians to count cases, then to calculate averages as a way of ordering the numbers they had produced. He did not bother to give examples of this, knowing that they were fresh in the minds of the academicians following recent discussions of the work of P. C. A. Louis and others. He and they were thinking of the kind of calculation that would count the number of patients in a given clinic suffering from pneumonia, then compare the average mortality rate of those who were treated by blood-letting with the average of those who were not. For Risueño, such calculation did not lead to worthwhile knowledge: "Counting cases and deducing averages is not how a man accomplished in the knowledge of the world gets to the bottom of the motives underlying human actions in their extraordinary diversity."[111] His own model of medical knowledge was that of the worldly, experienced man, and the fact of performing calculations gave no access to knowledge of that kind. Dealing in averages was a mechanical procedure that made the majority of cases equivalent in effect to the totality. "Numerists," he said, were not interested in healing any particular patient, only in healing the maximum number within a given group.[112] So they always found themselves "neglecting the minority" as they methodically set aside the possibility that there might be individuals or subgroups whose cases called for different forms of treatment.[113] By allowing the average to represent the group as a whole, then moving from average diagnosis to average treatment, they were making a double error, as they followed their "uniform, blind, mechanical routine."[114] They were compelled by their "numerical" logic to use "an identical treatment on all their patients."[115] This was a sharper version of the general point Cruveilhier had made earlier: "the variability of facts is not an exception in life; it is life's characteristic, essential—one could say its primary—rule," said Risueño.[116] For this reason, the answer for medicine was not to be found in the use of numbers since that was a refusal to know variability in itself: "it is because of this permanent variability of the phenomena that we must protest against numerical method, with its extraordinary pretension to bring about fixity."[117] The only kind of knowing able to grasp living variability was the "individual genius of the artist," who encountered and observed each patient as a particular case.[118] There was no mention here or elsewhere in the debate of the manner in which Étienne Geoffroy had used the concept of the normal state to speak of invariability as a distinctive stage in organic development, nor was there any consideration of how statistical mathematics might give an account of variability, rather than imposing fixity.

The question of types arose briefly at one point, as Risueño turned in his closing statement to nosologists and naturalists for a model of knowledge. What did it matter to naturalists how many birds or reptiles there were in a species? What reason could nosologists possibly have for counting the number of illnesses of a given kind? They were simply concerned with making clear distinctions of type. It might be in fact that there were illnesses or even species consisting of only two or three individuals, but that did not prevent them from being distinct types. The great Cuvier had certainly not had large samples of fossils at his disposal, but he had nonetheless succeeded in identifying "antediluvian species."[119] Here was a further point of polemical articulation with the discipline of statistics as it stood in 1837. The notion that the type might be known by calculating the average of a given population was central to Quetelet's conception of it: for him, the average man simply *was* the type. But that view was being rejected here, without being explicitly quoted, in favor of a typological model of the kind found in Cuvier's, and for that matter in Geoffroy's, natural history.

In the weeks-long debate that followed Risueño's opening address, there was some discussion of the calculus of probabilities, including a well-informed speech by François-Joseph Double, who was able to refer to discussions in the Académie des Sciences, but there was little or no discussion of the type. The theme taken up most vigorously was in fact one that claims a central place in our longer history: that of the average. What was the significance for medical practice, these leading doctors asked, of calculating averages? To what extent could averages make available a measured understanding of norms? The first person to intervene after Risueño had spoken was Frédéric Dubois d'Amiens, who pressed hard against the numerical method on this point. The leading exponents of statistical medicine, he said, were unable to point to any decisive outcome of their work to date. In spite of their parallel experiments, based on different hypotheses, no calculation of averages had led to an agreed method for treating typhoid fevers. They had actually produced competing accounts, each supported by its own set of averages.[120] Yet having failed so far in that regard, they might still lay claim to what Dubois called a more "modest" ambition, that of arriving "by *averages* deduced from very great numbers, at a kind of *norm*, whether it be in anatomy, pathology, or therapeutic practice."[121] If the numbers were great enough, Dubois was prepared to suppose, it might be thought that a norm could be determined simply by calculating averages. Here is how they might claim to do it: "Let us suppose that by dint of numerous dissections, carried out on thousands of individuals, one had deduced a series of *averages* for all anatomical variations worthy of note, *averages* such that one could form what I shall call the *average anatomical man*."[122] Dubois maintained that there would still be no value in such a calculation. All it would produce would be "a fictive being, a product of the mind. Far from having an exact idea of what will be found in dissections, one would have a general pattern [*patron*] that would never correspond exactly to naturally occurring beings because it would be an assembly of *anatomical averages*."[123] Dubois might have seemed close here to talking about the normal state that was opposed in physiological discourse to the pathological one, but he was not prepared to concede for a moment that valid knowledge about the normal state might be produced synthetically by accumulating averages: "Similarly in pathology, after having observed thousands of cases of a given illness to note either symptoms or organic alterations, if one decides to use those *averages* to organize an *average* pathological state for the illness in question, so as to have a norm, a type, a general pattern, one will have the unpleasant surprise in practice of never finding any actual examples that conform to the pattern, precisely because that *average pathological man* will be another fictive being, a product of the mind, a conclusion drawn from statistical tables."[124]

The key point was that the average man and the average illness were theoretical fictions that had no substance and no clinical value. When Double took his turn to speak, he followed that line of criticism relentlessly. The *Gazette médicale de Paris* commented a few days later that "the crushing arguments brought against the numerical method in these three sessions took on irresistible authority when articulated by M. Double."[125] Double was perfectly happy to allow that averages might be used to calculate the predicted outcomes of lotteries and games, maritime insurance, and, despite Cruveilhier, the probability of criminal judgments, but they simply did not apply to medicine: "Neither our successes nor our setbacks balance out over large numbers. We cannot be rescued by quantity, and each of our problems involves only one individual."[126] So the failure of numbers in the medical domain was not to be thought of as an accident, or a preliminary phase in the development of statistical science: it revealed something distinctive about the possibilities for medical knowledge. There was no such thing as average health, said Double, calling his colleagues to witness: "The healthy state, as everyone agrees, is more unified, more constant, more regular than the state of illness. But just think about it. There are around two hundred of us in this auditorium, all of us adults, all of the same sex, the same profession and the same social position. Yet even in the midst of these identical conditions, how many similar healths are there that can be gathered and added together so as to be able to say that one health plus one health makes two equal healths?"[127] The calculation of averages offered no algorithm that could be used to define health, and any doctor who set about his professional activity by doing sums of this kind was out of touch with a fundamental truth about the nature of health and sickness. There was in clinical experience no average anatomy or pathology. And there was, most important of all, no average patient: "I am put in mind, if you will forgive the comparison, of a shoemaker who, having measured the feet of a thousand individuals more or less, comes up with an average and makes shoes only according to that imaginary pattern."[128] The whole assembly laughed heartily at this gibe, as one of the defenders of statistical medicine felt himself obliged to acknowledge when speaking later.[129] The notion of the average was being held up to ridicule, and its normative value was being flatly denied. At this point, the statistical practice of the average was in danger of being laughed out of the medical academy.

IN DEFENSE OF "COUNTING"

On those occasions when historians refer to the 1837 debate, they speak of P. C. A. Louis as the central figure in it, even though he was not present for most of the Academy sessions during which the matter was discussed.[130] He was, it has to be said, the most prominent intellectual figure on the "numerist" side. The *Gazette médicale de Paris* referred to him as "the true dogmatic representative of the numerical method,"[131] but its editorialist was moved to comment rather condescendingly more than two weeks after the debate had begun: "His silence and his absence have led people to suppose that M. Louis will take no part in this discussion. Our own opinion is that he can hardly afford to refrain from doing so, since the fate of the unfortunate numerical method now depends exclusively on him."[132] Learned opinion appeared to be running so strongly against the "school" identified by Risueño that only its leading spokesperson could save it. But Louis, who was in fact being kept away from the Academy by a problem in his family, no doubt considered that his best response to the emerging antistatistical movement lay in his clinical work. Louis had published no theoretical essays about applying the calculus of probabilities. His publications were largely descriptions of his clinical work in which he insisted on the importance of "counting." That was the extent of his mathematical ambition and the measure of his contribution to the advance of medical statistics. In 1834, Louis had published a polemical text in response to an attack on his work by the bellicose F. J. V. Broussais. Louis made it clear there that he did not accept Broussais's claim that medical judgment should be based on experience. That had of course been the view advanced by Cabanis, one that was soon to be reiterated by Risueño in the Academy, but Louis was determined to put something more methodical in its place: "M. Broussais will probably say that his precepts are based on experience, to which I will reply that he is mistaken, that true experience in medicine can only be a result of the exact analysis of numerous facts, thoroughly noted, classified according to their analogies, closely compared, and counted. Experience acquired in any other way is not much more than imaginary."[133] Counting was thus, for Louis, part of a routine for the organization of clinical knowledge. It was simply a way of constituting clinical experience as data, and a way of affirming that all experience, insofar as it was professionally valuable, had to be underpinned by an orderly record of previous cases. One should never produce a general statement, he said, unless it was based on precise facts that were recorded in writing, analyzed, and counted.[134] Counting thus ensured the proper cumulative outcome of a process of observation and record keeping.

It must be said—and has indeed been said by other historians—that this hardly constitutes an elaborate or ingenious use of mathematics. Matthews comments that Louis's numerical method was not innovative from a methodological standpoint, and that the focus in his work remained on concrete empirical results rather than abstract or mathematical theorizing.[135] Indeed, when Louis invoked mathematics as a term of comparison for his work, he did so in a rather general way, as if he had been in search of a metaphor rather than a formula: "What I have done, it seems to me, has been to consider any particular series of observations as the givens of a problem to be solved. I have studied them, subjected them to rigorous analysis in order to draw from them general facts, as in mathematics one isolates the unknown quantity from an equation by means of a series of transformations."[136] Solving a problem, he was suggesting, had something in common with solving an algebraic equation: "I would wait for the end of the analysis in order to know the value of the facts analyzed, just as in mathematics one waits for the end of the work to which one is subjecting an equation in order to isolate the unknown and find its value."[137] But there were no actual equations to be solved, and thus no application of mathematics in the narrow sense. Louis's rather unspecific reference to mathematics was meant to do no more than characterize a style of clinical observation.

That style of knowing deserves to be thought of as modest, not because of any personal qualities that Louis may have had, but because he insisted that clinical knowledge could be brought together properly only if the task were carried out in painstaking fashion. Louis's colleague A. F. Chomel made just that point in the debate proper: "it is not a matter here of introducing mathematics into medicine, of imposing on medicine a precision to which it is unsuited; we are simply concerned with the modest addition of facts recorded in our archives."[138] Two very different conceptions of the clinician were in play in the debate. On one side, Risueño was a militant exponent of the set of knowledge practices that has come in our own time to be critiqued by Foucauldian historians of the "medical gaze." Whatever the practical virtues of that approach, it was undoubtedly a form of professional arrogance. By contrast, "counting," as Louis understood and advocated it, was the very intellectual style of modesty. The difference between them corresponds to the rhetorical asymmetry that obtained throughout this debate. Risueño and those who shared his view attributed to Louis's "school" a rather grand set of intellectual ambitions that involved the application to medicine of the calculus of probabilities as a mathematical principle, whereas Louis and his colleagues merely sought to use numbers as a code that imposed immediate disciplinary constraints on their day-to-day work.

When Louis did finally take his turn in the Academy debate, he offered a rather defensive version of the above points, but also addressed briefly the matter of the average man raised so tellingly by Double. The point of the numerical method, he said, was not to produce such an entity as that: "Thus numerical analysis applied to pathology, to pathological anatomy and normal anatomy does not have as its objective the fixing of the average, imaginary man."[139] As he saw it, the value of the numerical method was not to be found in averaging as such, but in "a complete enumeration of the facts."[140] Clearly no one in the Academy of Medicine, not even the leading "numerist," was prepared to defend the normalizing calculation of averages advocated by Quetelet.

There was, however, another issue that proponents of the numerical method could not easily resolve by dint of modesty: the question of large numbers. In his published response to Broussais, Louis had appeared once again to follow the dictates of modesty in that regard, declaring at several points that the number of observed facts to hand was not yet sufficient to warrant a conclusion.[141] But when he addressed the Academy, he was less cautious. He put forward the question of large numbers quite explicitly: "But how many particular facts will be required for the result in question to be definitive?"[142] He may have thought of this as a rhetorical question, but mathematicians like Poisson actually had an answer for it: they could calculate the number of facts required so that the likelihood of error would be held to a specified very low degree. Louis, however, showed no awareness that mathematical statistics had a way of addressing the difficulty: "Reasoning," he said simply, "cannot resolve that question, but experience shows that one does not require an infinite number of facts, so to speak, in order to determine a law."[143] This twist in his argument left him in an uncomfortable position, since *expérience* was being called on here to play a role that Louis had seemed to refuse it elsewhere. Experience in Cabanis's and Risueño's sense needed to be reined in, he had said in response to Broussais, by the regular practice of enumeration. Careful observation would presumably keep *expérience*, insofar as it persisted, closer to Pinel's sense of "experiment." But what kind of experience could serve to bring enumeration to an effective conclusion? Whatever it was, it was untheorized in mathematical terms or in any other. Louis would go on to cite a few instances in which studies of smaller samples had been confirmed by larger ones, but his conclusion was adventurous, not to say immodest: "one does not need a very considerable number of facts to reach the determination of a pathological law."[144]

A fellow numerist, Jean-Baptiste Bouillaud, addressed the Academy on the question of large numbers in a manner than was no more persuasive. Bouillaud had published a book in the previous year that addressed in one of its sections the central question now before the Academy. More specifically than Louis, he advocated the application of the calculus of probabilities to medicine, but appeared to do so on the basis of a misapprehension. He recognized that it was appropriate to "formulate a law after having used a method in a very great number of cases."[145] He recognized too, in keeping with Laplace's classic view, that "numerical results can only be exactly the same in two series of facts subject to calculation insofar as the circumstances of these facts remain exactly the same."[146] But he found in the very reference to probabilities an unjustified pretext for approximate calculation: "since facts in medicine very rarely present in such identical circumstances, it can easily be seen that *approximate* calculation or the calculation of probabilities is almost always the only kind available to us when it is a matter of *generalizing* a result."[147] This was a convenient misunderstanding of Laplace, who would not have accepted that "probability" signified approximate calculation. Laplace had in fact claimed to do quite precise calculations about probable future occurrences.

When Bouillaud took his turn in the Academy debate, he appeared to argue for both rigor and laxity at the same time. Aware that Risueño had made much of induction, rather than counting, as the proper path to medical knowledge, he invoked Laplace on the theory of large numbers, but did so in order to make of clinical medicine an exception to the rule of mathematics. How could clinical medicine possibly produce numbers so great, he asked, that Laplacian mathematics could be applied to them? Such numbers could never be reached in practice. As a consequence, said Bouillaud, the calculus of probabilities needed to be revised when it was applied to medicine; otherwise no difference would be made between medicine and games of chance. But Poisson's law of large numbers, following Laplace, claimed that the calculus could be applied analogously to a variety of fields, and it was precisely that claim that Risueño was contesting in the case of medicine. Bouillaud was effectively giving the game away. He was claiming to apply Laplacian principles to medicine while preserving medicine as a domain that had, or should have, nothing in common with games of chance. That was a somewhat desperate attempt to answer Risueño's taunt that medicine was in danger of being turned into a lottery, and it effectively conceded far too much to his opponent even as it reversed his proposition. Commenting during the following week, the editorialist of the *Gazette médicale de Paris* was scathing:

> We were delighted to hear M. Bouillaud say that medical probability was not the same thing as arithmetical probability, and that it would be folly to assimilate one to the other. That was music to our ears, but what then is the

use of figures, tables, additions, the extraction of averages, and mathematical formulas? If this does not have to do with the mathematical calculation of probabilities but merely with the estimation of philosophical probability, then everyone is in agreement. We ourselves will acknowledge, if called on to do so, that the practician's conjectures and the inductions he draws from the facts are only probable, that is to say fallible. But what does that kind of probability have to do with the kind that mathematicians claim to produce by rigorous calculations? Why are we even talking here about figures?[148]

Bouillaud had no position left to defend.

Two central tenets of medical statistics, the average as norm and the calculation of probabilities were being poorly defended and largely given up in the course of the debate. It was hard to see what, if anything, might now be saved for the cause of medical statistics. The numerists had lost the debate for at least three reasons. Their position was not of a kind that lent itself to the proclamation of new principles; their knowledge of mathematics was thin, much more so than that of their opponents Risueño and Double; and they were offering a view of clinical practice that, in its very modesty, was less gratifying for members of the medical profession than the model of the experienced, insightful clinician.

THE RANGE OF VARIATIONS

During the debate there was passing mention of another issue that hardly counted as decisive at the time, but would come to prominence later in the century. Louis himself adverted to it when discussing the possible application of numerical thinking to the analysis of "normal anatomy." He observed that at the "beginning" of medical science it had not been thought necessary to engage in any such analysis of anatomy in view of "the fairly great constancy of nature in the development of our organs." But it had since become clear that "regularity does not always occur." As a consequence, there had emerged a need to know "the extent of those variations." That knowledge, he asserted, could be arrived at by "counting the number of times they present in a given mass of subjects."[149] In other words, it was not simply a matter of locating the normal at a particular point by calculating the average. Normal anatomy could be identified as a range, and identified all the more reliably by aggregating observed variations. This was a way of getting beyond the rather dramatic impasse brought about by attempting to think conditions and patients in terms of averages, but it was scarcely taken up in the discussion. The professional academy was entirely preoccupied with the unresolved tension between the treatment of individual needs and the ordering of clinical observations.

The only participant in the debate to engage at any length in this kind of thinking was P. F. O. Rayer, who was relatively young in this company. He would publish a book on medical statistics in 1855, and was to be appointed to a chair in the Paris Faculty of Medicine in 1862.[150] Without being identified as a member of the numerist school, Rayer provided a cogent defense of numerical thinking where his more prestigious colleagues had failed. The first difficulty he addressed was the fact that an illness was not a "unique, fixed, invariable phenomenon." Bouillaud had been aware that this was a problem for numerical method, and had tried rather unsuccessfully to deal with it. If an illness was not an unvarying phenomenon, Bouillaud had asked, how could one go about the business of counting, given that one can only count things that are identical in a particular respect? His rather weak answer was that counting had to be approximate, but Rayer had a clearer view. It was patent, said Rayer, apparently conceding a crucial point to Risueño and Double, that there could be no "absolute, exclusive treatment" that corresponded in each case to a given illness. Yet it sometimes happened that "the phenomenon is similar enough to itself to allow the practitioner to take it as a unit. It is not fixed, but sometimes its inconstancy is not such as to alter the prognosis of the doctor and the results of his practice. It is not invariable, but sometimes its variations are contained within fairly narrow limits, so that they can be disregarded without any unfortunate consequences."[151]

This reformulation of the problem led Rayer to a defense of the average appropriately revised. He was only too aware, he said, that the notions of "average" and "probability" had proved uncongenial to many good minds, but people ought not to be frightened of words. In actuality, therapeutic practice often "took an average as its rule." And even though that average was not a rigorous mathematical concept, it was still of undeniable value to the doctor. In fact, posology in its entirety was nothing other than a succession of averages. Moreover, the "average image" of an illness in the mind of the doctor served as an aid in planning therapy. By refusing to recognize this, the adversaries of statistics had been led to deny the reality of this helpful resource.[152] Indeed, those who opposed statistics by asserting the uniqueness of each case were following a logic that led to a form of absurdity: "One must consequently uphold the view that nothing similar is ever seen, that no two facts ever resemble each other, that medical practice never takes the same form twice, and that theory is never entitled to assimilate one to the other. I must say, gentlemen, that when the doctrine of the adversaries of statistics is reduced to this, their position appears to me unsustainable."[153] It is unsurprising in this light that Rayer should have permitted himself shortly afterward to use the term "average man," which had become an object of ridicule in earlier sessions of the debate. Taking an example from the treatment of syphilis, he spoke of the need to vary treatment in cases where patients had "some condition that deviates [*s'écarte*] too visibly from the average man."[154] Nothing in the written record suggests that his use of the expression was greeted with derision on this occasion. The "average man" was able to stay in play precisely because it was a matter of considering deviations from the average as variations within a range.

AN ESTABLISHED ROUTINE OF MEDICAL STATISTICS

The Academy was undoubtedly in need of reminding at this point, such was the power of the antinumerist position, that certain statistical practices were in fact well established within medicine. That point was actually made by one of their number named Castel. Medical statistics proper, he said, needed to be distinguished from the numerical method. He had no quarrel with the standing of the former, which measured the influence on illnesses of place, climate, season, age, temperament, nutrition and profession.[155] This was one of very few occasions during the debate when anyone distinguished two kinds of medical statistics,[156] but Castel was perfectly correct to claim that statistics of certain kinds were already practiced in an uncontested manner across the medical field. In his view, the numerical method, for its part, involved only unworthy calculations—what he called "arid, speculative computation [*supputation*]." He then produced a derisory comparison that might have provoked as much laughter as Double's earlier gibe about the shoemaker who presumed to make all of his shoes in a single average size. Castel's comment was not about averages but about large numbers, or rather about the lack of them in clinical practice. "Including columns of bare, isolated figures within statistics would be comparable to the presumptuousness of a shopkeeper who used the term statistics to describe his stocklist or the statement of his accounts. Doctors in our time cannot be said to be doing statistics merely because they happen to use figures."[157] For all the methodical modesty of counting, doctors who dealt in numbers were still open to the accusation that they were merely tallying their daily transactions and then giving to those accounts the weighty significance, if not always the pretentious name, of statistics. The numerical method, said Castel in conclusion, was "a mutilation of medical statistics."[158] What, then, of the unmutilated version, and how was the gap between the two kinds established? How had one form of medical statistics been able to find acceptance in a professional context where the key principles of statistical mathematics—probability, averaging, and large numbers—continued to meet with firm, not to say militant resistance?

The fact is that population statistics had already claimed a central place in medical thinking about public health. As Theodore Porter observes, "vital statistics and calculations not only provided means of observation in public hygiene . . . but also had much to do with the direction taken by such hygiene after 1820."[159] In 1826, F. Bérard, who held the position of Professor of Hygiene at Montpellier, published an essay arguing that civilization did not degrade humanity, but brought benefits. In the first section, he made a qualitative argument. In the second, he offered statistical evidence in support of this view, beginning with statistics that showed mortality rates decreasing over time.[160] Bérard's work was enthusiastically expanded upon in Britain by Francis Bisset Hawkins, who made it clear even in 1829 that he was calling for the systematization of existing statistical practices rather than attempting to initiate brand new ones.[161] French physicians, as Bernard-Pierre Lécuyer has pointed out, "had as a professional group showed a comparatively high degree of interest in statistical studies."[162] The kind of statistics that interested them was typically the sort published in the influential *Annales d'hygiène publique et de médecine légale* under the leadership of such administratively minded figures as A. J. B. Parent-Duchâtelet and Louis-René Villermé. Those studies had to do with matters such as the quality of dwellings and the location within cities of refuse dumps and knackers' yards. As Desrosières points out, statistics of this kind allowed the medical community to intervene officially in public debates on matters that had to do with the organization of society.[163]

Public debates were never more strident nor the numbers more eagerly scrutinized than in critical times of illness such as the cholera epidemic of 1832. The *Annales d'hygiène publique*, for example, published statistics that set out death rates according to various criteria such as type of dwelling and level of income.[164] When the more general *Gazette médicale de Paris* began life in 1830, statistics of this kind were also on its agenda. The editor in chief, Jules Guérin, announced that the *Gazette* would report on matters of medical statistics, including "figures about the medical population of the kingdom."[165] This was, it should be noted, the same weekly that was to cheer from the sidelines for the antinumerist cause during the 1837 debate.

When cholera struck in March 1832, the *Gazette* began publishing tables of figures and gave them great prominence throughout the epidemic. Every issue contained a table for Paris showing the daily number of deaths in hospitals and at home, the number admitted to hospital, the number discharged, and the increase or decrease in those numbers from day to day. And when Villermé presented to the Academy of Medicine a report entitled *Statistique du choléra*, the *Gazette* quoted his address in respectful detail. That report contained tables about the effect of the epidemic on the Belgian population, and was coauthored by Quetelet. Villermé took the opportunity during his report to provide a full justification of population statistics. Statistical studies were of considerable interest to government because of what they revealed about "the social economy," but they were of particular interest to doctors as a group because they had to do with "the laws of our reproduction and mortality."[166] The value and the modernity of Quetelet's work, Villermé said with some insistence, lay in the fact of measurement: "We have had to wait until the present in order to see Mr Quetelet *measure*, mathematically speaking, the effects produced by age, not only on our mortality, but on our height, our weight, our passions, our tendency to crime, etc."[167] Villermé was one of those doctors of very high standing who were members of both academies, medicine and science. No one, least of all the *Gazette*, questioned the purpose or the validity of his engagement with statistics such as these. And when a new academy, the Académie des Sciences Morales et Politiques, came into existence, the *Gazette* supported the candidacy of Villermé and his fellow statistician Benoiston de Châteauneuf in the following terms: "Both of them have devoted their energies with great success to statistical research on population, mortality, epidemics, etc. Statistics is a new science that was unthought of thirty years ago. The work of these scientists is so satisfying, so rich in results, that they will be able to rehabilitate statistics in the thinking of serious men following many unfortunate efforts that have compromised its reputation."[168] This was five years before the great debate, but it pointed to underlying irresolution between "scientific" statistics of the kind undertaken by Villermé and uninformed or maladroit applications of the new discipline by medical practitioners.

One clear difference between Villermé's brand of statistics and that of Louis and his colleagues was of course the size of the groups on which calculations were based. It might be observed that population statistics were perfectly able to satisfy the mathematical requirement of large numbers, whereas data produced by medical clinics were not. But the fact is, as Foucault pointed out at the risk of overgeneralizing, that statistical tabulation had become routine in certain clinics. That was exactly the case with the clinics of Louis, Bouillaud, and Chomel. But it was equally true that clinical statistics were standardly published in the same places as figures on populations and epidemics. The *Gazette médicale de Paris*, for all its distrust of the numerical method, announced at its inception that it would publish "mortality tables of hospitals"[169] and subsequently did so with great regularity. To take but one example among many, it published in May 1830 a report on the clinic of Joseph Récamier at the Hôtel-Dieu hospital. Like others of its kind, the report contained tables listing the number of patients with a given illness, the number who had recovered over a period of several months, the number of those who had died, and the number of those still in treatment.[170] It is noteworthy that the numbers of patients in this particular report were quite small, as such numbers usually were other than in times of epidemic. In Récamier's clinic in 1830, there was one recorded case of colic and one of cholera. The largest number of cases, twenty-one in all, was of typhoid fevers. Statistical analysis in such clinical reports took the form of calculated proportions, mostly mortality rates for each illness. Everyone, it seems, was interested in such figures, and no one seemed inclined to observe during the week-to-week consumption of such tables that the numbers were so small as to offer no basis for a reliable calculation of probabilities.

It seems reasonable to conclude from all this that clinical statistics unsupported by mathematical theory functioned in effect as anecdotal currency in professional conversation. Hacking, in his history of probability theory, comments skeptically on their lack of scientific consequence: "The statistics of Paris were full of tables reporting the great hospitals. Would all these batteries of numbers lead at once to treatments of tests and cures? Not at all. When the numbers were used, it was more out of professional jealousy than in a quest for objective knowledge."[171] Castel had been right to say during the debate that these were not statistics in the strong sense of the word, even though they reflected a routine preoccupation with counting. There appeared in fact to be an unbridgeable theoretical gap between the practice of clinical tabulation, where the numbers were never sufficient to satisfy the exigencies of probabilistic mathematics, and the compilation of mathematically informed population statistics in the manner of Villermé and Quetelet, where they reliably were. But this gap became evident only when clinical statistics came to be debated in the Academy as a matter of method. To the extent that probabilistic thinking had played a foundational role in the birth of the clinic, as Foucault claims, this was a point at which the great majority of leading French doctors judged that kind of clinic to be a failure. In the first place, the style of generalized knowledge involved in the "counting" method was rejected by the majority of the French medical academy in the name of attention to the individual patient. The notion of the normalizing average bore the brunt of that critique. And in the second place, clinical statistics were deemed to be without mathematical or scientific significance. They failed the test of large numbers. In the majority judgment of the academy, "medical probability" was simply the intellectual condition of medical practice, and the use of numbers was no more, when it was all added up, than an attempt to mask that difficult truth.

CONCLUSION

The 1837 debate claims its place in our history of normality precisely because it was a high polemical instance of resistance to quantification. The medical academicians were comfortable enough with "the normal state" and "the normal anatomy," but they refused to allow in principle that those concepts might be translated into averages and probabilistic calculations. The debate was, as we noted at the outset, no moment of triumph in the rise of statistics. Even as it became thinkable that medicine might organize itself around a mathematically grounded understanding of the normal condition and the normal person, there emerged an intellectually organized discourse that confronted and rejected thinking of that kind. And this was no singular moment of resistance. It established a topos in which the opposing positions were still available to be occupied during the following decades. Twenty years after the 1837 debate, Paul Broca, who was later to play a leading role in French anthropology, wrote a series of two articles for *Le Moniteur des hôpitaux*, a biweekly publication aimed at the medical profession. The articles were structured as a dialogue between a first-person advocate of numerical methods and a person referred to as an "experienced doctor." Blows were traded figuratively. The numerist quoted a Hippocratic aphorism to the effect that experience is deceptive, and the experienced doctor retorted with the gibe about the shoemaker who made shoes in only one size. At the end of it all, Broca's speaker observed: "I have seen my friend since then. He assured me that our two conversations had only confirmed him in his thinking. I said that the same was true for me. That is how most discussions end."[172] A discursive routine had been established around medical statistics. From 1837 onward, statistics offered new means of talking probabilistically about long series of events and large classes of people, but ranged against statistics in the decades that followed was a strong body of professional opinion that considered such calculations improper to clinical medicine. The emergence of a numerically based discourse of normality in the medical field was cognate with the emergence of a counterdiscourse that refused to allow that mathematical forms of knowledge might properly be applied to living beings.

CHAPTER THREE

Rethinking Medical Statistics: Distribution, Deviation, and Type, 1840–1880

QUESTIONING "NUMERICAL METHOD" IN THE NAME OF STATISTICS

The average was able to be denigrated in 1837 by the opponents of numerical method in medicine because it was seen as an unthinking reduction of the quality of knowledge demanded of the clinician. The true doctor, it was said in effect, had the insight to understand each patient in their individuality, and the notion of the "average patient" was no more than a facile epistemic fiction that seemed to ignore the art of clinical medicine. The argument against numerical method could be forcefully made as long as the method was understood by members of the profession to be the aggregation of data (counting cases) that were then subjected to mathematical reduction and manipulation through the calculation of averages (average cases, average patients). In this chapter we move on to tell how the assumptions in play in the 1837 debate were called into question in the decades that followed. It should be clear from the outset that we do not consider the Academy debate to be an impasse in the course of scientific progress. Some questions raised there will continue to be in play throughout our history. But we will now move to expound a number of other issues that came into view after about 1840 as statistical notions were adapted, sometimes with considerable difficulty, to medical thinking. Medicine was emphatically not the primary domain in which data gathering and tabulation came to the fore, and we will acknowledge in passing the more straightforward development of statistics in other fields. But the very difficulty of applying numerical methods to medicine appears to have been a stimulus to intellectual and professional thinking, albeit of a contested kind. During the middle decades of the nineteenth century, that contest helped to shape what we now think of as the normal, as critical discussion of the place of numbers in medicine helped to produce some rather new propositions about such things as averages, deviation, and laws of distribution.

A leading figure in debates about the value of statistics was the Belgian Adolphe Quetelet, who published a number of widely read essays on the topic. Quetelet was a mathematician, and some of his writings were directed to a specialist audience, but much of what he wrote was aimed at an informed general public. In 1850, Sir John F. W. Herschel wrote about Quetelet's role in celebratory terms: "the centre of an immense correspondence, he has moreover succeeded in inspiring numerous and able coadjutors, not only in Belgium but in other countries, with a similar zeal."[1] Quetelet sought, and sometimes found, "coadjutors" among people in positions of political power, attempting to persuade them that the gathering and ordering of data were essential to the business of government. Statistics were after all, by etymological definition, a matter for the state. A key work by Quetelet published in 1846, *Lettres à S.A.R. le duc régnant de Saxe-Cobourg et Gotha sur la théorie des probabilités appliquée aux sciences morales et politiques* (*Letters to H.R.H. the Reigning Duke of Saxe-Coburg and Gotha on the Theory of Probabilities Applied to the Moral and Political Sciences*), enacts the rhetorical fiction of a teacherly scientific address delivered to a senior person in government. And although the use of statistics in clinical medicine was not primarily a matter for government, one of the letters was entirely devoted to that question.[2]

Quetelet began the letter by registering a history of difficulty and conflict: "Nothing has been more keenly contested than the usefulness of statistics in the medical sciences." But that, he said, was a consequence of the manner in which statistics were being used in the field.[3] One of his primary tactics was to answer skepticism with skepticism. Doctors viewed statistics with ironic reserve, and it was appropriate in Quetelet's view for statisticians to respond in kind. Medical practitioners, he said, had simply shown no patience in their use of numbers. When the results had not been as anticipated, they had resorted to mere speculation: "they suddenly give up statistics and launch headlong into the domain of conjecture."[4] The issue for him was not in fact whether statistics ought to be given a place in the field of medicine. He found them to be already in widespread use. The problem was that they were not employed as they ought to be. Our own history in the previous chapter certainly marked some inconsistency in this regard. We noted that, for all the self-conscious rejection of numerical method in medical circles, mortality tables were published with great frequency during epidemics. Historian Stéphane Callens notes in turn that "the keeping of hospital statistics means that medicine and statistics had to have some relation to each other whether that was desired or not."[5] Indeed, throughout the 1850s those tables were sometimes done with great care. In 1853, the *Gazette hebdomadaire de médecine et de chirurgie (Weekly Gazette of Medicine and Surgery)* commented on a statistical report on cholera deaths in England provided by the Registrar-General. The tables in this report were described as "admirably well done," and thus of great value to medicine as a whole.[6] But Quetelet saw the ongoing use of such tables, at least in the 1840s, as a contradiction in rhetoric and practice, insisting on the irony of it: "However, no one questions the usefulness of mortality tables in medical research and in speculations about the duration of life."[7] He went on to add: "They all do statistics in some way or other: some entrust their results to memory, others to paper; some produce statistics without realizing it, just as M. Jourdain produced prose."[8] Like Molière's character M. Jourdain in *Le Bourgeois gentilhomme*, who famously produced prose without being aware that he was doing so, doctors were engaging in statistics, but without the methodical self-awareness required of scientists.

In view of this general propensity, the notion that one might adhere to the genius of medicine by not practicing number was hardly credible: "Is it to be believed that people as prudent as this keep no account of their observations or that, if they keep an account so that they can make comparisons later, they prefer to entrust them to memory rather than to paper? Whenever observations are collected so as to be made comparable and allow deductions to be made, one is doing statistics, whatever one might say to the contrary."[9] All these doctors must have been doing statistics of a kind. They had to do so in order to organize their clinical experience, even if they did not admit it to themselves. A version of this point had already been made in the Academy debate by P. C. A. Louis when he spoke in defense of his own practice of numerical method, but Quetelet's irony was more pointed than that of Louis, who offered only the most general defense of counting. Quetelet had a rather more precise idea about what it might mean to do medical statistics. Doctors who were already counting, he said, must learn to count scientifically: "There is simply one difference between those who write down the results of their observations and those who entrust them to memory: the former are being true to the principles of science, while the latter are failing to do so in the crudest manner."[10] It should be noted in passing that Quetelet was preceded down this path by a lesser-known statistician who happened to be a medical graduate, Jules Gavarret. In two publications dating from 1840, Gavarret had revisited the 1837 debate and expressed the view that "the members of this learned society had only the vaguest idea of how to use mathematics in medicine."[11] Gavarret was particularly severe on "the advocates of numerical method" for their failure to apply in practice the theory to which they claimed allegiance.[12]

When Gavarret spoke of failure and Quetelet of crudity in medical practice, it was clear enough what they meant: a general lack of formal procedure in the handling of figures. But what forms might the scientific alternative take? What exactly were doctors failing to do when they counted "indiscriminately"? Quetelet did not provide a tailor-made answer to these questions, and certainly did not engage in a detailed critique of the errors he was characterizing in this way. His concern was to establish the principles that needed to govern medical statistics if they were in fact to be made scientific. The letter in question was not even addressed directly to the members of the recalcitrant profession, but Quetelet was taking the opportunity to spell out what it might mean for medical people to use numbers properly. Their profound unwillingness (*répugnance*) to count was particularly out of place when it came to dealing with things that "belonged exclusively to the domain of number."[13] Taking the pulse of a patient, for example, was in the most straightforward sense a matter of counting. But there was more to it than that. Pulse rates might be tabulated according to age, and scientific information gleaned from the ordering of those numbers. Yes, counting was essential to statistics, but there was further scientific work to be done with the numbers produced. It could now be shown by statistical method that medical writers in the past had been mistaken in asserting that pulse rates tended to slow with age. In fact, tabulation had recently shown that older people's pulses were actually faster on average.[14]

Other medical matters were more complicated. Key decisions about methods of treatment could be made scientifically, said Quetelet, only when the outcomes of competing procedures had been numerically compared. That ought to have been the way through the great controversy surrounding the treatment of bladder stones—whether to remove them by incision (lithotomy) or to crush them and flush them out through the urethra (lithotrity)—but the profession had generally failed to resolve this matter as it ought to have, by using the methods of statistics. Quetelet's account of what had happened was rather unfair to Jean Civiale, the principal advocate and practitioner of lithotrity. Civiale had sought data from a range of clinics in which either of the two procedures was used, and attempted to compile them for comparative purposes.[15] But Quetelet was unrelenting: whether lithotrity was to be preferred, all other things being equal, could only be decided by "the methods of statistics."[16] The debate among doctors had been undisciplined: "almost always, people neglect to investigate whether the number of observations is sufficient to inspire confidence."[17] What were indeed the dimensions of scientific "confidence," and how great were the numbers needed to "inspire" it? As it happened, that was for mathematicians a properly scientific question. Statistical mathematics offered a way of assessing the probability that a set of observations would be repeatable. The probability became greater, and measurably so, as the number of observations increased. And since the repeatability of observations was by mathematical definition accuracy, accuracy itself could be assessed from the outset in statistical terms. But answering the requirement for large numbers was a difficult matter for clinical medicine.

Quetelet's main attack on medical uses of number bore on what he saw as the failure of informal practices to provide genuine knowledge: "they needed to avoid generalizing merely on the basis of particular cases. Each of them had knowledge of only one face of the dice that turned up several times in succession, and had not had occasion to know the other faces for the lack of a sufficient number of trials."[18] Beyond the significant technical matter of the number of observations, he was drawing attention to something he considered more important still. While using a dice metaphor in the time-honored manner of probability mathematics, he was giving the metaphor a rather new significance. The dice with its multiple faces now stood for a plurality of causes at work in a given medical circumstance. By accumulating clinical experience in mathematically uncontrolled ways, doctors were failing to identify proper forms of treatment because they were not establishing predictable links between causes and their effects. They might appear to be successful in identifying a cause and treating it, but that involved reading only one face of the dice that happened to turn up more often in a particular circumstance. Other causes might be in play, and some of them might happen to be observed and treated by other doctors. All this led to unreliable diagnoses and rather haphazard patterns of treatment.

Medical claims to have found the cause of an illness were thus to be thought of as largely speculative. They fell into the logical fallacy of post hoc, ergo propter hoc, making unjustified inferences as to causes and failing to establish therapeutic regularities on which others could rely: "They do not limit themselves to saying, 'I have saved a lot of patients.' They go on to say, 'That is because I have identified the cause of the illness and have succeeded in applying the true remedy.' But they in no way prove the link between the effect and the so-called cause, which is what was required."[19] Doctors were likely to fail in scientific terms, not because they had no hypotheses about causes, but because the conclusions they drew as to etiology were not properly constrained and supported by their data, which were poorly organized: "They even take the further step, after assigning a so-called cause to an illness, of treating in the same manner all new patients who present, most often without taking into account their constitution, their age or their sex. This is where one finds the abuse of statistics, insofar as it is proper to speak here of statistics at all."[20] This criticism could well have been applied to other debates in the French Académie de Médecine in the 1830s during which leading doctors such as Louis and Bouillaud presented competing treatments for such high-profile illnesses as typhoid fevers. Each did his own "counting," citing numbers from his own clinic as proof of the effectiveness of a favored treatment, but their counting was not properly tabulated and the differences in therapies remained unresolved.[21]

For Quetelet, all that was no more than pseudostatistics. It did not take account of differences between kinds of patients. Doctors tended to suppose a standard illness whose etiology called for an unvarying treatment: "If identity existed rigorously in all men, it would therefore suffice to observe only one illness followed by healing in order to have the same success every time that the illness recurred in other individuals. But that perfect identity will probably never exist."[22] In the 1837 debate, the "average patient" had been made an object of derision and a sufficient reason for rejecting the use of number in medicine. Quetelet was not attempting here to defend that notion, although he had a strong theoretical commitment to the average. His first move was simply to counterattack by pouring scorn on the logic of inference practiced by doctors. They might have disdained the notion of the average patient, but Quetelet was suggesting in reply that they were effectively committing to the clinically absurd notion of a standard, unvarying illness.

It has to be said that some other statistically minded writers were less severe about the use of numbers in the medical profession. Writing more than a quarter of a century after the 1837 debate, André-Michel Guerry was still prepared to side with Louis against Risueño and other critics of numerical method. Those critics, he said, had been mistaken in implying that numerical method was designed for therapeutic application to individual cases. In fact, Louis himself had expressly said that the general treatment of a disorder did not reliably dictate the treatment of any particular case.[23] Numerical observation, said Guerry, could make a valuable contribution to the study of pathological anatomy and physiology. It had become possible by this method to ask whether different categories (according to sex, age, etc.) were equally prone to a given illness. The answer to those questions could be found by counting, and Risueño ought not to have objected to that. Guerry referred to contributions made in this regard by Rayer, Civiale, Villermé, and others, although he was compelled to acknowledge that even in 1864 the place of numerical method was far from assured: it was "still the subject of acute controversy in the medical world."[24]

Quetelet was less easily placated than Guerry because, in a sense, he expected more from medical statistics. He was insisting on the relation between cause and remedy, and thus giving a quite particular inflection to the study of probabilities. The classical calculus, it should be remembered, was focused on combinations rather than causes. It served to assess the chances that certain events would occur in combination or in simple succession. Events were considered as matters of calculable chance rather than in terms of consequence. Quetelet, on the other hand, was bringing to the fore the question of causality, even as some other statisticians continued to pursue a theory of combinations.[25] For Quetelet, the true scientific purpose of statistics was to point to causes, and that was precisely what he meant by "statistical laws." Laws implied recurrent patterns of causation, discoverable in principle through statistical inquiry. Herschel, in his generally positive review of Quetelet's *Lettres*, refused to follow this line of reasoning: "Into the delicate and refined system of mathematical reasoning, now generally known as the 'calculus of Probabilities,' the metaphysical idea of Causation does not enter. The term *Cause* is used in these investigations without reference to any assumed power to effect a given result by inherent activity. It simply expresses the *occasion* for a more or less frequent occurrence of that result, and may consist quite as well in the removal of an impediment as in any direct agency."[26] Indeed, while Quetelet was undoubtedly for his fellow Europeans the most prominent statistical thinker of his time, it can reasonably be said that the discipline turned away in its subsequent development from his concern with causality, focusing instead on correlation. As many later statisticians saw it, the task of statistics was not to suppose that one event or phenomenon was the cause of another, but merely to take account of the fact that they occurred together.[27] Yet while Quetelet's commitment to an understanding of causality based in probabilistic mathematics and his in-principle application of it to medical statistics may not have proved decisive for the long-term development of statistics as such, they mattered greatly for the mid-nineteenth-century development of normality as an object of medical knowledge.

When one looks back from the *Lettres* of 1846 to the Academy debate of 1837, there is a nice historical irony in the position adopted by Quetelet. Unlike Guerry, he was in no sense following in the steps of those who had advocated numerical method. The great European champion of statistics was taking as his primary target precisely those who had advocated counting nine years earlier. He was critiquing and indeed dismissing their numerical method as scientifically trivial, arguing that their mathematical ineptitude was characteristic of the whole profession. So it is quite misleading to suggest, as one historian does rather blandly, that "the numerical method of Louis continued to gain ground" after the 1837 debate.[28] Quetelet's severe criticism of it shows that there was in fact no gradual capture of contested territory during the decades immediately following. And as if to compound the irony, Quetelet took numerical method to task, not just for its lack of mathematical precision, but for its failure to give due value to clinical *expérience*—with its double sense of "experience" and "experiment"—the very notion that had been mobilized to such effect in the Academy by the opponents of medical uses of number. Quetelet spoke, much as Risueño had done, of the unrepeatability of clinical events: "In the entire course of his life, a doctor will perhaps never act twice in circumstances that are absolutely the same."[29] That is why the clinician was required to display the finest discernment: "this examination can only be carried out by men of exquisite tact and certain judgment, by observers endowed with that aptitude for patience to which Buffon gave the name of genius."[30] The requisite quality appeared precisely to be clinical genius as Cabanis and Risueño understood it. The doctor's *expérience* was held up for admiration for the very reason that it allowed him to perceive each case distinctively.

But while valuing clinical tact as the capacity to perceive the distinctive and even the unique, Quetelet was thoroughly integrating his mathematics into the theory of medicine by accounting for *expérience* itself in broadly statistical terms. He did not believe it possible to practice a generally exhaustive method of observation that allowed for all possible combinations and causes. That would be a vast, abstract exercise that could not be accommodated within the space of clinical medicine: "If one were required to provide for all the cases that may present themselves and gather enough observations to verify all the possible combinations, there could be no hope of reaching a satisfying result. One would have to give up not only on the use of statistics, but on observation itself."[31] Medical science could never hope to trace a path from every possible cause to every possible outcome. The number of observations could never be large enough, and there would be no point in speaking of *expérience*: "Experience would be an empty notion because one single illness could undergo an infinite number of modifications under the influence of all the causes that give rise to it."[32] In other words, it was necessary in practice to ensure that the number of observations would be sufficient to justify a limited number of inferences about causality. Managing that statistical exigency was, at least in part, the role of "experience." That was what had been lacking in the debates about bladder stones. It was not in fact that doctors had failed to carry out an ideal number of observations, since the ideal itself was something of an abstraction. Rather, they had not properly assessed what was required because they had not integrated statistical findings into their clinical practice. By claiming to build knowledge without the support of statistics, they were "launching into blind empiricism and rejecting the enlightenment of experience."[33]

When statistical thinking was integrated into clinical experience, it tended, however, to produce equivocation in Quetelet's account of things. If medicine were to be constrained in theory and in practice by the need for statistical methods, how might that apply, if at all, when dealing with individual cases? In the *Lettres* and elsewhere, he stated as a matter of principle that individual cases could not be judged according to statistical tables,[34] but he did sometimes offer a statistically based account of how the clinician might deal with an individual patient. That he did so is significant for our genealogy because he came to speak then, perhaps for the first time in history, of the average and the normal as proximate notions. That occurred in *Sur l'homme et le développement de ses facultés ou Essai de physique sociale* (*On Man and the Development of His Faculties, or an Essay in Social Physics*). In that 1835 essay, Quetelet put forward the concept of *l'homme moyen*, the average man. He offered it mainly as a way of organizing knowledge about populations, but he also used it to make a theoretical incursion into clinical medicine. That venture called for him to use a key term that informed physiological discourse at the time, that of the "normal state." While the term had become available for medical usage in the 1830s and 1840s, Quetelet's statistical thinking gave it a rather new inflection.

In *Sur l'homme*, Quetelet was attempting to show the theoretical power of his concept of the average man, and the medical sciences seemed to him an area in which it could be usefully deployed. He offered a more specific version of the claim made in the *Lettres* that conscientious doctors were of necessity engaged in statistics, arguing here that some notion of the average man must be in the doctor's mind whenever he examined a patient: "Taking the average man into account is so important in the medical sciences that it is almost impossible to judge the condition of an individual without relating it to that of another fictive being who is considered to be in the normal condition. That condition is effectively the one we are discussing here."[35] To assess the presence or absence of a pathological condition in a patient was effectively to measure discrepancies between the patient's current state and that of the average man. Quetelet told it as an exemplary story:

> A doctor is called to the bedside of a patient and, after examining him, finds that his pulse is too rapid, his breathing agitated beyond measure, and so on. It is clear that to make such a judgment is to recognize, not only that his characteristics observed deviate from those presented by the average man or man in the normal state, but even that they are beyond the limits that can be reached without danger. Each doctor, in making such an assessment, refers to the documents currently available to science, or to his own experience. The assessment he reaches is tantamount to the one we wish to make on a larger scale and with greater precision.[36]

According to Quetelet's description, when the doctor examined his patient he always carried in his mind a point of reference that was a representation of the average man or the normal state.

Quetelet was aware that his position was rather adventurous. He could not have failed to recall his own dictum that the proper calculation of the average man could only be done on the basis of a large set of observations, which did not seem to apply in the case of a single patient. "It is proper to recall here," he said as if to remind himself, "that general laws in relation to masses are essentially false when applied to individuals."[37] But he found his way through that difficulty by hypothesizing multiple observations of the individual patient. The normal state for a given patient could be understood as the average of his conditions at various times. To know the patient well was to know those conditions as a set of variations. Since it was "not ideal" to do this from memory, the doctor should really have at his disposal, for a given patient, a table showing the limits of the variations or deviations (*écarts*) that the patient might safely experience. In the absence of such a table, "the doctor finds himself obliged to bring the patient back to the common scale and assimilate him to the average man."[38] The physician was compelled in practice to refer all diagnosis and treatment to the standard measurements of the average or normal man. That made all doctors practitioners of implicit mathematical knowledge, and left little or no room in principle for Cabanis's conception of medicine as an art that disdained counting.

Having advanced a claim that tended to make the average man and the normal state into equivalent notions for clinical purposes, Quetelet then withdrew onto more secure terrain. He restated his position that the calculation of averages ought not to be applied to individual cases. But he withdrew only part of the way. If the notion of the average man were to be applied in a particular instance, he added, it would be best for each patient to have a table of personal statistics. In that way, "it will be possible to draw up a table deviating to a greater or lesser degree from the one that relates to the average man. That will allow the patient to recognize anything about himself that presents a greater or lesser anomaly and calls for urgent attention."[39] The table imagined by Quetelet would set out the characteristics of the individual patient, displaying them as averages, while at the same time allowing these personal qualities to be compared statistically with those of the average man. The doctor could then consult this table "in order to assess the extent of deviation from the normal state and determine which organs were particularly affected."[40] Here, at least hypothetically, the calculation of averages was to be used to measure ill health precisely as deviation from a calibrated standard. The clinical norm, in this more refined version, would not simply be that of the average man, but would be specifically established for each patient. Illness would be marked by deviation from a personal rule, itself known through averages, which could then be assessed in relation to a more general standard identified as that of the average man.

As it happened in practice, however, given the lack of an individual scale, the "average man" provided the only point of reference: "the doctor finds himself obliged to measure deviations according to the common scale, assimilating the patient to the average man."[41] By making this acknowledgment, Quetelet was effectively conceding a key point to those who had attacked numerical method in the Academy debate of 1837. To refer to the average man as the basis for calibrating the symptoms of a given patient was to confront great practical difficulties: "It is obvious that a doctor called to the bedside of a patient he is seeing for the first time is bound in certain circumstances to commit errors in subjecting him to the common rule."[42] "Grave misapprehensions" were likely to result, he quickly acknowledged: "it is apposite to point out here that general laws applying to masses are essentially false when applied to individuals."[43] Nonetheless, Quetelet went on, as if to ratify and accommodate the contradiction in his argument, "that does not mean that differences with respect to the common scale cannot be fruitfully consulted; deviations are always worthy of consideration."[44] Things would in fact be more straightforward—and the tension between theory and practice made less acute—if people were to consult their doctor in times of health: "People generally call the doctor when they feel unwell. I believe that it would be very useful to see him when one is in a state of health. He could then study our normal state and collect the information that would allow him to make comparisons in cases of anomaly or indisposition."[45] The logic was compelling once the premises were accepted. Indisposition, statistically understood, was always identifiable in principle as an anomaly, and the measurement of anomaly against a personal norm could not fail to serve as a guide for the clinician.

Whereas in the 1837 debate the emphasis had been on the demonstrably absurd notion of the average patient, Quetelet was claiming rather to calculate, at least in principle, an average condition that would constitute the normal (nonpathological) state for a given individual. And because the normal state varied from one person to another, there could be no clinical circumstance in which it was strictly proper to treat a patient as average. But since no refined data about individuals were available in practice, doctors were obliged to refer their treatment to a "common scale" that was in fact the average of all the normal states. That scale amounted to a notion of average health, summed up in Quetelet's "average man." "The constitution of the average man serves as a type for our species," he wrote succinctly. That the average was able to be represented as the "type" helped to constitute a normalizing view of good health. For Quetelet, the close theoretical relation between average and type functioned at the level of whole populations as well as within individuals: "Each people has its own particular constitution that deviates to a greater or lesser degree and is determined by the climatic influences that characterize the average man of the country in question. Each individual in turn has their own particular constitution that depends on their bodily economy and their manner of being."[46] To know the normal state of a population as a whole was to know the average set of qualities found across the population. To know the normal state of a given patient was to know that individual in the aggregating and averaging way one knew a type.

MORAL STATISTICS

The contest about the place of statistics in medicine did not typically arise with such intensity when statistics claimed a place in other fields. Quite often the gathering of statistics simply became an unquestioned and unquestioning routine. The *Journal des travaux de la Société Française de Statistique Universelle* (*Journal of the Proceedings of the French Society for Universal Statistics*), for example, reveals a learned society simply getting on with the job even as the Académie de Médecine was riven by controversy. The Société Française de Statistique Universelle was, as Zheng Kang has pointed out, a place in which the practice of political economy grounded in statistics served to reduce the distance between science and government.[47] Libby Schweber observes that the society was divided between administrators led by Alfred Legoyt and moral and medical statisticians led by Louis-Adolphe Bertillon, but the division to which she refers was a professional rather than an intellectual one.[48] The journal, published from 1832 to 1849, bespeaks an undisturbed sense of purpose, offering a series of reports on countries around the world. From time to time, it made room for general affirmations about the value of statistics such as the following: "statistics, like nature for the physicist and the naturalist, are an inexhaustible source of facts and positive truths."[49] When members of the society did reflect about the field of statistics as a whole, they most often did so in order to map the terrain of current and future activity. A standard description identified three broad kinds of statistics: physical and descriptive, moral and philosophical, positive and applied.[50] The first of these corresponded approximately to what might now be called geography and trade, and the third to politics and public administration. Nearly all of the journal's content, as it happened, belonged to these two areas. The second category had little or nothing to do with morality in the sense that has become current in our time. It included in principle such matters as criminality, judicial decisions, and marriage rates. This was by far the most contentious area of the three in wider discussion, not because it was ambiguously defined but because there was disagreement about the validity of statistics in this field.

In 1837, at the same time as debate was raging in the Académie de Médecine, the mathematician Siméon-Denis Poisson laid a strong claim to the "moral" domain, asserting that it was entirely appropriate to build statistical knowledge of that kind. The calculus of probabilities, he said, applied just as well to "the things that are called moral" as to others. Any belief to the contrary was just "prejudice."[51] It was true, he acknowledged, that some moral things were affected by "variable chances" and therefore did not display the kind of regularity that lent itself to calculation.[52] But such matters as jury decisions could be studied in probabilistic ways, and he offered equations to show how the calculations could be done. He was able to demonstrate that, at any given level of the French legal system, the proportion of condemnations to acquittals varied little from year to year,[53] so that one might in fact wager with confidence on the overall outcomes of jury decisions.[54] That was the occasion for a further high-profile academic debate that was compared at the time to the great debate between Cuvier and Geoffroy Saint-Hilaire.[55] Since our history does not call for fine polemical detail on this question, we will simply represent the position of Poisson's opponents by quoting Alexandre Moreau de Jonnès, who was director of the general statistics section in the French ministry of commerce. Moreau was unequivocal, expressing strong disapproval on both technical and ethical grounds.[56] Moral and intellectual statistics were, he said, "a vain attempt to make the mind and passions submit to calculation, and to compute the movements of the human soul and intelligence as if they were defined units able to be compared with each other."[57] The only statistics worthy of the name, said Moreau, had to do with "social Facts," insisting that the word "fact" be spelt with a capital to emphasize the distinctness of the social as a source of data and to ensure that it was not confused with incalculable human qualities.[58] As a contributor to the *Journal des travaux de la Société Française de Statistique Universelle*, Moreau can only have approved of the fact that the journal had little to say about moral statistics or about clinical medicine. Both of these were, in his view, outside the domain of government and of the properly factual. The *Journal*, when it did actually have something to say about medicine, simply concerned itself with the administration of hospitals, publishing a study of the average number of days spent by each patient admitted to a hospital in Paris.[59] These were "social facts" of a straightforward kind. By focusing on such data, Moreau and his colleagues were generally able to avoid the theoretical issues raised by Quetelet.[60] No one writing in the journal worried openly about the definition of average patients, average illnesses, or indeed average hospitals.

In an address delivered in 1846 to the Royal Academy of Belgium and published as "Sur la statistique morale et les principes qui doivent en former la base" ("On Moral Statistics and the Principles on Which They Must Be Based"), Quetelet took direct issue with Moreau de Jonnès and defended moral statistics in principle.[61] He was, as Lorraine Daston observes, extending the practice of statistics beyond the "traditional areas of demography and mortality rates" in which some earlier French statisticians had distinguished themselves.[62] But there was more to it than that. He certainly undertook an ethical and philosophical defense of moral statistics against the charge that it would "degrade man by taking no account of his moral and intellectual qualities, reducing him to a machine whose every movement could be calculated."[63] But he also promised to make a strong positive contribution to the discipline, arguing that different forms of statistical inquiry could work in convergent ways: "whenever one is working on a large number of men, the same thing applies to their moral qualities as to their physical ones: one can suppose that there will be an average term around which all the observed elements will be organized as a group, some being greater and some lesser."[64] So when he offered his own tripartite division of statistics in a later essay, it was eminently clear that the point was not to quarantine certain human qualities and activities as unsuited to quantification and calculation, but rather to claim the full range for probabilistic inquiry. In *Anthropométrie, ou Mesure des différentes facultés de l'homme (Anthropometry, or the Measurement of the Different Faculties of Man)*, published in 1870, Quetelet focused on the mensuration of human physical features while announcing two volumes to follow, one devoted to moral qualities and the other to intellectual ones.[65] For each of these categories, he affirmed, the intensity and range of qualities would follow a comparable pattern. It was not, of course, that the qualities would be distributed in such a way that any given individual would be located at the same point on the scale for physical as for moral or intellectual ones. Rather, the formal pattern of distribution of each set of qualities, expressed mathematically, would prove to be the same. The mathematical principle of regularity would apply to every quality. At the end of his statistical work, he would present consolidated evidence of "the homogeneous nature of men taken together," not just their physical proportions, but "their moral and intellectual properties."[66] If that was indeed to be the outcome, Quetelet's statistics would make it possible to define the full range of human qualities, making knowledge about all of those qualities available in comparable, tabular ways.

PATTERNS OF DISTRIBUTION

Quetelet's concept of the average man served to sum up what he called "general facts" about a given population. The average man was not a single person existing in the world, but a figure of scientific knowledge produced by calculating the average of various features, and indeed various sets of features. That consolidated average would serve quite properly, Quetelet claimed, as an index of the population under study. As he saw it, this was a mathematically sound means of considering a human society as a whole. Here as in other contexts, he noted that the average could fulfill its representative function only if it was drawn from an appropriately large set of numbers. So the "average man" was in the first place a technical term, and whatever Quetelet might have said in 1835 about the average man as a potentially useful reference for clinical practice, it was in the ordering of knowledge about populations that this figure was called on to do most of its work. It is noteworthy that reference to the average man was regularly attended in Quetelet's writing by celebratory metaphor. Turning aside from the study of individuals, who could not in any case be well known by statistical means, he wanted to "examine the social body."[67] Calculating the average man would allow him to know that body by locating its center of gravity. The very centrality of the average man was thus held up for attention: "The man I have in view here is, within the society, the analogue of the center of gravity in the body. He is the average around which the social elements oscillate."[68] Alain Desrosières rightly observes that Quetelet was committed to "thinking society as a stable whole, rather than analyzing and hierarchizing its parts,"[69] and the figure of the average man helped to achieve that, ordering everything around it without actually constructing a hierarchy. Far from being a cipher nestled in some nondescript middle ground or indeed an undistinguished figure lost in mediocrity, the average man was for Quetelet the center around which meaning, statistical and social, took shape.

For this reason, when the average man was described in *Sur l'homme* as "the ideal type," that expression was for Quetelet productively ambiguous. The average man was certainly "ideal" in the sense that he was the product of mathematical calculation, but he was also a figure to be admired by virtue of his qualitative equilibrium. He represented not just the sum, but the summum. As Daston says, "Quetelet held up *l'homme moyen* as a literary, artistic, moral, and intellectual ideal . . . the great man was he who subsumed his individuality in 'all of humanity, nature, and the universal order.'"[70] Epistemic and aesthetic values converged here, as *l'homme moyen* became the standard of beauty defined mathematically: "if the average man were perfectly established, one could consider him as the type of beauty."[71] And while the "average man" may have been less important in the long-term history of statistics than some other mathematical innovations, this was a notion that had a strong influence on cultivated public discourse in the decades that followed. Hacking puts it simply: "there is no doubt that the 'average man' stuck, even though almost no one had favorable things to say about the concept when taken literally."[72] By its very quotability, the contested notion of *l'homme moyen* played a role in the genealogy of the normal. We will see in later chapters how the average and the normal eventually came to function together in the measurement of human bodies and in knowledge about how populations could be represented.

Quetelet was keenly aware of the attacks on his statistical model, and defended it at every opportunity,[73] but it eventually became clear to him that the concept of the average man could not of itself suffice to organize statistically based knowledge about whole populations. The problem, put simply, was that the calculation of a set of averages gave no account of the spread of qualities across a large group. It surely mattered for the description of a population how broadly other qualities were disposed or dispersed about the figure representing the average. In *Sur l'homme* (1835), Quetelet had appeared unconcerned about this set of questions. It was true, he acknowledged, that there were likely to be "exceptions" to the general rule summed up in the average, but such was the case with all laws of nature. The summary figure itself would "best express what takes place in the society, and that is the most important thing for us to know."[74] Once that figure was established, the qualities of the average man could be identified by definition as "a nice balance, in a perfect harmony that is equally distant from excesses and deficiencies of every kind."[75] But how great was the distance separating the average from the excesses and deficiencies? How excessive were the excesses? Quetelet made no particular attempt in the first instance to measure such distances or calculate their significance. If the average man was indeed the ideal type of beauty, Quetelet wrote in 1835, then "everything that was furthest from resembling his proportions or his manner of being would constitute deformities or illnesses; anything that was so different, not only in its proportions and its form, as to stand outside the limits observed, would constitute monstrosity."[76] At that time, his mathematically based aesthetics could only characterize those far distant figures as monstrous and the somewhat closer ones as deformed. There was no properly quantitative account of the distance separating them from the center, and therefore no promise of a qualitative account interpreting that distance in detail. Anomalies were known to exist, said Quetelet, but he treated them at that time much as he treated exceptions. The task of the statistician was simply to focus on the average man: "He is the one we need to examine, without being distracted by particular cases or anomalies."[77]

That is how in *Sur l'homme* we come to find the historical beginnings of a statistically grounded theory of the normal unaccompanied by anything that resembled a concomitant theory of the nonnormal or the abnormal. More narrowly, Quetelet had little or nothing to say at that time in mathematical terms about the anomalous or the exceptional. They were apparently no more than "particular cases" to be methodically set aside as the statistician went about the task of computing large numbers. But that was not the position defended three decades later in *Anthropométrie* (1870). There Quetelet reminded his readers of something that statisticians had tended to forget: "Let us note right away that in their concern with the idea of the average where quantities are subject to variation, people have rather lost sight of the *limits* within which these variations occur. Wherever one can say *more* or *less*, one is required as a matter of necessity to take account of an average condition and two limits."[78] The reminder about limits may have been all the more apposite as, in his earlier writings, the author himself had undoubtedly helped his readers to forget them. Now, as he referred to measurements of height in human populations, he revisited and rethought the very terms he had used in *Sur l'homme*: "According to a commonly held view, giants and dwarves are anomalies, monstrosities of the human species. But from the point of view of my own theory, which is, I argue, the point of view of science, they are necessary links in the human chain. They complete the range of measurements over which height in humans must necessarily extend."[79] Whatever was to be said and known scientifically about a human population was now to involve not just the calculation of an overall average, but an assessment of the range as such, marked by outer limits, in which every variation needed to be given a place. Only when the ideal type and the former monstrosities were taken together would a complete description become available.

It is important to note that computation, in this revised view, did not just apply to averages. The limits of the range were themselves subject to probabilistic calculation. In *Anthropométrie*, Quetelet spelled out the statistical work to be done in giving a full description of height in a human population: "We can no longer restrict ourselves to observing the average height of the individual. The respective number of heights corresponding to each age must be known. For that to be done, all that is required is to arrive at the law of agreement that applies across them, and as we shall see shortly, to know the two extreme heights as well as the average."[80] There was, he was suggesting, a law governing the distribution of heights at any given age.

We should note before concluding here that the limits of which Quetelet speaks were themselves within the reach of mathematics. Statisticians were actually able to "establish the limit that will probably not be surpassed."[81] Poisson had in fact given a more detailed account of such calculation than Quetelet himself, but the mathematical procedure was known to both. Poisson had drawn attention to a method of calculating limits that could be used at the beginning of a process of observation, making it possible, not to determine limits directly, but to calculate the probability that a result would fall between two chosen points.[82] So while these limits were not absolutely fixed, they were certainly not arrived at in an arbitrary manner. Without being determined empirically—by finding, for example, the tallest and the shortest person in a given population—they could be arrived at through a calculation of probabilities. It was unlikely, to a calculable degree of unlikelihood, that a measured observation would be found beyond a certain limit. And it was even less likely, again to a calculable degree, that a measured observation would be found beyond a more extreme point still. So it was a matter of deciding what degree of probability was called for, and determining the limits accordingly.

THE CURVE

For distribution as such to be adequately represented, the task of statistics was to provide a mathematical model that made the average salient while at the same time making the limits available to calculation. Quetelet found a way to do this by applying the binomial law, which, as he noted, "establishes intensity and limits."[83] This law describes what happens in a series of trials with only two possible outcomes, such as coin tossing. Imagine a trial in which a coin is tossed n times. It will happen in this trial that heads turns up a certain number of times. Let us call that variable k. Then imagine another trial where the coin is tossed the same number of times (n). The number of times heads appear is likely to be different. That gives a different value for k. If n trials are themselves repeated many times, there will be a set of different values for k, which, if plotted on a graph, will show a predictable pattern of distribution. Very low values for k (close to 0) will be rare if n is reasonably large, as will very high values. The results will tend to cluster around the average.[84] When represented side by side as vertical lines on a graph all with the same base, these results will make a shape in such a way that a curve can be drawn through their upper ends. They do not of themselves constitute a curve, but the drawing of a curved line smoothes out the representation and makes them appear continuous when they are in fact discrete.

Theodore Porter observes that "the history of this curve, now known as the Gaussian or normal distribution, is practically coextensive with the history of statistical mathematics during the nineteenth century."[85] Readers with even a passing acquaintance with statistics are likely to recognize it as the normal or bell curve, but it is important for the sake of our intellectual history that it not be recognized too quickly. Quetelet did not refer to it in those terms. On the rare occasions when he used the word "normal," he typically meant by it the opposite of "pathological," thus following the pattern of medical usage that had become established in the 1830s. There is in fact every reason to believe that he would have found the expression "normal curve," had he ever encountered it, to be a misleading and unhelpful ellipsis. The curve he described was initially called "binomial," in recognition of its mathematical derivation: it was "this simple curve to which we have given the name 'binomial.'"[86] His fellow statistician Antoine-Augustin Cournot, writing in 1843, spoke rather of a "curve of probability," but there appeared to be no great stake in the choice between the two terms.[87] Indeed, Quetelet's own terminology was not constant, even if it was broadly consistent. He sometimes spoke of the "general curve,"[88] and that term was entirely apposite. The curve represented a pattern of distribution that seemed to be generally found for any quality across any population: "The analytical explanation of the same curve can be done not only for height, but for weight, strength, speed and any other quality that can be measured in man. Moreover, the same law can also be observed for moral and intellectual qualities. And not only does it apply to man, it is also suited to animals. So in general it relates to everything that belongs to the law of growth: to man, to animals, and to plants."[89] "That the curve could be drawn and its properties observed in all such cases led Quetelet to speak of it as the graphic representation of a law. He was affirming the law-governed nature of this particular pattern of distribution. Its mathematical shape he took to display a general truth.

Quetelet found an additional means of conceptualizing the use of the curve when he talked about it as if it were a distribution of errors. He offered the example of repeated measurements of the height of a mountain. Such observations could not be expected to coincide perfectly with each other. Rather, they would be distributed in a known pattern, the very same that had been revealed by the analysis of binomial coefficients we have just described: "Thus when one measures a height, between the greatest and the smallest value observed the other values stand at different distances from the average, but the majority of them fall in the neighborhood of that average. This is because the further one strays from the true number being sought, the probability of a result falling at that point diminishes more and more."[90] That it should be so was to him a source of wonder: "Is it not wonderful that errors that are made by accident should take their place in such a perfect order? And that slipups should occur, unbeknown to us, with a symmetry that appears to be the result of the most carefully reasoned combinations? The curve that represents the order of succession and the number of results would be exactly the same as the curve of possibility."[91]

So here was another name for the curve, the curve of possibility, but the very multiplication of terms was itself significant. While the mathematical principle governing the production of errors might prima facie have appeared different, in practice it happened to display the same regularities as a growth pattern. Quetelet believed he had found in the binomial curve a law of remarkable generality. When observations of various kinds were regularly found to follow the shape of the curve, it seemed to him that the possible was equivalent to the probable and the probable to the general.[92]

FINDING LAWS OF NATURE THROUGH STATISTICS

What had Quetelet found, in fact, when he adapted the binomial curve to the statistical study of populations? He was claiming to find comparable regularities in the distribution of virtually all measurable human qualities whenever they were recorded as numbers and organized statistically. Once the familiar curve appeared, with a balanced distribution on either side of the average, he did not doubt that there were natural laws in play. The mathematically predictable order of the numbers served for him as the sign that a law was at work. We have noted Quetelet's identification of the "social body" as an object of knowledge, with the average man functioning as its center of gravity. That it was a single, unified body had to be due to the sustaining presence of "remarkable laws that govern the unity" of human properties.[93] Quetelet wanted to use statistics, by contrast with many of his colleagues, in order to identify causes rather than mere combinations. His claim to do so was supported by the belief that his calculations, while not revealing specific causes, were displaying the effects of law-governed causality. Every statistical investigation he conducted, insofar as it reproduced the binomial pattern, seemed to confirm a belief in the existence of underlying laws. Ian Hacking makes the point with his usual vigor: "Quetelet's extraordinary hypothesis about the law of errors—that it is the standard curve for the physical and moral attributes of people—was a foisting of law on to humanity and free choice."[94]

The claim that laws were at work in such domains as the choice of partners and the timing of marriages was considered by quite a few of his contemporaries to be ethically reprehensible, as Hacking implies. These were "moral" questions, and any talk of underlying laws at work in such decisions as the choice of a partner could only serve to aggravate the controversy that already surrounded moral statistics. How could it be, as Quetelet claimed, that in deciding when and whom to marry "man instinctively follows prescribed laws that he carries out, unaware of what he is doing, with the most exact regularity"?[95] In response to that question, he published an essay in about 1846 entitled "De l'influence du libre arbitre de l'homme sur les faits sociaux, et particulièrement sur le nombre des mariages" ("On the Influence of Man's Free Will on Social Facts, Particularly on the Number of Marriages").[96] He presented two quite different kinds of argument in defense of his position. The first was a mathematical explanation. It was that all decisions taken by individuals acting with a sense of their own free will, once they had been counted and averaged, were in fact subsumed into overall social regularities: "We can see first of all that, in an important class of social facts in which man's free will plays the strongest role, everything down to the finest detail proceeds from year to year with such constancy and regularity that the effects of individual wills can be regarded as almost completely neutralized."[97] The population as a whole behaved in such a way that the effects of individual intentions were "neutralized." It was not that free will did not exist in the minds of those people who were choosing to marry. It is just that, at the scale of the population as a whole, individual intentions produced no recordable effects. They were lost in strongly recurring overall patterns. Statistical work of this kind succeeded effectively in neutralizing the individual through its mathematical procedures. When decisions and actions were tabulated, it was exactly as if free will did not exist.

The second argument adduced by Quetelet was of a quite different order. It was a broadly philosophical one that took the regularities he observed to be the sign of a greater intelligence at work in the world: "When one sees that from year to year almost identical numbers are reproduced, it will never be possible to believe that chance presided over arrangements like this. Something mysterious is going on here that defeats our intelligence."[98] To deny in this way that chance was at work might well seem a little surprising on the part of a mathematician whose work owed so much to a tradition of probability theory, but this rather new philosophical inflection brought Quetelet's understanding of the average and its significance closer to a recognizably modern notion of the normal.

Before addressing that development, we will mention two issues raised by this account of general laws underlying distributions. The first point, made astutely by Alain Desrosières, is that Quetelet's preoccupation with unity led him to turn away from any principled investigation of diversity.

When speaking in his earlier work about the unifying function of the average man, he tended, as we saw, to relegate any markedly different qualities to the uncharted regions of the deformed or the monstrous. In his later work, he might have seemed to rectify this oversight by taking the distribution of errors of measurement—in our example, measurements of the height of a mountain—as a model for the spread of qualities in a population. But he was still supposing that there was a singular principle of truth—the actual height of the mountain—underlying the range of deviations. As Desrosières says succinctly, "the diversity inherent in things to do with man is reduced to the inessential spread of errors of measurement. It is thereby denied as an object worthy of interest in itself, and in fact Quetelet is unconcerned with it."[99] There is an irony here that was inherent in his practice of law-governed statistics: distribution could only be thought of as the recurring accidental manifestation of an underlying unity. The "social body" could not accommodate otherness.

Moreover, the logic of inference that took Quetelet from the binomial curve to natural laws was hardly inexorable. Poisson had presented his own law of large numbers as a mathematical law that simply served to constrain statistical calculation: "This law of large numbers can be observed in the events that we attribute to blind chance, in the absence of any knowledge of causes, or because the causes are too complicated."[100] It made no difference to Poisson whether blind chance or unknown causes were at work: the outcome of the calculations was strictly the same. All that was needed was to proceed with the business of statistics while ensuring that these unknown and unknowable elements were bracketed out. Nothing in the mathematics called for talk of "mysterious laws." It may even be the case that many of the regularities that provoked in Quetelet an experience of wonder were in fact produced by his own method. Historian of statistics Stephen Stigler certainly thinks so. He makes the point that Quetelet's inference about the existence of natural laws was not justified by the data he was examining. He was, says Stigler, simply reproducing the mathematical regularities that were built into any such calculation, rather than revealing natural or social laws: "Quetelet had been too successful—he found nearly all distributions to be normal, and he thus lost the war by not being able to isolate homogeneous aggregates in the way he had wished to."[101] The problem according to Stigler was that Quetelet's mathematical method was too powerful, nearly always finding a way to confirm itself: "too many data sets yielded evidence of normality."[102] It may well have been that Quetelet was finding law-governed regularity every time he looked at a pattern of distribution because a standardizing mathematical procedure was built into his own method of investigation. It was certainly not clear to his contemporaries whether laws of nature really could be discovered by studying binomial curves.

Wherever the calculation of averages was established as the principal algorithm of statistics, Quetelet's place in the discipline was assured, even if his conclusions about natural laws attracted only intermittent support. But the practice of averages was not accepted in principle by many outsiders. The opposition encountered in the Académie de Médecine had been an acute phase of disparagement and rejection, but that had not been the end of controversy. Although, as we have noted, the Société de Statistique de Paris was not given to internal polemics, its members knew only too well that principled criticism of the use of averages threatened to undermine the standing of statistics as a discipline. That is why, a full thirty years after the medical debate, L. F. M. R. Wolowski, who happened also to be a member of the prestigious Institut de France, felt called on to address the society in defense of the average: "People have made fun of *averages*. They have said that statistics did not express real facts, but were rather abstractions arising from arbitrary calculation. However an *average* precisely reveals the exact law, the law freed [*dégagée*] from accidental pressures, variable elements and capricious oscillations. If you want to know how social phenomena really work, you have to proceed by *averages*. Only *averages* are true." Wolowski went straight on to speak of the "average man." Without naming Quetelet, who was known to all statisticians as the author of that concept, he was calling on his colleagues to close ranks against the mockers: "*the average man*, who has given rise to so much facetious raillery, is really the expression of humanity, of whom Pascal said that it is like a single man who goes on living and learning."[103] In its appeal to a kind of timeless humanism, this defense was true enough to Quetelet's thinking, and the record of proceedings tells us that the speech was greeted with "prolonged applause."[104] It is, however, somewhat ironic that Wolowski should have cited Pascal, one of the early contributors to the calculus of probabilities, without raising any mathematical issues. The success or otherwise of statistics as a scientific enterprise called for more than rousing moral speeches. What appeared to be needed was a careful theorization of the average.

REFINING THE CONCEPT OF THE AVERAGE

There can be no doubt that averages were the core business of nineteenth-century statistics, but a little more needs to be said here about just how that business was transacted.[105] In practice, the work done on averages by statisticians was more complex than the opponents of statistics allowed. Proponents of statistics continued to work on the concept of the average, refining it as they did so. They argued, for example, that what was meant by the average should change according to the kind of group being studied. Quetelet was one of the first to make a key distinction of this kind, and the examples he offered were recycled and revised by others. His first example, and his first kind of average, involved measuring the height of each building in an unregulated street, then calculating the average height of the buildings. The product of that calculation would be an average in the arithmetical sense—the total of the heights divided by the number of buildings—but it would not, he said, be a statistically valuable outcome. Risueño and the other early opponents of numerical method might have been surprised to learn that this was not regarded by a statistician as significant because that was exactly what they took averages to be, but the number produced by averaging the heights of buildings in a street was simply declared by Quetelet to be of no scientific interest. That was because the set on which it was based was not held together by any "law of continuity."[106] No binomial curve could be drawn were the numbers to be represented in graphic form. The true average for statisticians, said Quetelet, corresponded to a second kind of grouping, which he exemplified by imagining twenty attempts to measure the height of a single building. The numbers produced in that way would constitute a different kind of set. The average would be effectively constrained by the unchanging height of the building and the set of measurements could be expected to correspond to a binomial curve, with the average standing at the middle and deviations ranged around it symmetrically.[107] In this second case, statistical method would have produced a worthwhile outcome. It would have contributed to the management of error in measuring a physical object, and the standard pattern of distribution would be testimony to that.

But repeated measurements of the same object, while theoretically exemplary, did not provide for Quetelet the most telling example of the average at work on humans. He used a set of measurements that had been conducted on Scottish soldiers, and noted in tabulations of them that the binomial pattern was repeated for every quality measured. This he took as irrefutable evidence that there was in this population an "average man," despite the criticism to which that concept had been subject.[108] Such patterns of distribution pointed to what Canguilhem and most thinkers of the twentieth century would have called a "normal" set of physical features. But Quetelet did not use the word "normal" in this context. His key word was "type":

"this all takes place as if there were a man, the type, from which the others deviated more or less."[109] The mathematical indicator once again was the symmetrical distribution of frequencies: "this symmetry of the results exists only insofar as the elements that enter into the calculation of the average can be brought back [*ramenés*] to a single type."[110] Across this population, and indeed wherever the average height of men in any given country was concerned, there was "a regularity that one does not find when one looks at buildings."[111] In the case of buildings, there simply was no "shared law" to be brought to light by statistics.[112]

It is important to see that probability theory was fully integrated into Quetelet's theoretical understanding of the (normal) type and of its function within a distribution. It was as if everyone in the population had been "made, so to speak, from the same mold." The differences between individuals could in fact be understood as if they were "purely accidental," and precisely because those differences were accidental or quasi-accidental, "their numerical values, according to the theory of probabilities," would be "subject to preestablished laws."[113] In other words, individual qualities would be distributed around the type in the predictable pattern found in errors produced by repeated attempts to measure the same object. It was as if all the individuals in a population corresponding to a type obeyed a probabilistic law of errors, deviating from the average man by predictable degrees. If the type was perfection, then the distributed set of individuals functioned like a graded set of imperfections. The average man had thus become "a mathematical truth": "The human type for men of the same race has become so well established that the deviations between the results of observation and those of calculation, in spite of the many accidental causes that can give rise to them and exacerbate them, are hardly more than those that would result from slipups in a series of measures taken of the same individual."[114] This was a strong answer to give to Risueño and his colleagues. It amounted to saying that, at least in the construction of knowledge about populations, there was ultimately nothing to be gained by thinking of individuals as unique. Quetelet's theory effectively began with the average in order to understand individuality, by a potent theoretical fiction, as an iterative, although somewhat inaccurate, reproduction of the normal type.

Quetelet believed that he had found what Michel Armatte describes as an "internal test" for statistical data. Insofar as a binomial distribution appeared around the average, he was prepared to declare that the calculated average was a "true" one pointing to a "law of continuity."[115] The key thing from a mathematical point of view was that the differences being measured should have arisen by accident. It may seem paradoxical in the light of Quetelet's continual references to laws that accident should have played a key role in his thinking, but that was precisely where he leaned on the theory of probability. The kind of law that corresponded to the binomial curve was a probabilistic one predicting the pattern of distribution that would result from a long series of accidental events: "Another consequence of the theory [of probabilities] is that, the greater the number of observations, the more the fortuitous causes cancel each other out, allowing the general type that might otherwise have been masked by them to predominate. In the same way, with the human species, if one considers only individuals, one finds them to have all heights, within certain limits at least. Those who are closest to the average are greatest in number; those who are furthest away from it are fewest in number, and *their groups numerically follow a law that can be assigned in advance.*"[116] While Quetelet was drawing on a mathematical theory of accident, he was doing so in order to show how, in the long run, causes producing chance events could be expected to cancel each other out. At the end of a long series of observations, statistical inquiry would find in populations where a law of continuity existed "a character by which one recognizes whether individuals belong to the same type, and are only differentiated by fortuitous causes."[117] When chance and error were distributed, cancelled out, and subsumed by the curve of probability, the type would be fully visible as the average.

It could be said that during the middle decades of the nineteenth century Quetelet's concept of the average man was influential without being persuasive. It was, so to speak, theoretically assertive, promising to provide standardized knowledge about human populations by purely statistical means. For the strength and relative simplicity of its claims, it remained topical throughout this period. But it was hardly ever accepted by Quetelet's peers, and one quite fundamental question was posed with some regularity. Antoine-Augustin Cournot formulated it clearly in 1843: "The average man defined in this way, far from being what might be called the type of the species, would be quite simply an impossible man. At least, nothing justifies the claim that he is possible."[118] Quetelet responded vigorously to this criticism, but his response was hardly more than a reiteration of his initial claim. He summed up his achievement in *Anthropométrie* (1870) by saying that he had now established each of the physical proportions of man "in our climates." No one was entitled to say that "these averages must constitute an impossible man." Proof of the contrary was to be found in the very manner in which "the numbers obtained for each part of the body are grouped around the average and obey the law of accidental causes."[119] But that was not the point of the initial criticism. Louis-Adolphe Bertillon's version of it, published in 1876, was more explicit: "The synoptic table of all these average measures could not be considered to define an individual who was the type, nor even a possible individual."[120] By their criticism, Cournot and Bertillon meant, not just that the average man was a mathematical construction — Quetelet readily conceded that he was "ideal" in just that way — but that mathematical reasoning did not support bringing the various qualities together "synoptically." There was no justification, they were saying in effect, for Quetelet's claim that the averages of various features and qualities could be appropriately aggregated into one quasi-personal figure. Perfectly average height need not be accompanied by perfectly average weight, and so on. Summing up all the averages in one perfectly average person would not reflect the diverse ways in which the range of qualities worked across a population. There was no single binomial curve for all the qualities, and there could be no single average man.

While remaining skeptical about the possibility of aggregating measurements to produce a singular average man as the ideal type, Bertillon was fully persuaded that the measurements of Scottish soldiers, when taken severally, revealed the existence of an ethnic type. Each average measurement was a "natural" one, and the fact that they could be arranged in a curve provided graphic proof of that: "This is a natural average of the kind that one would find if one were dealing with a group of men descended from the same ancestral couple, or at least with a human group whose elements have been mixed and combined over such a long time that they have finally constituted a single type. In that case one finds a succession of terms that, when classified according to size, yield a series in which the values are closer to each other as they come nearer to the average. And if one represents these values graphically, they produce a convex curve of remarkable regularity."[121] Here was an application of the average that was to have considerable significance for anthropology. The singular figure of the "average man" could not stand for the naturalness of an ethnic group, but results displayed in a curve could. The same pattern emerged, Bertillon added, when one grouped the height measurements of French conscripts. But that method would not be appropriate in Sweden, where there were held to be two racial groups, the Swedes and the Lapps. For this reason, the average measurements of Swedish men would resemble those of buildings in a street: they would be only an "artificial average."[122] Bertillon went on to give another example of an artificial average that could be produced in the study of populations, that of the age at which people died: "People might say that in France the average age of death is thirty-six, but it is not in fact around that age that one finds the greatest number of deaths. The greatest number of deaths occur at the two extremes of life. Thirty-six years is purely a mathematical abstraction, serving the purpose of relieving memory and attention [of the need for effort] by summing up in a single term a number of diverse values."[123] The average produced in this case was not so much artificial as facile and misleading. It was important, Bertillon was saying, for statisticians to know when *not* to calculate averages.

BERTILLON AND MEDICAL STATISTICS

Bertillon shared Quetelet's critical preoccupation with medical statistics, and in fact intervened more publicly than his Belgian colleague in debates on the topic. Through the Académie de Médecine, he advocated the collection by the French administration of statistics regarding cause of death, and was eventually successful in bringing about change on that score.[124] He regularly addressed members of the medical profession through their journals, seeking to draw their attention to the value of statistics.[125] In one of his essays, he stated that he wanted doctors to "give up their prejudices" and familiarize themselves with "this powerful new instrument, which will very much extend the limit of their observations."[126] Like Quetelet and Gavarret, he saw the medical practice of number at the time as part of an ongoing problem rather than the beginning of a solution. It would hardly suffice to eliminate "vague language" by introducing numbers into medical talk: true statistics as he saw them had to do with "establishing *average values*, studying these averages, noting their oscillations, and so on."[127] Bertillon advocated a particular understanding of medicine in which the practice of number represented, and acted as a vehicle for, a materialist approach. In a series of articles published in 1857 in one of the professional journals aimed at doctors, *Le Moniteur des hôpitaux,* he responded to an attack on medical statistics by a doctor named Pidoux. Medical statistics were no more, Pidoux had said, than *numérisme* (numberism), that is to say, the use of number for its own sake.[128] Bertillon's response in this instance was not a narrowly mathematical one, but a self-consciously philosophical rejection of what he saw as Pidoux's "spiritualism."[129]

According to Bertillon, medicine had a lot to gain or lose by its response to the requirements of statistics. Whereas Quetelet tended to say that doctors were all doing statistics while being unaware of it, Bertillon was more inclined to say that none of them were doing anything worthy of the name. It was well known, he said, that statistics had found a place in astronomy, in political economy, and sometimes even in public hygiene. But had they been applied to medicine? Bertillon was doubtful that they ever had, declaring that he could not find even one example of a proper statistical inquiry in the field.[130] Not only was numerical method failing to carry out the task, but certain questions that might have been raised by statistics were not even being asked. Analysis that focused entirely on individual cases, said Bertillon, would not be capable of revealing the influence of climate or sex, for example, on the incidence of certain illnesses.[131] No standard medical texts talked about the difference between the sexes in the incidence of early childhood illness, even though one-fifth of boys died in the first year of life, as against one-sixth of girls. This fact had escaped medical observation because doctors had been considering only individual patients, whereas questions ought to have been asked whether some illnesses were more severe for boys than for girls, and if so which.[132]

Conflict arose during the 1850s around the ongoing effects of "the vaccine," as smallpox vaccine had long been called in etymological recognition of its derivation from cowpox.[133] Opponents of the vaccine fought a kind of guerilla war. Some argued for a return to the earlier technique whereby people were inoculated with secretions from an infected patient in the hope of inducing a mild case that would subsequently confer immunity. The editor of *Le Moniteur des hôpitaux* argued forcefully against that practice, declaring first that it was risky for the individuals inoculated and second that it would be likely to abet the spread of smallpox in cities.[134] A doctor by the name of Bayard made a different kind of case, arguing that any apparent decrease in the number of deaths from smallpox was simply due to the fact that typhoid fever was a variant of that disease and was continuing the same deadly work in a somewhat different pathological form.[135] To rebut that claim, a number of doctors presented medical arguments in such professional organs as the *Gazette hebdomadaire de médecine et de chirurgie* (*Weekly Gazette of Medicine and Surgery*). Dr. Barth, for example, pointed out that if it were true that typhoid was merely a variant of smallpox, patients who had not been vaccinated and who bore the scars of smallpox should have immunity to typhoid. His observations in a hospital in Lyon showed that that was not so: a number of patients there who had already had smallpox were suffering from typhoid. After describing eight such cases, he conceded that this was not a sufficient number to draw an "absolute conclusion," but if such observations were to be "multiplied," one could conclude that those who were opposing the vaccine were behaving most imprudently.[136] A similar case was made in the same journal two months later by another doctor also based in Lyon.[137] Senior medical figures were able to respond to Bayard's argument by citing their clinical experience.

But there was another, more intense, battle being waged against the vaccine, and a telling weapon in that battle was statistics. Professional disquiet about this issue came to a head in 1853 when a leading member of the Académie de Médecine, Dr. Malgaigne, asked the question: "Is it true that before the discovery of the vaccine a greater proportion of individuals lived to maturity?"[138] Malgaigne's question had been provoked by a campaign led by Hector Carnot, who claimed that when the nineteenth century was compared with the eighteenth, a larger proportion of the mature population was now dying, and that this was the fault of the smallpox vaccine. The medical profession as a whole was not well equipped to respond to such an attack. Even as doctors from Lyon and elsewhere put forward their medical arguments, they had to declare themselves incompetent to fight on the statistical front. Dr. Leudet, one of the regular contributors to the *Gazette hebdomadaire de médecine et de chirurgie*, summed up the state of the debate in 1854: "This question is naturally subdivided into two parts, one purely statistical and one belonging only to the domain of doctors."[139] For all their willingness to give battle on the medical front, doctors were not generally armed to respond in kind to an attack that was "purely statistical." What was called for were doctors with sufficient mastery of statistics to answer Carnot's criticism in kind.

The striking thing from the viewpoint of our history is that so much had changed since the 1837 debate in the Académie de Médecine. It was not that the medical profession had since undergone steady progress in the application of statistics: everything we have seen so far in this account shows that that was far from being the case. But a crucial question about medical policy and practice was now being raised in statistical terms from outside the discipline, and was finding at least marginal support within it. Far from decrying the use of numbers in medicine as Risueño had done less than twenty years earlier, Carnot was actually relying on statistical evidence to make his case. This was not a debate for and against the use of numbers, but one in which the authority of statistics was claimed by the opponents of the vaccine, and needed to be mobilized in turn by its defenders. The epigraph to the first chapter of Carnot's *Essai de mortalité comparée avant et depuis l'introduction de la vaccine en France* (*Essay on Comparative Mortality before and since the Introduction of the Vaccine into France*), published in 1849, was a quote from Laplace, perhaps the greatest authority in the history of French statistics, and the arguments made throughout were based on mathematical interpretations of historical population data. The essay's drastic general conclusion was announced almost at the outset, but it was hedged about with the protocols of statistical reasoning: "all the probabilities come together to indicate that the introduction of the vaccine into France was the sole cause of the social disorder that is revealed to us today by statistics."[140]

Bertillon had a ready-made place in this debate because of his rare combination of intellectual skills. He happened to be one of the very few qualified doctors in France with a sound grasp of statistics, and this enabled him to lead the campaign in defense of the vaccine, providing his medical colleagues with the applied statistical knowledge they needed in order to respond to this professional emergency.[141] From 1853 to 1859 he published many short pieces on and around the topic in the *Gazette hebdomadaire de médecine et de chirurgie*, *L'Union médicale*, and *Le Moniteur des hôpitaux*.[142] One of these was a review of a text by Achille Guillard intended to serve as an introduction to statistics. Guillard, who happened to be Bertillon's father-in-law, had nothing to say about the attack on the vaccine, but Bertillon insisted on the usefulness of his work for doctors. The problem in 1855 was that "for nearly ten years, M. H. Carnot and his supporters have squarely laid down a challenge, and no one has taken it up directly. They talk statistics, and people reply by talking medicine. In short, those who are in the wrong have been very bold, while those who are in the right have been at a loss as to what to say." It was imperative that doctors learn to use the "necessary method" made available by experts like Guillard so that they could "reduce these pretensions to nothing."[143] That they were reluctant to do so was no doubt due in part to the unconvincing efforts of the old champions of numerical method: "Before doing statistics, one has to know its rules. This requirement of science, which seems so natural, has sometimes not been properly understood by men of the first order. That is how M. Louis and M. Bouillaud, in their *therapeutic* research, have compromised the standing of statistics in the eyes of the uninformed. They have worked with numbers that were too small and have been unable to avoid the effects [on their data] of accidental disturbances. That has led them to contradictory, hollow conclusions."[144] The failings of numerical method were being widely held to blame, Bertillon believed, for the current lack of statistical skill, and indeed of interest in statistics among doctors. The old problem of large numbers was still in evidence: the numbers required for proper calculations could hardly be attained in the practice of individual clinics. But

Bertillon went on to list a set of technical matters that ought to have been part of the professional practice of medical statistics: "Moreover, some men [he meant Carnot, but also Bayard] who are ignorant of the importance of averages and natural series, of homogeneous couplings and the laws that govern changes in the size of the population, imagine and proclaim that statistics condemns the vaccine."[145] A better command of averages and of "natural series" was urgently required by the profession.

The aggressive use of statistics against the vaccine constituted a particular threat to established medical practice because vaccination against smallpox was arguably the greatest success story in population health at the time. Quetelet himself had taken it as exemplary, saying that the first use of Edward Jenner's vaccine in Britain had been founded on a statistical study.[146] In fact, as Baxby's detailed 1981 history shows, that was not strictly true. The first extensive trials were carried out by William Woodville in 1799, three years after Jenner had introduced vaccination to the public on the basis of very limited trials.[147] But that was no longer the issue in the 1850s. What mattered for Bertillon in particular was that statistical method itself was being tested by this attack. In 1857, he published a monograph, *Conclusions statistiques contre les détracteurs de la vaccine* (*Statistical Conclusions against the Denigrators of the Vaccine*), with the intention of "making people realize the power of this method of research and analysis in professions where it has most to offer." It was now not just a matter of saving statistics from "neglect" as might once have been the case,[148] but of wresting back the weapon of statistics from those who were misusing it against the vaccine.[149] Here Bertillon set about the business of undoing Carnot's argument in narrowly statistical terms.[150] Carnot was comparing mortality in the eighteenth century with mortality in the decade leading up to 1850, but the terms of his comparison were misleading. Since infant mortality in the eighteenth century was measurably higher than during the nineteenth, the actual number of people remaining alive out of the cohort born in any given year had been markedly smaller during the earlier period. It was true that the number of people who died at a given mature age during the eighteenth century was proportionately smaller than during the nineteenth. That was simply because a greater proportion of the eighteenth-century population had already died before reaching that age. In other words, during the eighteenth century, the number of deaths at earlier ages meant that there were fewer people left to die at mature ages, so that deaths at mature ages were indeed a lower proportion of the whole at that time. The calculable difference between the eighteenth and nineteenth centuries was a result of reduced mortality overall. The mathematical argument was presented by Bertillon as an "utter condemnation" of his opponent's, whose work could now find credit only with those who were not part of the scientific world and did not understand its principles.[151] The Académie de Médecine and the Académie des Sciences concurred with Bertillon: both awarded him prizes for his essay.[152]

THE AVERAGE AND THE NORMAL

We have seen the close attention paid to the significance and definition of averages, but it remains to be seen just how the term "normal" first came to be used by statistical thinkers, and what relation that usage bore to their talk about averages. Where Quetelet was concerned, Stephen Stigler's assertion was that he simply did not use the word "normal."[153] That is not strictly true, however. It would be more helpful — and a more generous reading of Stigler's apparent oversight — to say that a modern reader looking for the term with its current statistical overtones would be unlikely to find it. We have noted that Quetelet did not use the term "normal curve," and Stigler can be taken as making the stronger point that he did not use "normal" when talking about distributions. But the word did occur from time to time in Quetelet's writings, and we are about to discuss his use of it. That prompts an observation about our own historical method. It might seem as if a discursively focused genealogy of the normal were required to part company at this point from the history of statistics: whatever "normal" might mean in Quetelet's texts, it does not appear to be closely articulated with the practice of that discipline as he understood it. But this is a complication that can often be found in discursively minded histories like ours. Our task is to work with the difficulty and around it. We must accept that we cannot carry out our genealogy simply by tracing the history of a single discipline or by mapping the usage of a key term — even less by merely following the use of a key term within a discipline. We have examined what Quetelet had to say about general laws at work in human societies and concluded that he was committed to a normalizing intellectual practice. His writings did not, however, provide us with the term "normalizing," still less with a definition of it. For now we will undertake the more straightforward analytical task of talking about the word "normal" as it was actually used by Quetelet, before coming to discuss how the word did eventually find a place in statistical discourse after his death. While discussing *Sur l'homme*, which was published in the 1830s, we found Quetelet using the word in what was then its standard medical acceptation. The question that opens up for us now is whether any other senses of the term were also in play in his work. We will be asking in particular whether the standard medical usage was being extended or inflected by Quetelet in a manner that brought the medical term closer to the broad sense of the word current in our own time.

The simple answer is that the 1870 essay *Anthropométrie* did just this kind of discursive work. Quetelet's focus there was not just on the dimensions of the human body: that would simply have been a matter of careful measurement followed by averaging calculations to minimize the effects of error. In *Anthropométrie*, he was interested in the ratios of human dimensions, which posed a different set of questions for statistical inquiry. And when he came to speak in general about those dimensions, he often used the word "normal." The first use of it in his essay occurred in a reference to studies of human skeletal structures conducted in order to describe patterns of deformity produced in children by severe work conditions. This was mensurative science, the proper terrain of statistics, with long-term therapeutic significance: "others have studied, in the normal makeup of the human body and in the balance between different parts of the skeletal structure, the means of straightening limbs that present an impaired disposition [*organisation vicieuse*]."[154] The opposition between the normal and the pathological was still at work here, with *vicieux* standing in, as it quite often did at the time, for *pathologique*. But Quetelet was not quite speaking of the "normal state" in the standard medical way. He was talking about the normal structure and the normal proportions of the human body. When "growth defects" appeared in the limbs of children doing factory work,[155] those defects were by definition measurable departures from normal proportions. The normal appeared then as a kind of hypothetical: "a set of favorable circumstances would be required for all the physical faculties to develop in a perfectly normal manner."[156] The task of statistics was principally to establish just what those normal proportions were, and the job of a therapeutic medical science was to do all that it could to reduce the gap between the actual and the normal/ideal.

Quetelet's interest in the measurement of human proportions extended well beyond orthopedic applications such as these. As he moved on to other topics, the signification of "normal" seemed to broaden beyond its medical sense. He examined archival sources for records listing human proportions and sought to make comparisons across space and time. Some ancient Sanskrit texts, for example, provided data that called for explanation. One key proportion, the height of the head in relation to that of the body, was compared with the ratio established on the basis of European examples.

Quetelet found "nothing disconcerting" about the difference between the two. The yardstick here was Raphael. That ratio "can even be found in several paintings by Raphael, the artist who had the finest knowledge of relations between the different parts of the body." But the Indian proportion was nonetheless outside the range of "normal": "one cannot consider it as belonging to the normal state, unless it be admitted that the Indian type is different from that of Greece or Italy."[157] As we can see, it was perfectly possible in Raphael's work for a given proportion to have a place within the canon of Western art without being judged normal. And it was even possible, at least in principle, that the normal proportions might vary from one place to another, thereby attesting to a variety of "types." But the normal state was being taken here as identifiable, and it bore no particular relation to the question of health or illness. As in Étienne Geoffroy Saint-Hilaire's comparative anatomy, the term was not functioning in simple binary fashion: it was not defined in practice as the contrary of "pathological."

In his earlier essay *Sur l'homme*, Quetelet had used another word that was not derived as directly from medical discourse as "normal" was. He did so in order to speak about such things as height and weight compared across a population. The word in question was "regular"—a word that had its own history in the discourse of comparative anatomy, as we have seen. This word served to indicate in practice that the subjects of whom he was speaking were located within a certain range. At one point, he examined maximum and minimum height for "individuals of regular build."[158] At another, he spoke of "the extreme limits of weight for regularly formed individuals."[159] Elsewhere, it was "fully developed women who are of regular build."[160] In each case, the use of "regular" was a sign of method, ensuring—or at least affirming—that his sample did not include the deformed and the monstrous, as he was wont to call to them at the time. This usage begged the question of how "regular" was to be defined in the first place. How could the range of regularity be reliably delimited? In the later *Anthropométrie*, for all Quetelet's talk in the intervening decades about the importance of limits, that fundamental question remained unanswered. When doubts arose about whether the human type might vary from place to place, he answered them by declaring emphatically that "the human type, in our climates, is identical with the one that can be deduced from the observation of the most regular ancient statues."[161] But that did not explain why certain ancient statues rather than others should be chosen for the purposes of comparison. Why were their particular proportions to be considered "regular" to the exclusion of the others?

Quetelet may not have been quite certain whether there was a single type of human being, but he did not doubt that every time he found symmetrical dispositions about a set of average qualities, he had found a type. In the identification of the type, statistical analysis of a population was produced as a summary figure. And that was where Quetelet's claim about a mathematical law found its vindication: the type was quite precisely the incarnation of the law. The fact of symmetrical distribution provided "the most indisputable proof of the unity of his kind and the existence of a type. I have named this type the *average man* in order to express its main property. The average man characterizes the nation to which he belongs."[162] In this case, the type was that of the "nation." Elsewhere it appeared to be the whole of humankind. Was it appropriate, in fact, to speak of a single human species identified as a type? Anthropometric statistics answered the question in the affirmative: "By reasoning that is more or less conclusive, a great number of naturalists and philosophers have striven to prove the *unity of the human species*. I believe I have succeeded in demonstrating, not only that this unity exists, but indeed that our species allows for a type or module whose different proportions can be determined with ease."[163] Statistical science might thus find itself establishing local types, particularly those of a nation, while all the while helping to identify something like the type of types, that of humanity as a whole: "Our highest priority is to know the *general type* of man. All our attention will be directed initially toward that study. Subsequently, we will endeavor to recognize the characters that may be regarded as secondary and that represent certain human faculties."[164]

This was how Quetelet undertook the conceptual and practical work of studying groups of humans in order to identify a type. What is more, he was confident that statistical analysis of this kind would not be defeated by the requirement of very large numbers. That was because the type would appear readily once statistical inquiry had begun: "When considered as individuals, men are so diverse that it seems pointless at first to look for a type or module of what they might be in the normal state. And yet the type exists. To find it, one does not need to go so far as to examine a large number of individuals. The exact observation of a limited number of them is sufficient to eliminate the particularities that characterize them and distinguish them from one another. Furthermore, among the variable elements that nature presents, there is perhaps none that is more characterized [*caractérisé*] than man."[165] The type was quite easily found, we must suppose, because most humans were naturally grouped close to the average. As a consequence, man was more easily averaged and characterized than other elements found in nature. There was, Quetelet was saying in effect, a law of convergence revealed through the practice of statistics, a law that allowed the average, the type, and the normal state of man to become readily apparent—and effectively equivalent—once an inquiry of this kind was undertaken.

When the "regular," the "normal" and the "type" came to function in Quetelet's writing as near equivalents, the circularity of his thinking was reinforced and its effects compounded. The conceptual strain showed itself in the method by which he chose subjects for measuring bodily dimensions and calculating averages. While insisting that his subjects were not selected in any way, he added that "deformed" individuals had been excluded: "The subjects measured were not subject to selection. Nonetheless, it was considered necessary to set aside those who presented deformities. Those who were of excessive or defective height were studied separately, and are not counted in the general tables."[166] It is predictable enough that a distinction should have been made between the regularly formed and the deformed in view of Quetelet's focus on the average as type. What is striking, and the sign of an unresolved difficulty, is that such decisions should have been taken even before the measurements were constituted as a set and the type inferred from them. So it was not in fact the case that "deformed" subjects were defined statistically because they showed great degrees of deviation from the average: in their case, deviation as such was not measured, or plotted on any curve of distribution. This ought to have been a problem for Quetelet's understanding of laws. To arrive at a knowledge of general laws, he noted, one had to constitute large sets that would serve as a basis for calculation. Yet even as he constituted those sets he considered it "appropriate to set aside accidental and inevitable anomalies."[167] The danger of circular reasoning was patent. How was an anomaly to be defined if the law was not yet known? How could deformed individuals be excluded when no distribution had yet been done that put them at a great distance from the type? It is hardly surprising that, in contemplating his own results, Quetelet should have been moved to comment about the narrow range of the measurements collected: "since I was only able to collect my numbers with individuals *of regular build*, the differences between the *maximum* and *minimum* numbers are very slight."[168] Excluding irregular individuals may have made the average more readily calculable, and may indeed have made the supposed type more salient, but the narrowness of range had become a cause for concern in itself. Was this narrowness indicative of the natural tendency of humans to be quite closely grouped around a type, or was it an artifact of Quetelet's method? He appeared to be claiming as characteristic of humankind something that was in fact implicit in his own mode of inquiry.

Other judgments that had no demonstrable basis in statistical method tended to inform Quetelet's interpretations of his anthropometric data. Having noted a significant difference in structural proportions between men and women—the midpoint of men's height being on average higher than that of women—he went on to celebrate the natural propriety of men's and women's body shapes for their social roles: "One can only admire the fact that these proportions are so appropriate. Man, as the defender of woman, needs agility, and he appears to have proportions that are better suited to ease of movement. Woman, on the other hand, is the hope of the family's future, and her build seems to provide a stronger guarantee of fixity and poise."[169] The difference between the bodily structures of men and women could only be thought of here as "admirably" confirming their suitedness to established roles. Gender difference was inscribed in their body shapes in such a way that mathematically normal proportions and "natural" behaviors were held together in a self-confirming circle that made the normal something more than a state of health. Underlying gender laws were at work in the sexed structure of human bodies. These qualities were not only being included within the range of the normal, but were serving to reinforce its authority.

Moreover, the normal was taken by Quetelet to constitute a kind of material perfection. By a general tendency said to be inscribed in human nature, Quetelet had no doubt that every population group recognized the normal as a model: "All peoples have reached agreement about seeing the human body in its normal state as the most perfect type."[170] The word "perfect" had a variety of meanings here that were in play at the same time, much as it had had in the work of Georges Cuvier. Perfectly normal proportions were precisely those that coincided with the average, while the full set of average proportions served to define perfection and make it knowable. Health was included within the normal state, certainly, but so were beauty and indeed typicality itself. The perfection of the normal was made by the coincidence of all three. And perfection of form was reiterated and distributed throughout the body as a law of proportion applied to every part: "One of the first corollaries one can deduce from the foregoing research is: not only must the entire body of man respond in its development to the law we have formulated, but the very parts of the body must follow that same law, and obey the same compelling authority [*puissance*] that nature invests in all of its works."[171] A natural law worked thus to reproduce in the normal body, at every level, a set of perfectly average proportions. The normal state of the body was an orderly disposition of well-ordered parts. Typicality went all the way down.

When Georges Canguilhem, in his essay *Le Normal et le pathologique*, written in 1943, came to consider Quetelet's work, he had no difficulty finding a fully developed view of the normal, although Quetelet used the word "normal" only on rare occasions prior to his last major work. Canguilhem understood just how theoretically convenient it was for Quetelet to consider the norm and the average as "two inseparable concepts."[172] He recognized too that the assimilation of one to the other constituted a genuine attempt to address the need for a norm in scientific thinking about living organisms, acknowledging that "an average determining deviations that become rarer as they become greater is a norm in the proper sense."[173] Quetelet's average functioned as a norm, Canguilhem was saying, precisely because it provided the midpoint of organization for a binomial distribution. Quetelet had not made the "mistake" of regarding the average as an empirical basis from which some kind of norm might be induced, but had actually pointed to an ontological reality that expressed itself in the average.[174] This was indeed what Quetelet had meant when he said that the human type as the standard of beauty stood beyond the taste of individual artists: "this type, so strongly desired and yet always left to the arbitrary decision making of artists, exists in actuality. It is only a matter of apprehending it in nature, not according to the taste of such and such a man, but by following properly conducted observations."[175] So while Quetelet conceded readily that his average man was "ideal" in the sense that it was a figure based in mathematics, this was not to be taken to imply that he was not real. First, there must be real individuals resembling and representing the average man: "it is not possible that a given large group, such as a people, having a common type in the way we have shown, should not contain individuals who represent the type to a greater or lesser degree."[176] More generally, the type existed in (human) nature: "the type of beauty is a singular entity, in men as in women."[177] To arrive at the type through statistical calculation was to find something that was already there. Quetelet's repeated affirmation that the average corresponded to the type as a matter of law reassured Canguilhem that his statistical approach would not fall into the collection of shapeless and ultimately pointless data. But it was nonetheless a mistake, according to Canguilhem, to define the normal in narrowly mathematical terms. Canguilhem argued that the logic by which one arrives at the understanding of a norm should be discrete from the mathematical reasoning that produces an average.[178]

This was an argument about the significance of the curve. The binomial curve stood for Quetelet as a figure of knowledge pointing to fundamental regularities in the world, but Canguilhem insisted that distribution on either side of the average did not properly address questions of value:

"Establishing one of Quetelet's curves does not resolve the problem of the normal for a given character, for example height. There need to be guiding hypotheses and practical conventions that will make it possible to decide at what values on the scale, both upper and lower, the change from the normal to the abnormal occurs."[179] Quetelet had of course addressed the question of limits in a mathematical way that allowed him to determine the range within which the great majority of individuals would fall, but that involved no judgment of principle about where the normal ended and the abnormal began. It was simply a matter of finding a point beyond which the numbers of individuals became very small to a calculable degree, thereby ensuring that they would have a negligible influence on the averaging pattern. It is noteworthy that Quetelet almost never used the word "abnormal," even in *Anthropométrie*,[180] and made no attempt to define a liminal point between the normal and the abnormal. One might say in general that for Quetelet, the "problem of the normal" referred to by Canguilhem simply did not exist. The formal requirements of his mathematics allowed this to be so, and that is precisely why Canguilhem was questioning their function. According to Canguilhem, decisions needed to be made about normative values, but Quetelet was using mathematical procedures to make those decisions and avoid them at the same time. Some initial exclusions were indeed normative in untheorized ways, and others simply rehearsed the outcome of his calculations. They were not held up for examination in normative terms of the kind Canguilhem was seeking to analyze.

Quetelet's elision of the "problem" of the normal is in many ways characteristic of this phase in our history. As we have seen, a great deal of intellectual effort was invested in discussions of the average. In medical and moral statistics, the problem of the average called for urgent attention and received it, often in strained and controversial ways. The normal, in the medical sense that was current in the middle third of the nineteenth century, hardly seemed to be problematized at all. But, as Canguilhem suggested in his analysis of Quetelet, statistically minded work on the average acted as a vehicle for implicit judgments about the normal. In that sense, talk about the average was helping to produce normalizing effects. The conceptual history of the normal was already bound up with that of the average, and indeed of the typical.

CONCLUSION

Even as statistics became an intellectual and administrative routine in some fields, attempts to practice statistical analysis in medicine were fraught with difficulty and attended by controversy. The source of difficulty is not best understood as a matter of professional tardiness: it was not simply a failure on the part of doctors to fall into step with scientific progress. Indeed, those doctors who did seek to practice numerical methods were often the ones most severely criticized by statisticians for failing to understand the requirements of mathematical discipline. The primary requirement of statistical analysis was for data in sufficient quantities collected according to unvarying procedures, and clinical "counting" was almost never adequate from that point of view. It was, however, generally unclear to statisticians just what statistical thinking might actually contribute to clinical practice. Quetelet's own attempt to apply it to individual cases was unsustained after the 1830s, and it could be argued that he had unwittingly conceded the general point of those who claimed in the course of the Academy debate that true clinical insight had nothing to do with number. During the middle decades of the nineteenth century, statisticians focused their attention on population health rather than on clinical matters. They found their preferred terrain in the study of such things as epidemic mortality rates and the effects of vaccination on whole populations. As this work was going on, there appeared to be a comparable shift in the understanding of the "normal." Whereas medical usage of the term had for decades consisted in using "the normal state" to refer to the state of health, the "normal" now began to undergo a certain mathematization. In the work of Adolphe Quetelet, the average and the normal came effectively to coincide, although that shift was never fully owned as a matter of statistical method. We will see in the course of our history that thinking about the normal was often marked by the theoretically unresolved cohabitation of such notions as the average and the typical. That very looseness came to characterize the normal as a figure of knowledge.

CHAPTER FOUR

Measuring Bodies and Identifying Racial Types: Physical Anthropology, c. 1860–1880

INTRODUCTION

In 1861, during a meeting of the newly formed Société d'Anthropologie de Paris, Jean-André Périer gave a talk entitled "Sur l'hérédité des anomalies" ("On Inheriting Anomalies"). Périer was a senior doctor at the Invalides hospital for returned soldiers, but he made no reference to his clinical experience, which could hardly have been pertinent to his topic. Like the many other medical people in the Society, he was addressing what he and his colleagues saw as a properly anthropological question, and he was doing so with the intellectual means made available to him by medical thinking.[1] When Broussais had studied anomalies, it had been to read them as signs of a pathological state. When Isidore Geoffroy Saint-Hilaire had offered detailed descriptions of anomalous organs, it had been for the purposes of a classificatory teratology. Physiology and comparative anatomy had thus both treated anomalies as objects for scientific study, to be examined wherever they became available. But Périer was now taking the extra step of asking whether anomalies produced by accident could persist across generations. In other words, could anomalous formations be expected to endure through time? Insofar as the anomalous and the normal were mutually self-defining for medical science, that necessarily involved a question about the normal state. Did it too persist through time, and from one body to another? The conclusion reached by Périer was that the normal state endured beyond the life span of individuals, with the result that "natural deformities" that had arisen by accident or artifice tended not to persist: "in man, natural anomalies and deformities, when they are transmitted by heredity, do not persist beyond a certain number of generations, and they eventually disappear without trace as the normal state takes over again [*reprend ses droits*]."[2] This was the "normal state" much as Étienne Geoffroy had understood it when the term first appeared: a biological capacity serving to produce and maintain order. And now, with little fuss, Périer was adding a temporal dimension that had not been present when the concept of the normal state had come into circulation during the first half of the century. This historical shift was not received in the Société d'Anthropologie de Paris as any sort of grand innovation. The preoccupation with hereditary normality simply appeared from the first to deserve a place in discussion. Anthropology was required, like physiology and anatomy, to take full account of the normal state, except that here the normal state was understood to maintain order throughout the bodies of human beings as they lived in hereditary succession. When anomalies occurred in one generation, the normal state could be expected to reassert itself in generations that followed.

Some kinds of anomalies, said Périer, were particularly likely to disappear over time. That was notably the case with deformities that appeared in the skull: "artificial anomalies or deformities, those of the cranium in particular, even though they are sometimes transmitted, never fail to disappear in the early generations that follow, with no disturbance of the ethnic type. . . . It follows as a consequence that, with all these kinds of deviations, the return to normal conditions [*conditions*] is a law of nature."[3] Anomalies were "deviations," and the workings of hereditary normality could be expected to bring them back into line. Périer was supposing that each race or ethnos had its own normal state, and indeed that the persistence of a normal state was precisely what constituted an ethnos as such. The study of race understood in these terms was from the outset a mainstream line of inquiry for the Society, which five years later elected Périer as its president. In this chapter, we will examine how anthropology was built during the 1860s and 1870s on the premise that race was a normalizing tendency inherited by all humans and that it was best studied through the close observation of crania.

This is a step away from our discussion of the decidedly complicated business of applying statistical thinking to medicine—but only a single step. As we made clear at the outset, we have largely set aside the study of governmental normalization pointed to by Foucault as a subject of investigation. Population statistics certainly did play an important role in the business of government, helping to shape the questions that might be asked and the answers that might be given to them in administrative terms. That was notably the case with birth rates and matters of public hygiene, and we commented in passing on those matters. But the primary business of our study continues to be elsewhere. Rather than governmental normalization, what interest us most here are emerging knowledge practices that allowed individual humans to be assessed according to the extent to which they deviated from an identified average. Statistically minded science was bidding to make the assessment of normality and abnormality a matter of numbers and averages. It was sometimes thought possible to arrive at a standard by a process of calculation, and to locate each individual according to the extent of their deviation from it. Insofar as that seemed possible, the comparative measurement of human bodies now became for many a privileged mode of inquiry. We have seen how Quetelet made anthropometry the field of his final major work, but Quetelet's version of the discipline was not in fact the most sustained and most influential. That place was taken in France and beyond its borders by physical anthropology of the kind practiced by members of the Société d'Anthropologie de Paris.[4]

Physical anthropology offered a quite narrow way of describing and classifying humans in numerical terms. For Étienne Geoffroy Saint-Hilaire, the unit of knowledge had been the "composition" that gave each organ a structural place. For Broussais, it had been the physiological state, continually subject to dynamic transformation through pathology and healing. For physical anthropology as it came to prominence around 1860, the basic unit of knowledge was the character. "Character" is not to be understood here in the summative sense that is still current in psychological and moral discourse: it was for physical anthropology a material feature of the human body. A character was a quality that could be measured, such as hair color or the distance between the tip of the nose and some anatomically defined point at the back of the skull. The self-appointed task of physical anthropology was to identify characters for which measurement could be carried out with repeatable precision. Yet accuracy of measurement alone did not suffice. If in the long run the numbers produced did not lend themselves to the differentiation of individuals, the character was not considered to be of enduring scientific interest. Moreover, the measurement of difference was itself a means to an end. The desired outcome was a set of orderly data that revealed a degree of stability throughout the variations themselves. The point of measuring human bodies was to identify natural groups and locate individuals within them. Physical anthropology found its scientific outcome in "the identification of types," and those types were deemed to be most natural, most fundamental when they corresponded to race.

This form of anthropology did not lend itself to philosophical disquisitions about humanity. Its leading exponents did not wax lyrical about the perfection of the human form or wonder at the heights attained by human intelligence. Physical anthropology as they practiced it was a tough-minded zoological study of the human animal. In the course of its development, physical anthropology found methodological convenience and a promise of epistemic stability in the study of human bones, especially crania. Skeletal characters, particularly the skeletons of dead humans, were subjected to precise measurement. Whenever those measurements presented a certain statistical order, that could be taken to reveal an underlying pattern in nature. Individuals could thus be arrayed according to the manner in which they spread and clustered, allowing differentiation and (racial or ethnic) categorization to occur as part of the same process. As we now focus on the use of skeletal measurement to organize knowledge about humans and the scientific debates that it provoked, we will look closely at the development of anthropometry in France after 1859 under the leadership of Paul Broca, Louis-Adolphe Bertillon, and Paul Topinard, who came together in the Société d'Anthropologie de Paris, and we will analyze some of the debates that took place in that society.[5]

Our approach will be much the same as when we discussed the 1830 debate about comparative anatomy that occurred in the Académie des Sciences between Cuvier and Geoffroy Saint-Hilaire and when we analyzed the 1837 controversy surrounding numerical method in the Académie de Médecine. Here we will examine some exchanges that took place in the Société d'Anthropologie de Paris between 1859 and 1880. None of these exchanges attracted the same public attention as the great debates of the 1830s, but their net effect was to inflect the course of physical anthropology in decisive ways.[6] We will focus here on craniometry because Broca and his colleagues did so, even as they were considering "characters" that called for other modes of inquiry. French physical anthropologists considered the human skull to be the locus of hereditary normality because it was less subject to variation and deviation than any other part of the body.

Craniometry was not brand new in 1859. It had played a role in a late eighteenth-century study of the differences between humans and animals conducted by the Dutchman Petrus Camper.[7] Since then, cranial measurement had at times been practiced in the field of phrenology, which had attracted a great deal of attention during the first half of the century in spite or because of its lavish hypotheses and uncertain methods.[8] Phrenologists sought to identify psychological qualities via a reading of skulls. Their key assumptions were that brains were differently shaped according to the greater or lesser development of local areas within them, and that irregularities on the surface of the skull could be read as an index of personality traits informed by the shape of the brain. In 1861, Broca offered a historical reflection and a summary judgment on the whole phrenological enterprise. Its leading proponents, he said, had been right to "proclaim the great principle of cerebral localizations," but their applications of it were "largely mistaken." That was because they had not had the patience to bring together a broad set of observations. Instead, they had presented themselves as the champions of an "ingenious system," a general theory that could now be declared a scientific failure: "there is no longer a single physiologist with sufficient faith in their doctrine to believe that it is possible by feeling the bumps on the skull to determine the degree of development of each personal tendency [*penchant*] and each faculty."[9] Phrenology had failed, he was suggesting, precisely because it did not accept the arduous requirements of true scientific method, preferring systematic and doctrinaire interpretations. But the primary thing phrenology was attempting to know, even as it maintained a laudable interest in cerebral localization, was itself of no interest to physical anthropology. Reading distinctive individual qualities was not the point for Broca. "The true goal of craniometry," he said, "is not the study of individuals; it is the study of groups."[10] For phrenology, each personality was different and worthy of being examined; for physical anthropology, an individual observation became a properly constituted object of scientific knowledge only when it was located in a series. Individual qualities did not deserve to be studied discretely: they were simply "variations" on something that required to be known. And for that very reason, while phrenologists might occasionally have shown an interest in measurement as a way of identifying or confirming individual traits, they had had no general need, as anthropologists did, to engage in the statistical management of distributions.[11] Physical anthropology had to deal with the fact that it was variations as such that constituted a series, even as each variation made the series a little harder to know. So it was only possible to know "groups" of this kind by carrying out calculations that would serve, in Broca's words, to "eliminate the uncertainties that result from individual variations."[12] Only when the variations were organized statistically as a spread could a series of observations be properly known.

CRANIOSCOPY

When discussing anthropological craniometry, Broca and some of his closest associates tended to mark off the scientific craniometry they were attempting to build from a disparate set of lesser practices to which they often gave the general name of "cranioscopy." So there were in fact three terms in play: the general term "craniology," referring to any study of crania, the term "craniometry," naming the developing practice of careful measurement, and "cranioscopy," study based merely on looking. Phrenology was in this sense a cranioscopic practice, although it tended to recede from view after having been largely discredited. Some other forms of "cranioscopy," however, were more persistent, and some were actively defended within the Society.

In examining debates and discussions of this kind, we have steered away from any suggestion that one single intellectual formation might have been dominant at a given point in history. Accordingly, we will eschew a term that is much used by Anglophone historians whose allegiance to Foucault is often unspecific and sometimes uninformed, that of the "medical gaze." Within the Société d'Anthropologie de Paris, many of whose members were indeed doctors, there was an enduring contest between different ways of understanding what it was to examine crania. In that sense there were competing "medical gazes." In a model of clinical perception expounded by Cabanis, true medicine was the art of the coup d'œil, of knowledge at a glance. The expert clinician, he claimed, could recognize in a trice the overall condition of a patient, determining from that first moment what treatment was called for. Cabanis's model had its advocates in medical discourse throughout the nineteenth century and also served at times, albeit anonymously, for physical anthropology. Some anthropologists recycled and reclaimed the metaphor of the all-knowing glance, using it to characterize their own observation of crania. A single coup d'œil, they said, could identify natural patterns and natural types. Instantaneous perception, when practiced by experts, might bring to light an overall truth that would not have appeared in the course of painstaking analysis.

A noteworthy proponent of the cranioscopic glance was an erudite German named Pruner-Bey who served as president of the Society for a number of years, presenting research findings almost as regularly as Broca himself. On one occasion, Pruner-Bey claimed to have identified the skull of a Lapp among a group of Ligurian skulls that had come to light during an excavation. If, as Broca and his associates regularly asserted, the business of anthropological craniometry was to examine series of crania in order to determine just how they constituted a series, Pruner-Bey's claim to have found a typical Lapp among a group of Ligurians deserved to be considered an aberration of method. For Broca, it was only possible in principle to talk about Lapp skulls if in fact one were dealing with a series either found in Lapland or transported elsewhere by circumstance. Finding a skull that had no demonstrable place in a purportedly Ligurian series would have meant for Broca dealing with an eccentric or anomalous individual. His statistical method had its own questionable way of dealing with such difficulties. But when pressed by Broca to give an account of his findings in serial terms, Pruner-Bey could see no methodological emergency, and felt no compulsion to respond to Broca's concern. For him, the anomaly was an observational fact that, for the present, must simply be allowed to stand. That this was in fact a Lapp skull, he insisted, could be seen at a glance. It was typical of its kind. That affirmation reflected no insouciance on his part about the business of craniometry: Pruner-Bey knew perfectly well how to carry out measurements and was known to perform them with care. But he was practicing and defending a form of synthesizing, quasi-metaphorical knowledge that in his view went beyond the analytical and the metonymic: "In order to make this comparison [with a Lapp skull], it is not enough to carry out measurements; one has to have an overall assessment of the skull's architecture."[13] At a point where measurements in their plurality were likely to fail, the singular expert look, he was saying, could ensure a truthful outcome.

The divergence of these two approaches to craniology was underlined when Pruner-Bey himself offered a contrastive account of his method and Broca's. Discussing some skeletal remains found in the Dordogne that had been characterized in general terms by Pruner-Bey, Louis-Adolphe Bertillon, whose position was close to Broca's, made a strong polemical declaration about method: "Until such time as we have learned to analyze our impressions about forms and translate them into mathematical formulas, we will not have done anything scientific."[14] Not so, said Pruner-Bey. There were, he said, two ways of going about scientific work in this field because there were two different kinds of project. It was true that Broca had just produced a list of measurable differences among the remains being examined, but that was simply Broca's way of doing craniology. Broca was committed from the outset, said Pruner-Bey, to finding difference. Indeed, he was so good at doing so that he was capable of finding differences between two skulls belonging to the same family. And anyone with his methodological bent could attempt to do the same thing: "If anyone, even without the talent and experience of M. Broca, were to set about examining two skulls belonging to the same race or the same family with the clear intention of finding differences, he could be expected to find even more than M. Broca has done with those I have presented here." Finding differences was not the point of Pruner-Bey's own undertaking: "The key thing here is to grasp [*saisir*] the characters inherent in a cranial type."[15] As Pruner-Bey summed it up, Broca was hunting for differences, while he himself was attempting to grasp the type.[16] If Pruner-Bey was right, these two kinds of inquiry simply took divergent paths, and there was no reason to believe that fastidious measurement could ever produce a clear perception of the type.

Broca could never have accepted this apparently benign account of separate lines of inquiry. Pruner-Bey was speaking as if difference could be for Broca a scientific object in its own right, and further implying that the measurement of differences could be pursued indefinitely without yielding knowledge of the whole. The point for Broca was not, however, to study differences for their own sake, but to identify variations within a series, organizing them so as to yield a statistically grounded perception of a type. For him, the type was something to be arrived at eventually — not instantly — as a natural regularity manifesting itself in a series. While holding the floor in meetings of the Society, Broca took the time to warn his colleagues against what he saw as the scientific inadequacies of the coup d'œil. No one was so experienced or so clever as to know crania properly in that way: "However different cranial types might be in the human races, there are so many gradations linking them and they are moreover so variable within one population that the most experienced observer cannot reasonably claim to perceive all those nuances merely by looking at them [*à la simple vue*]. The most skillful of writers could not hope to describe them with utter precision."[17] That was why craniological research as he saw it called for the use of numbers. Numbers actually made observation more precise, registering as they did a finer level of detail. It then became possible to "express all the degrees [of an observation] in numbers and take note of all the differences, even those that are too slight to catch the eye."[18] Numbers were finer and more accurate than any look.

When Topinard, a pupil of Broca, read to the Society a paper entitled "De la méthode en craniométrie," he too felt entitled to refer to some of Pruner-Bey's work as a counterexample of the method he was describing and advocating.[19] Since craniometry had now become a precise, methodical science, he said, there was no place in it for the "feeling [*sentiment*]" that had led some colleagues to declare that they could perceive the truth without the aid of measurement: "it becomes obvious that the most practiced eye will be incapable of arriving unaided at the aggregate outcome [*résultante*]."[20] According to Topinard at least, practitioners of the coup d'œil, even the most experienced of them, no longer deserved to hold authority as scientific observers. Looking outside the Society, Topinard criticized an article by the Italian anthropologist Paolo Mantegazza for adopting, at least on one occasion, an approach he considered unscientific. Together with two unnamed colleagues, Mantegazza had put together pell-mell skulls "of both sexes and all races" and reached the conclusion that "the measurements given by craniometry were not in keeping with their aesthetic ideas," that is to say that the most pleasing skulls to look at were not those of Caucasians. This aesthetic look, said Topinard, had nothing to do with craniometry properly understood. Anthropologists should leave that kind of thing to artists: "The illustrious anthropologist regrets the fact that craniometry does not demonstrate the hierarchy of races as he likes to think of it. But what if craniometry refuses to support that hierarchy? Everyone should play their proper role. Let us leave feeling to the artists. That is their essence. And let us keep for ourselves strict observation, without which there would no longer be any science."[21] For Cabanis, medicine itself was an art, summed up and performed in the coup d'œil. For Topinard, anthropology could be properly scientific only by turning its back on practices of that kind.

Yet for all of this resolve within the Society, the march of progress was not quite so assured or so relentless. In Broca's own *Mémoires d'anthropologie zoologique et biologique* of 1877, we find him speaking in passing of a group of skulls using a term that one might have expected him to eschew. Having just listed a number of groups of skulls, he comments: "these points of resemblance give them what one might call a family resemblance that can be recognized at first glance [*au premier coup d'œil*]."[22] So the glance of recognition did appear after all to have some place in Broca's practice of physical anthropology. Even Topinard, the assertive champion of mensuration and calculation, made a surprising concession on this score: "Admittedly, one cannot deny oneself the use of a simple look [*la simple vue*] in order to conceive of an immediate, approximate opinion. One does not always have the necessary instruments to hand and there are, for example, curves that will be always difficult to render numerically. But it should not be forgotten either that that system is full of illusions."[23] Here was an admission that some sort of knowing glance did in fact occur from time to time in properly scientific practice, and that the simple act of looking might allow the perception of curved cranial shapes that could not be well characterized using numbers. Furthermore, Topinard even conceded that some people were particularly good at overall perception of that kind: "Some people, it has to be recognized, possess an admirable synthesizing glance." Yet, he insisted, any knowledge produced by that means was not to be trusted: "try as they might, their judgments are subordinated to their current state of mind and to the most recent impressions that have been made on them."²⁴ That was how the knowing glance could remain in play even while being regarded with suspicion. There were indications that a statistically based anthropology could not manage entirely without it.

EXCLUDING ABNORMAL INDIVIDUALS FROM CRANIAL SERIES

Craniometry in the rather strict sense expounded by Broca was devoted to the study of series, but if a set of skulls happened to become available to the Society, for example through the relocation of a cemetery, that offered no guarantee that they would constitute a proper series. There might be skulls that did not belong. Yet what exactly did it mean in practice and in theory to declare that they did not belong? To put it in Broca's terms, what characters by their presence and their relation to the other characters in a series would justify the exclusion of particular individuals?²⁵ Given that the whole point was to study variations, what kind of variation deserved to be considered excessive, and how could it actually be that some variations were so great as to deserve exclusion? Broca made caution a scientific virtue, and attempted wherever possible to include in the same series all skulls found in the same place. When he came into possession of a set of Basque skulls from a cemetery in southwest France, he posed the question explicitly. Ought some of them to be excluded from the series because they were deformed? On balance, he thought not: "there are indeed some that are not normal, but their degree of deformity is too slight for them to be eliminated."²⁶ We see here that even as Broca was setting about the statistical business of identifying the normal, he was also using the term rather casually, with the medical sense of "not pathological," "not deformed." For one so scrupulous about method, there was a potentially worrying circularity here. The point of studying the Basque skulls as a series was to identify qualities about them that were normally characteristic of Basques. But if a number were eliminated at the outset because they were judged to be "deformed," how could the observer claim to be sure of that judgment before any calculation began? Broca was acting as if he knew from the outset just what it meant for Basque skulls to be well formed. This is an important question for our genealogy. The point of carrying out observations in series was to determine what was normal for a given ethnos or race, but prima facie judgments were sometimes being made about what was to be considered abnormal before any calculations were actually performed.

The Society had the opportunity to ponder a version of this difficulty when one of its members read out in French translation a paper by the English doctor and craniologist Joseph Barnard Davis. Davis raised a question about a skull that had come to light in excavations at Neanderthal: "What is the real meaning [*valeur*] of the form of the skull from Neanderthal? Is it normal, abnormal, or pathological? Does it show the characters that belong to a race?"[27] It was clear that once again medical thinking was informing a question in physical anthropology. While medically trained anthropologists from Périer to Broca had been able simply to transfer their assumptions about the normal state to the field of human anthropology, they were struggling here to do the same with the abnormal. Davis was allowing in principle for two kinds of abnormality: divergence from a healthy norm and divergence from a racial one. Was the Neanderthal skull abnormal in either of those senses, or was it in fact normal for a largely unknown race to which this individual belonged? A German anthropologist had examined the marked development of the frontal sinuses in the skull and declared them to be "a character that was typical of a race," but Davis's own view was that the character was pathological: "its large dimensions must be regarded mainly as an individual particularity that can be found in living individuals."[28] With the evidence available at the time, the matter was hard to resolve, but in any case it pointed to an ongoing difficulty for Broca's method. When was it appropriate to declare an individual pathological or otherwise eccentric, and what was the theoretical basis of such a declaration? How could serial normality be studied without first eliminating abnormal individuals on the strength of an overall assessment of their characters and thereby begging the question of what was normal within the series?

Broca's first preoccupation was to constitute a series that allowed the calculation of an average. In statistical terms, individuals whose characters lay at a point distant from the average were a problem. Instead of facilitating and clarifying the comparison of groups, they "only confused and obscured matters."[29] That was because their presence could be expected to distort the average. Exceptional individuals, when their measurements were added to those of others close to the mean, had a markedly greater effect for strictly mathematical reasons, pulling the average toward them, so to speak. Extreme measurements counted more, and thus threatened to disturb the identification of the type being pursued via the fine calculation of an average or set of averages. In that sense, extreme individuals were an impediment to the study of the type, and a proper anthropological corpus could be produced only by "excluding the exceptional cases from series."[30] That included any individuals with pathologies, but also children, who would almost by definition provide extreme measurements when included in a series with adults.[31] The same rule of thumb applied to those with pathologies and to those who were "manifestly abnormal."[32] But it was only a rule of thumb, for all of Broca's close attention to method. Just how and when was the abnormal "manifest"? What kind of observation justified that conclusion?

Statistical methods were not utterly bereft in the face of this difficulty. In a later presentation, Broca pointed to a properly statistical way forward, although that did not actually resolve the fundamental difficulty we have just highlighted. In the case of a particular series, he said, there were "subjects that seem to us to be normal." This observation, we note, could only be an overall perception arrived at before analytical study of the series proper had begun. But as the study proceeded, some anomalies might become apparent via measurement and calculation. At the outset, it was appropriate to include any doubtful but apparently normal individuals: "we must therefore begin by giving them a place in our list." As the business of statistical analysis proceeded, however, anomalies might appear: "Only when the reckoning [*dépouillement*] of the series has been carried out will the unexpected anomaly be revealed to us."[33] So Broca also proposed to exclude those individuals who, once the series was in place, stood furthest from the average. He would close off the series at what he called the "second maximum."[34] If between the first (greater) maximum and the second there was a significant interval, "it is appropriate to ask if the first one might not be due to some anomaly, and the likelihood of an anomaly is all the stronger when the series is larger."[35] Thus, in setting up a series for the purpose of statistical study, Broca envisaged two forms of exclusion. Those who appeared from the first to be clearly abnormal were not to be measured at all, and those who, having once been measured, occupied a markedly extreme position in the range were eventually to be eliminated. Guiding Broca in his initial decision making about abnormal individuals was his scientific interest in "the cohesion of the ethnic type."[36] The focus on ethnic type, he said in a precise statistical argument about the probability of error, allowed a series to be relatively limited, but this was so only if the observer had first taken the step of "excluding abnormal individuals."[37] The ethnic type, defined in principle as hereditary normality, had to be known in some overall way so that abnormal individuals could be identified and excluded accordingly.

We find ourselves in a position here to contribute to a history of discrimination, as long as the notion of discrimination is understood quite narrowly. By this we mean that our history of physical anthropology is not in the first instance a history of prejudice understood broadly, but rather an account of how science of this kind was deemed by those who practiced it to require the exclusion of certain kinds of individuals from the series under study. To discriminate was to engage in fine measurement, but it was also to make decisions about who did not deserve to be counted. For reasons of statistical method, it was not appropriate to lump everyone together and wait to see how the mathematics came out. The statistical procedures followed with such care by Broca and his closest colleagues did not therefore result in anthropological knowledge that gathered all humans into a single group. The point was not to study *anthropos* as a single entity. Precisely because of its preoccupation with groups, physical anthropology was not inclusive in any generally humanistic sense. We have already seen that, according to Broca, "deformed" individuals were to be omitted from series, as were children. And most consistently of all, so were people of different races. It would have been a great mistake in Broca's view to treat all humans in the same way. That would involve bringing them together in one vast series that would prove not to be a proper series at all: "But one would then be committing a grave error in considering all men as if they made a single group. That would mean giving up on the search for the craniological characters of the different races. The extent of deviations must be studied, not in humankind in general, but within each population in particular."[38] Broca's logic did not necessarily lead to the verbal disparagement of deformed people, of children, or of non-Europeans, but it did lead systematically — as systematically as it was able — to the separation of humans into groups and to the exclusion of atypical individuals from each group.

We will refrain here from asking baldly whether this kind of physical anthropology was racist — or indeed sexist, ageist, and dismissive of people with disabilities. In the sense of those terms current in our own time, it was patently all of those things, but the point of intellectual history as we understand it is not to denounce thinkers of the past for failing to see issues as they are generally seen today.[39] It seems more helpful to say that Broca and his colleagues were utterly committed to discriminating between groups of humans, and that they put a great deal of mensurative work into normalizing those groups, particularly in racial terms. We can then ask exactly what that entailed for the development of knowledge practices. Scientific leaders like Broca sought to make strictly quantitative distinctions and usually refrained from making qualitative judgments on the basis of them, but there was undeniably an easy slippage from establishing craniometric differences to producing disparaging categorical statements. Gustave Le Bon, to take an inglorious example, slid easily from the one to the other. A busy craniometer with relatively little authority in the Society, he moved enthusiastically from technical discrimination to dismissive judgment. Presenting a paper entitled "Sur la capacité du crâne d'un certain nombre d'hommes célèbres" ("On the Cranial Capacity of a Certain Number of Famous Men"), he excluded many kinds of people from his studies: "women, criminals and pathological subjects" were all "classified separately."[40] There were in fact technical reasons for excluding the skulls of women from craniometric series because their heads were smaller on average, but it has to be acknowledged that any historian wishing to dismiss the whole craniometric enterprise on ethical grounds might look no further than Le Bon's pithy résumé of the significance his own work. He summed it up all too neatly in a final disparaging reference to "inferior beings: women, savages and children."[41]

Work like Le Bon's can be simply cast aside by modern historians, but the ethical strain becomes greater for those who seek to understand how thoroughly committed physical anthropologists were to race as a category of knowledge even when they did not engage in disparagement. Of particular interest to our own genealogy is the fact that anthropologists usually held to the idea that zoological normality constituted and defined a race, and were thus able to give an account of each race according to its own norms. Considering each race in its own right involved no particular moral effort, no special practice of tolerance. It was simply believed to be necessary for the zoological classification of humans. So when dealing with the vexed question of comparative brain sizes across the full range of humans, Pierre Gratiolet, one of the Society's founding members and intellectual leaders, was able to insist that what was normal for one race was not likely to be normal for another. Some argued that brain size was a clear indicator of intelligence, but Gratiolet demurred from that view on two counts. He pointed first to anatomical research, some of it conducted by Broca, tending to show that in actuality the number and complexity of folds in the brain mattered more than the volume of the whole. But his second point was a more fundamental one. It was that configurations of the brain ought to be understood as normal *within* particular races. Certainly, a contrast could be made between Caucasians and Bushmen, he said, as if to quote a standard view of hierarchical difference: the average size of Bushmen's skulls and brains was known to be one of the smallest among humans, and, on average, post mortem examination showed that they had fewer folds in their brains. But this was not an occasion for Gratiolet to talk about any supposed inferiority of Bushmen as a race. Instead, he offered his colleagues an interpretive grid for making sense of the differences that had been observed: "This simple form of cerebral folds in Bushmen is normal. It suffices for the kind of perfection that is proper to them, and is consistent with intelligence. But a white man with this shape would not have acquired the normal type of his race. For him it would be an imperfection, a degradation, and would necessarily result in idiocy."[42] This statement was discriminatory in the narrow sense in which we have begun to use the term here, and could easily have been interpreted as marking the superiority of Caucasians. But insofar as it was presented as a description of biological normality, a form of scientific relativism was necessarily entailed. Gratiolet's statement was reminiscent of Cuvier's view that perfection must exist by definition wherever there was a type. No ethics of tolerance were called for on Gratiolet's part. He was merely affirming as a zoological principle that to each race corresponded its normal type: "So with brains that are equal in weight and complication, two men who belong to different races will be as different as it is possible to be. That is because intelligence does not have as a necessary condition a certain brain weight, nor even a particular shape of folds in the brain. What it requires is its typical perfection, which functions as the harmonious expression of an accomplished creation."[43] It could still be argued, of course, that Gratiolet's position deserves to be called racist when viewed in a modern perspective, but it matters for our history that his view of racially specific normality actually served at the time to forestall or at least question affirmations according to which one race was generally superior to another. For physical anthropology at its most disciplined, there could simply be no general superiority because there was in effect, as Broca said insistently, no general set of humans and no general range of characters that might serve to distribute all individuals, however they might be clustered, along the same scale.

A clear contrast can be made between the view of race advocated by Broca and the one put forward by Arthur de Gobineau in his *Essai sur l'inégalité des races humaines* (*Essay on the Inequality of the Human Races*), which had been published six years before the Société d'Anthropologie de Paris was founded. Gobineau had put forward a theory of racial inequality and interracial hostility, arguing that mutual "repulsion" was a natural consequence of the need for races to defend their blood from the terrible damage that would necessarily be wrought by hybridity.[44] This was not just a theory of race, but a theory of natural racism, and Broca gave it short shrift. Against Gobineau's claim that hybridity was bound to result sooner or later in "physical and moral degradation," Broca adduced the example of the

United States, a country built on a mixture of races that was currently thriving.⁴⁵ Moreover, the people of France presented a long-established example of racial diversity, with no evidence whatsoever of diminishing fertility as a consequence of their hybridity.⁴⁶ And while Gobineau had seen fit to claim that in colonial contexts dangerous physical effects had been brought about by the cohabitation of races, Broca simply stated that colonial conquest was "an act of pure violence" that consisted precisely in displacing people of different races and forcibly creating for them difficulties of adaptation. Racial difference was not in Broca's eyes a matter of superiority or inferiority, but of suitability to natural environment. So there was no natural right of conquest founded in supposed racial superiority. On the contrary, there was a natural requirement to respect racial difference, and that made colonialism violently unnatural. Clearly, Broca's disagreement with Gobineau was not an in-house argument between physical anthropologists. Gobineau's essay was never discussed in the Society, and was not regarded by Broca and his colleagues as a contribution to the scientific study of race. Even Broca's criticism of Gobineau was made in another place.⁴⁷

We can gain a fuller understanding of what race meant in the work of the Society by observing that, by contrast with Gobineau, physical anthropologists were not preoccupied with racial purity. The point of a scientific study of racial types was simply to identify the stable elements that persisted through heredity despite migrations, invasions, interbreeding, and the like. In practice, racial purity was found to be vanishingly rare. To the extent that a particular group of skulls, for example, came from a well-defined area, that was valuable from the point of view of observational method, but not because it was a promise of purity or of racial identity in the narrow sense of strict similarity. "Purity" was no more than a hoped-for convenience of method that almost never arose. Hybridity was an empirical given, and in that sense a proper object of knowledge, but physical anthropologists sought a knowledge of race as something more profound and more stable. They were looking for the serial particularity of racial groups through data that were presumed to be complicated by the presence of diverse racial characters. When they worked along the standard axes that were thought to define cranial types, such as the length of skulls from front to back, they came up with mixed results—numerically and literally. Almost no sample of any size was marked by strict similarity of the individuals. Yet that was not taken as a defeat for a scientific theory of race precisely because race understood scientifically was thought to require series of painstaking observations, the final outcome of which continued to be deferred.

These anthropologists had no scientific interest in or commitment to racism as such. By contrast with Gobineau, they offered no account or justification of its causes or effects.

The broad theoretical frame in which Parisian physical anthropology was conducted differed significantly from that of Quetelet's anthropometry. Quetelet was in touch with the Société d'Anthropologie de Paris, as he was with so many scientific and administrative bodies in which statistics played a central role. Moreover, his *Anthropométrie* was published in 1870, when the Society was flourishing. But he could never have been enrolled in the research program led by Broca because he could not have accepted Broca's way of constructing race as a scientific object. Quetelet's aim was broader, and by that fact alone decisively different. He referred quite simply to "the whole race," declaring that "humanity in its full extent" was the proper object of anthropometrical study.[48] He was unpersuaded that measured "deviations" between human bodies were sufficiently great to justify the hypothesis that different groups of humans might have different origins.[49] On those occasions when he did have the opportunity to measure individuals from outside Europe, he found no reason to give up this general view. He observed two Indian chiefs from the Ojibwa people who were visiting Belgium, comparing them with one of the finest individuals he could find among his compatriots. The Indians proved to be bigger and stronger, being outside the range usually found in Belgium. Their broader chests he attributed to the "exercise" that necessarily accompanied the use of bows and arrows. Apart from that, it could be seen that "their proportions are approximately the same as ours."[50] And when he measured three black Africans, he drew conclusions of the same kind. From such observations, Quetelet felt entitled to conclude that "the broad outlines of the human species appear to be more or less the same for different countries and for different races. The characters that separate them are found in parts that are harder to assess [*apprécier*]: the facial angle, the width of the nose, the thickness of the lips, color, hair, beards, etc."[51] When, in *Anthropométrie*, he came to compare his statistical tables with tables from a variety of countries, he believed that he had now shown well enough that human proportions were "fixed," thereby "demonstrating the type of our species."[52] There might, he conceded, be perceptible racial differences among humans, but these had "much more to do with relations of forms than relations of size," and so belonged "in the domain of the arts rather than that of the sciences."[53]

That did not prevent Quetelet from speaking on occasion of a plurality of races among humans. He spoke at one point about dividing humans into five races according to color, but went on to note that it was difficult to make and maintain that set of distinctions in practice. Broca, we should note, had no patience with what he called the five-color classification: "as if, for example, all individuals of the Caucasian type resembled each other sufficiently to constitute a single race."[54] For Broca, the number of racial groups had to be far greater than five, and could not be limited in principle, but Quetelet regarded the five groups produced eighty years earlier by Blumenbach's classification as an acceptable way of ordering the observable varieties of a single species that could be taken to include all of humanity. Racial differences might indeed be perceptible, but they would doubtless be best perceived by a singular look: "Broad proportions vary only slightly in man. The real differences presented by races have to do with characters more easily grasped by the eye than by the compass. They can be established only by dint of very fine judgments using the kind of tact that is developed through the sustained practice of such research."[55] So Quetelet did not deny that anatomists might perceive differences of race among humans. He simply declared that any such differences were "impossible to assess [*apprécier*] by numbers alone."[56] Where Broca was claiming that the characters reflecting differences in race were so subtle that they could be brought to light only by measurement, Quetelet, the best-known statistician of the time, was asserting that they were so slight that whatever "research" was carried out on them could not be the business of a statistically based science. They were just too small to be described through measurement and calculation. Broca and those who followed him bet their whole scientific careers on the opposite view. Certainly, they said, the differences between races were tiny, but that was precisely why the task of identifying them was so demanding. A well-practiced glance could not be subtle or precise enough to yield the truth in all its fine differentiation.

CRANIOMETRY AND RACE

Physical anthropology, when settled into its scientific routine, involved a search for characters that offered some form of "fixity." Insofar as a character was found to be stable in the sense of being normal across generations, it was usually taken by definition to indicate the presence of a race. If it recurred from generation to generation, that was precisely what it meant to be normal through heredity, and the persistence of characters allowed race to be identified by—and as—that fact alone. It might be any character at all as long as it offered both difference with respect to others and stability within itself: "To differentiate two races, all that is required is a single character, no matter how slight, as long as it is hereditary and sufficiently fixed," said

Broca.[57] When making this comment he took hair color as his hypothetical example, but the great bulk of his scientific effort went into the study of skulls. That was not in his eyes an arbitrary choice. Quetelet, in his *Anthropométrie*, had observed that "the noblest parts of the body are those that are least subject to change. For example, the head varies very little in adults."[58] It mattered even more for Broca's anthropometry that head measurements should vary only slightly, but, as an anatomical scientist, he had not the least interest in distinguishing parts of the body according to degrees of nobility. Broca's mentor P. N. Gerdy, in a book published in 1830, had already pointed in general terms to the property of skulls that would later be at the heart of Broca's thinking. Gerdy described the skull as one single resistant piece of bone: "the skull resists like a hollow sphere or ovoid. It is very solid, although of uneven thickness, and is made of a single piece."[59] Solidity and resistance were likely indicators for physical anthropology of a capacity to endure across generations, and Broca developed that theme. Comparing the skull with the nasal opening, he remarked that the skull was "much more fixed. It is closed on all sides; the bones that compose it are constrained by each other [*solidaires*]; their edges touch; no single one of them can extend beyond its limits without being stopped or restrained by its neighbors. So the chance events [*caprices*] of ossification that produce individual variations run into obstacles on all sides. The causes of variations in the form of the skull, leaving aside pathological cases or mechanical acts, are the overall conditions that govern the general development of the brain and the skull that contains it rather than local, partial conditions."[60] Not just solidity then, but "solidarity." That was why the human skull appeared likely to reward close study. It was not just a set of bones growing together, but a set of bones whose very closeness provided a guarantee of structured stability. Within the human skeleton, the skull could thus be seen as the locus par excellence of normality.

The very qualities that made skulls a valuable object of study for physical anthropology were at the same time a source of persistent technical difficulty. Crania might well be stable, but they were certainly not geometrically regular, and it was no easy matter to translate their shape into numbers. In response to talk about supposed general differences in brain size separating racial groups, Broca declared that crania did not lend themselves to the establishment of a single comprehensive scale because they were so irregular: "The differences cannot be expressed in figures because the techniques used to measure them have been so diverse and, because of the irregularity of its shape, the skull is not suitable for geometrical evaluation."[61] So here was, so to speak, a concrete paradox: the skull was an inscription of—and a semiotic vehicle for—racial normality, but it carried that normality in ways that were resistant to precise observation. While the in-principle anatomical solidity of skulls appeared to guarantee that they would reward scientific observation, it had to be recognized that they were in actuality subject to remarkable variation: "The great variability of cranial shapes constitutes the main difficulty for craniology. Whatever part of the skull or face one considers, one always finds that, by its arrangement and its dimensions, it offers oscillations that are extensive enough to make descriptions uncertain and comparisons difficult." There was consolation to be had in the thought that "not all conditions were equally variable" and that, while there were none that proved to be "absolutely fixed," some did "reveal relative fixity."[62] Craniometry involved a search for characters of the latter kind in particular, but variation was not in itself an insurmountable problem for Broca. The important thing was that the variations, when measured, should constitute a proper series, thereby revealing normality in the form of a patterned distribution. A certain part of the brain, he noted, could occur in a variety of shapes and sizes, but the full range of those shapes and sizes was present "in normal man."[63]

A great difficulty faced by craniometry arose not from variation in the size of crania, but from the fact that it was difficult to arrive at a (numerically) summative account of each individual. Skulls might have been relatively stable, but the characters they revealed did not stand in an unchanging relation to each other: "In any race, in any population, no matter how pure, each cranial character shows fairly extensive individual oscillations, and these variations, which can be thought of as oscillations with respect to the essential type of the race, are quite unevenly distributed over the diverse characters of the same cranium."[64] Oscillations there might be, but there was no single axis of oscillation. So when characters were identified that could be ascribed to race, it did not follow that whole individuals could be reliably sorted according to full sets of racial characters. For the same reason, it was not possible by statistical calculation to arrive at the conclusion that a given individual had the full set of characters belonging to a race. The different characters could not themselves be set out as an orderly distribution: it was "impossible" to establish a "set of multiple statistics that would pertain successively and particularly to each of the racial characters."[65] There was thus no final sum of mensurative knowledge, no significant average of averages, and no single mean allowing all cranial characters to be synthesized in one formula. That is how the inviting general promise of craniometry continued to be attended by apparently unending difficulties of measurement and tabulation.

USING OTHER CHARACTERS TO IDENTIFY RACE

A variety of characters presented themselves as possible objects of serial observation, and while craniometry was by far the most discussed and the most resolutely pursued, a number of others were considered in the Society from time to time. The most obvious of those in every sense was color: of the skin, of hair, and of the eyes. In 1864, Pruner-Bey reported on a study of skin color, asserting that it was the character that served undeniably to differentiate the human races.[66] It is unsurprising that Pruner-Bey should have taken this initiative. He was a great advocate of the knowing glance, and color seemed to lend itself well to that form of inquiry. But Broca did not simply cede that terrain to his methodological rival. He set about the anthropological use of colors in a quite different way, making various attempts to standardize them so that they could serve as characters. He initially envisaged three chromatic scales, one for hair, one for eyes, and one for skin.[67] When discussing the observation of eyes, Broca asserted that it was not a matter for scientists of noting fine nuances within the iris, but of studying "the *average* shade of the iris."[68] In a later presentation, he spoke of developing a scale that showed "types of colors."[69] The logic he was following required colors to be reproduced in a form that could be consistently identified by all. Only then could they provide a stable set of references.

It became clear from time to time that unease was being generated within the Society by attempts to build a research program around color. Charles Rochet was provoked by Pruner-Bey's presentation into declaring that the importance of color in the study of human races had been greatly exaggerated. In his view, skin color did not deserve to maintain a central place in physical anthropology. Color was in fact "a fragile, secondary character."[70] People who were said to belong to "races" standardly classified by color did not even have skin of the corresponding shade. "White" people were not white, "yellow" people were not yellow, and "red" people were not red. "What could color possibly tell us about the value of an individual?" he asked. One might as well judge individuals by their clothing or their tattoos.[71] In other words, color was unworthy of scientific attention because it was superficial. Rochet might have found further support for his view in a paper delivered a few months earlier by Louis-Adolphe Bertillon in which Bertillon pointed out that when people were exposed over time to hot climates, their skin tended to darken.[72] Here then were two reasons to disqualify color. It was widely and carelessly used in unscientific talk about races and it was in any case subject to change in a way that made it an uncertain racial character. But Broca was not quite prepared to discard it and actually spoke up in defense of Pruner-Bey when he was attacked on a later occasion for some work he had done on hair. It was true, said Broca, that one should not expect to find "fixity" in characters drawn from hair, but this research was still of great importance.[73] Broca spoke only in general terms as he defended Pruner-Bey, who happened to be absent because of illness. When one considers the sum of Broca's many contributions, this in-principle defense of work on hair color appears to be little more than professional courtesy. His own practice reflected the belief that, whatever the technical difficulties presented by craniometry, human skulls offered to scientific inquiry the great advantage of material solidity. They were of far more interest than color because they were not exposed, as color was, to superficial change and superficial observation.[74]

Within anthropology, craniometry was also required to compete for a place alongside historical linguistics. Broca was quite prepared to refer to linguistic practices as characters, and on more than one occasion asked a general question of his colleagues about "the anthropological value [*valeur*] of characters drawn from language."[75] As he moved at one point to answer his own question, he emphasized the fact that the study of language made it possible to reach back into a relatively distant past. He was referring in particular to the work being done by linguists, including Pruner-Bey, on language families, work that involved the reconstruction of proto-Indo-European. "Language," said Broca, "is almost always the most ancient monument of a people."[76] That very metaphor served as a measure of his respect for the discipline: the linguistic genealogy of peoples offered access to a character that could be referred to in terms of monumental solidity. However, this was not in Broca's eyes a sufficient reason to make historical linguistics the preferred method of anthropology: "But we have available characters of a quite different kind that will allow us to group and classify the human races by determining the analogies and dissimilarities between them. Those are drawn from the physical organization of races. The natural method that we must adopt requires us to take account of all characters, and the ones provided by linguistics are certainly worthy of our attention. But this same natural method requires in addition that we give pride of place [*la primauté*], in keeping with the principle of *the subordination of characters*, to those that present the greatest fixity."[77] The natural method to which he was referring was that of the great naturalists, from Linnaeus to Cuvier and Étienne Geoffroy Saint-Hilaire. As interpreted by Broca, it called for characters to be organized hierarchically, with greatest importance being attached to those that were most fixed. In natural history, he reminded his colleagues, "the characters of first importance are the most permanent ones, and it can easily be shown that the [anatomical] organization of man is more permanent than his language."[78]

The impermanence of language could be demonstrated by historical examples. Africans forcibly taken to Haiti had lost their native languages and now spoke French.[79] More generally in situations of conquest, the development of racial characters over time revealed no reliable parallels between language and anatomical organization: "after a certain number of generations, when mixing has occurred, the crossbred race tends to draw closer to the physical type of the majority race, while sometimes the language of the minority supplants and replaces that of the majority."[80] Ancient Gaul invaded by the Romans was taken to reveal that pattern. That was how it was possible for Broca to resist what looked like a straightforward racial characterization of Polynesian peoples. It was true that they spoke much the same language over a vast spread of islands, but there was every reason to withhold judgment about the existence of a common race until a sufficient number of anthropometric measurements had been conducted. It had to be remembered that "man can change his language as he changes his customs and beliefs. His physical characters are better conserved. In many cases, they survive the influence of crossbreeding."[81] This was principled obstinacy on Broca's part. It might seem obvious that Polynesians belonged more or less to the same race, just as it might have seemed obvious that, say, northern Europeans with blond hair were part of the same race or set of races. But obviousness of that kind was of little scientific interest to him and to his close colleagues. The self-appointed task of this kind of anthropology was to find an enduring truth about race, to find the truth of race as that which endured in human skeletons even when more strikingly visible characters were subject to change: "observation proves that languages always become slowly extinct, and that most of the peoples of western Europe have changed languages several times, even while keeping their type, despite the crossbreeding they have undergone."[82]

QUESTIONING THE VALUE OF CRANIOMETRY WITHIN THE SOCIETY

In order to sustain its central place within the Society, Broca's craniometry needed not only to assert its place against rival research programs, but also on occasion to repel direct attacks. Charles Rochet thought it regrettable that some of his colleagues should spend so much time observing skeletons. It would be better, he said, if anthropologists were to conduct their observations on the living. The study of "animated man" ought to be "preferred to that of man without life."[83] Cranial measurements might have some value, he allowed, but they offered less to anthropology than the study of the human face and those "characters of expression presented by the physiognomy of individuals."[84] Broca, who was usually quite zealous in the defense of his own work, did not even rise to respond to Rochet's criticism. He could only have seen it as a historically regressive appeal to the old art of physiognomy. But when Pierre Gratiolet, a founding member of the Society and its first president, challenged the central place of craniometry, it was quite a different matter. Gratiolet, himself a comparative anatomist, posed a direct question about method. A quite small Mexican skull had proved on observation to contain a considerable cerebral mass, and this led Gratiolet to declare that it was far more valuable to measure fresh brains than to measure skulls.[85] When craniometers set about measuring the volume of a skull, Gratiolet remarked, they usually filled it with a substance like millet, then weighed the millet. But that, in his view, gave only an approximate indication of the actual volume of the brain. He went on: "the procedure that involves measuring the brain in its fresh state after removing the membranes is much more exact."[86] This was, so to speak, a rather idealized discussion of method in view of the difficulty of gaining regular access to fresh human brains, but Gratiolet was attempting in the first instance to raise a question grounded in theory. He referred to a skull said to be that of Descartes, noting that it was in fact quite small and therefore, according to the standard view that correlated cranial capacity with intelligence, most unlikely to be that of a genius. But the skull in question was "admirably shaped [*conformé*]." This served for Gratiolet as a reminder that "the form and not the volume constituted the dignity of the brain."[87]

It was not Broca who replied first to Gratiolet, but a colleague named Auburtin, who quickly perceived the challenge to one of the standard assumptions underlying the work of the Society. Was Gratiolet really saying that there was no relationship between intelligence and cerebral mass? Surely he could not mean that? There were so many geniuses who had been shown through autopsy to have big brains, said Auburtin.[88] To that Gratiolet responded in a properly statistical manner that could only have compelled the respect of Broca and Bertillon. There was certainly a limit of brain weight below which a man ceased to be intelligent. That was about

900 grams. But among eminent men who were almost equal in intelligence, the weight of the brain had been found to vary across a range from 1,200 to 1,900 grams.[89] Gratiolet was making some strong affirmations. First, measuring skulls was not a good way of determining the size of people's brains. Second, brain size was not in any case correlated with intelligence as long as it remained within the broad range of normality. And to these he now added a third, that cranial capacity should not be read as an index of the capacity for future cultural development inscribed in a race. Greater cranial capacity did not indicate the possibility for a given race of reaching an advanced stage of "civilization."[90]

These criticisms required Broca to reply at length, and he did so with his usual resolve. As often in debates of this kind, his reply serves the purposes of our genealogy by making a key assumption explicit. He began in the classic manner by conceding a point to Gratiolet. It would indeed be better to weigh the brain directly rather than to follow an indirect method. But since that was not usually possible, it was better to know the capacity of the skull rather than to find oneself with no idea of the brain's volume, not even an approximate one.[91] But when during the next meeting of the Society Broca returned to speak at length, he presented craniometry as something rather more than an ersatz for the weighing of brains. There had been, he said, a long history of inquiry dating back to the late eighteenth century during which the study of the human races—always a plural term for Broca—had pursued two paths. Some people had paid particular attention to anatomical characters while others had been preoccupied with intellectual and moral characters. But these two kinds of inquiry had found "common ground" in craniology: "On the one hand, studying the shape of the head provides precious elements for the anatomical parallel of races. On the other, the cranium contains the brain, which is the organ of thought whose disposition could be expected to influence intellectual and moral phenomena as well as influencing the external configuration of the head."[92] Craniology, presented thus, could be said to address a problem that we considered previously: the difficulty of bringing together different kinds of knowledge about humans, namely the physical, the moral and the intellectual. By focusing on crania, Broca was claiming, it was possible to study all those things together. The cranium deserved to be closely examined not only because it was the locus of normality in the human body, already identified as such for its stability, but also because craniology was a field in which different scientific programs could converge. The skull was the part of the human body in which the physical, the moral, and the intellectual could all be observed together. This was Broca's own contribution to intellectual history, offered as a diagnosis of why craniology now claimed such a prominent place in anthropology: "The great importance of craniology has had such an impact on anthropologists that many of them have neglected the other parts of our science, devoting themselves entirely to the study of crania. This preference is probably legitimate, but that would not be the case if the examination of the bones in the head only made sense in anatomical terms. There is also the hope in this work of finding some data about the intellectual value [*valeur*] of the different human races."[93]

We noted earlier that Broca had not deigned to respond when Rochet suggested that it would be better for anthropologists to practice the art of physiognomy on the living rather than run slide rules over the dead, but there was a point in Rochet's remarks that was not forgotten. While hardly anyone showed an interest in physiognomy, there was in fact ongoing discussion within the Society about the relative merits of conducting anthropometry on living persons rather than on skeletal remains. Broca was called on from time to time to provide a full rationale for the study of skeletons. In 1868, he presented to the Society a paper entitled "Comparaison des indices céphaliques sur le vivant et le squelette" ("A Comparison of Cephalic Indices on the Living Person and the Skeleton"). This was, like so many of his contributions, a discussion of the technical business of measurement. His claim, put simply, was that the measurement of skeletons produced more accurate results than that of living people: "Cephalometry, that is, the measurement of the head of the living man, offers only an approximate idea of the true shape of the cranium."[94] On a later occasion, he was more specific: "How many errors one risks making if measures are taken on living subjects! The hair, the thickness of the skin and muscular insertions are all causes of error. One sometimes finds more than a centimeter of difference in the semi-circumference of a head if one of the temporal muscles is more developed than the other."[95] That is why cephalometric measurements were "far from maintaining fixed relations with the corresponding craniometric measures."[96] Freshness worked against fixity, and was of no value to craniometric investigation, as Broca observed on a third occasion while addressing the same point: "One does not practice craniometry on fresh skulls. It is necessary to wait until they have been stripped of their flesh, cleaned and dried."[97] Humidity was in itself a potential source of error: "the capacity of crania varies a great deal according to the hygrometric condition of their inner walls."[98] In order to fulfil its scientific promise as an anatomical representation of hereditary racial stability, the skull had to be made into a dry, clean object. For Broca, its very dryness constituted an opportunity for exact method. Thus reduced and desiccated, the human was at its most measurable.

Anthropometry continued all the while to be practiced on the living, not least in the measurement of conscripts, and the Society moved to address worries about consistency of method, establishing a set of procedures that would serve to standardize the measurement of living persons. A special committee had been charged with "drawing up instructions aimed at making anthropological observations and measurements carried out on living man as regular and uniform as possible."[99] In 1865, it was announced that, this task having been completed, "the study of living man can now be conducted with the help of rigorous, uniform procedures, so that the results obtained by different observers will be able to be compared with confidence."[100] Following that announcement, Broca commented that the same thing now needed to be done for craniometry: "The measurement of crania, the location of points of reference, the choice of diameters, curves, lines, angles, in short all the elements of craniometry, have not so far been properly fixed. Most of these elements can be measured in several ways, and that results not only in many examples of confusion, but also in serious contradictions"[101] Fixity, already claimed as an anatomical property of skulls, was also being envisaged as an artifact of craniometric method. In effect, measuring the living according to a strict code of practice was approximately equivalent—but only approximately—to working on skulls in the laboratory. Measuring large numbers of individuals with the greatest of care, then organizing the measurements serially would produce knowledge about human bodies that was able to be used by science and by government. And the close measurement of well-prepared skulls, so Broca hoped, would allow for the exquisite measurement of individual differences and their interpretation in series.

A book-length study could be written about the mensurative practices developed by Broca, but that would not materially advance the history of normality. It will suffice for our purposes to note that he was remarkably ingenious in conceiving of new techniques, and was ably supported by a skilled instrument maker.[102] Germane to our history, however, are the principles that guided and constrained his inventiveness. The first of these was that the cranial bone should always be treated as a singular object. It was not acceptable to cut into it or do anything that might disturb its wholeness: "The only craniometric procedures that can be generalized are ones that make it possible to respect the integrity of the skull," said Broca.[103] So the skull was not something from which samples were to be drawn for the purposes of material analysis. Only when envisaged as a whole did it serve as a vehicle of anatomical truth. That was, he noted, a reason to be wary of making plaster casts. One colleague, while seeking to strengthen a skull so that it could be examined more closely, had filled it with plaster. Unfortunately, the plaster had expanded and the skull had fallen apart. For Broca, that was the end of it.[104]

Broca also paid a lot of attention to the surprisingly difficult business of standing a skull on a workbench for observation. It was not enough to have a dry, clean, whole object. That object had to be positioned in such a way as to allow the consistent measurement of angles because facial angles had been considered ever since Camper's discussion of them in 1770 to be variables that might serve as racial characters. Johann Friedrich Blumenbach, in his 1795 discussion of Camper's work, spoke of the measurement of facial angles with respect to the *norma verticalis*.[105] This was in fact the classical geometrical usage of the Latin word *norma*, signifying perpendicular. By a striking redundancy of history, the vertical axis was now serving Broca's geometrical fastidiousness and functioning quite literally as an anatomical norm. In the laboratory, the horizontality of the workbench had to be made to correspond to a natural horizontality inscribed in the skull. That could be done only when the natural horizontal had been determined by an experiment conducted on living people as they looked straight ahead.[106] When Broca had established that axis, it became a practical problem of maintaining it in laboratory conditions. The earlier practice of simply placing the skull on the bench was now to be regarded as a source of error. If, for example, the alveola was used as a resting point and the teeth happened to be worn away, that would change the Camper angle, which ran through the bones of the face.[107] Broca had an apparatus built that made the natural and laboratory horizontals coincide, thereby producing once again "a fixed point of reference."[108] There was here a kind of epistemic mimesis that Broca undoubtedly saw as proper to craniometric science. At every point, he sought to reproduce a practical fixedness that would make available for observation a fixed character deemed to exist in nature.

QUESTIONING THE DEFINITION OF RACE WITHIN THE SOCIETY

We have seen that craniometry, while holding a prominent place in the Society, could hold that place only by dint of assiduous self-defense. Everyone seemed prepared to suppose that there was a natural relation between anatomy and intellectual qualities, but the modes of inquiry proper to investigating that relation were by no means agreed on. And while a set of debates continued around observational methods, there was also ongoing disagreement of a more fundamental kind. It had to do with a theoretical notion that seemed actually to constitute the fundamental for physical anthropology, the notion of race. Even as members of the Society were committed to a scientific project that found its purpose in the identification and observation of racial characters, they were far from agreeing about how race was to be defined. On those rare occasions when a definition of sorts was proposed, disagreement ensued. So these anthropologists found a sense of collective purpose in an idea that they did not properly share. That circumstance is hardly unprecedented in the history of science and is not a reason for intellectual historians to disregard their work, however unworthy and incoherent the notion of race may seem to us now. We are interested here, as we have been throughout this study, in the manner in which ways of thinking about normality came to prominence while at the same time being contested and placed under strain. It has become clear in our own time that there are many ways to contest the notion of race, and more recent critiques of it have largely undone its intellectual influence. But it should be noted that the disagreements that took place in the Society did not constitute critique of that kind. No one stood up to argue that it was ethically improper or scientifically unsound to make racial divisions among humans. So our interest here, once again, is in assessing the state of thinking at the time and identifying the presuppositions in play. Just as in the debate opposing craniometry to cephalometry some key assumptions remained uncontested, in the debate about race certain things were taken for granted even as people were disagreeing.

One member who took every opportunity to make an issue of race was André Sanson, whose scientific activity involved experiments breeding domestic animals. On one occasion, Armand de Quatrefages, a long-serving president, spoke of the development of new races of domestic animals, thereby provoking Sanson to a polemical response. Sanson said in effect that there were no such new races. He began by drawing the attention of his colleagues to what he saw as a general habit in the Society: "By the way, may I point out that we often speak here of race and races, but that we have never defined them?"[109] He then went on to observe that he and Quatrefages did not have the same definition, spelling out his own as follows: "For my part, here is the one I adopt: a race is a group of individuals displaying a set of similar forms that are able to be transmitted by generation. Homogeneity of typical characters and the power to act through heredity are the necessary attributes of race."[110] The key elements in this definition would have been familiar to his colleagues. They were the attributes referred to by Périer in the statement with which we began this chapter: the presence of a recognizable set of characters deemed to be typical and the generative capacity to pass those characters on through heredity. These were in effect the two dimensions of racial normality.

Sanson made two quite lengthy presentations to the Society on this topic. On one occasion, he did not merely elaborate on his own definition, but attempted to provide an account of how his colleagues had been able in various ways to elude the difficulty of defining race. There were, he said unexceptionably, two groups of scholars contributing to physical anthropology: naturalists and linguists. The naturalists did not have "an idea in common about the set of characters by dint of which race can be determined," while the linguists simply had "an ingenious idea" for dividing humans into distinct races according to the languages they spoke.[111] Ingenuity was required by the linguists because their kind of knowledge gave no access to a "natural set of characters."[112] The naturalists, on the other hand, ought by trade to have provided just that, but were not up to the task. The problem, said Sanson, was that they did not have a proper theory of race. Their theoretical focus was on the species, and race was for them no more than a "conventional" name given to a variety within a species.[113] The net effect was that "race and species were often confused in their accounts."[114]

Sanson then went on to outline a position he considered to be a quite new one, although it is noteworthy that for all its professed novelty this view was supported by assumptions that had been circulating in the Society from the first. Disagreement here was enfolded in broad agreement. According to Sanson, individuals in the state of nature were grouped "first of all" in determinate races recognizable by the presence of anatomical characters belonging exclusively to individuals in that group. The essential property of the characters—and this lay at the heart of his definition—was that they were "infallibly reproduced by heredity." It followed logically from this assumption that "the permanence of characters" was "the fundamental attribute of race."[115] Moreover, the anatomical characters whose hereditary permanence defined a race were limited in number and location. They were found in "the bones of the skull and the face." Those configurations, when observed, "gave the absolute type." Other parts of the body might happen to be comparable to those found in individuals of a different type, and thus could not serve to define race. Characters of the latter kind might alter in response to a new environment, and were therefore not "essential."[116] There were in each race, Sanson asserted, "essential, permanent, typical characters and secondary, variable ones."[117] Naturalists, when constructing their taxonomies, had failed to make this distinction. They had focused on the species, defined biologically by the fact that members of the same species who were of different sex could reproduce other individuals who could do the same in turn.[118] As a consequence, naturalists were unable to identify with clarity those particular groups of individuals whom they ought to have defined from first principles as races, affirming instead that "racial types had nothing fixed in nature."[119] Sanson conceded that races always existed within species, but argued that a naturalist analysis of racial groups needed to take full account of the reproduction of essential characters. Within the species, each individual reproduced itself with all typical characters if mating with another of the same race. If not, they could still procreate as members of the same species, but would then produce offspring with the mixed characters of a crossbreed. It might be also that adaptation to environmental conditions led to a change in secondary characters, but the essential, typical characters would not change. Even if they did appear to change temporarily through crossbreeding, individuals with different racial characters would revert to a dominant racial type after a few generations. It was not enough to recognize a biological law at work in the perpetuation of the species. The "permanence" of race was itself the "expression of a natural law, absolutely like that of species."[120] Sanson was engaged here, like many of his colleagues, in a struggle to establish fixity, not just as an artifact of investigative method, but as a law of nature. Understood thus, race named something that was anatomically inscribed and was passed on through "permanent" heredity.

Sanson was rehearsing some of the Society's favored themes, but he was taking them to disturbing conclusions. Charles Rochet, for one, was ill at ease. He saw no need to drive on relentlessly toward unequivocal definitions of species and race. Yes, they did correspond to something in nature, but they were also conventional notions that allowed naturalists to converse.[121] It had to be recognized that "the character of the species or the race is not a character of absolute fixity or stability."[122] Anthropologists should attempt as far as possible to reach agreement about them, but while the characters of a species were not subject to question, the characters of a race certainly were: "the characters referred to as racial cannot be understood by everyone, and that is true even for my learned colleagues, and indeed for most men who have written on the topic."[123] Colleagues were free to make whatever contribution they wished to the definition of race precisely because the whole field remained entirely open to uncertainty and doubt. By contrast, even the most ignorant of people were able to recognize the characters of the human species.[124] Rochet's key point was that what Sanson saw as a failure to define scientific terms was not a mere oversight, but reflected a genuine, widespread uncertainty about the nature of race. Given that individual characters were perpetuated through heredity in any case, there was no particular need in Rochet's view to make racial heredity an object of study.[125]

Sanson was unmoved. He reiterated a view of racial permanence that may well be the high-water mark of nineteenth-century theories of anatomical normality: "race is a fixed, permanent thing thoroughly determined by essential or typical characters owed to anatomical forms."[126] At the heart of his theory lay the concept of reversion. In the event of crossbreeding, he conceded, it might seem for a generation or two that new anatomical forms had appeared and that a new race was emerging, but biological processes would ensure that these new forms did not in fact constitute new types: "There is as yet no properly conducted observation known to us, no rigorous observation showing that more than three generations have elapsed without the offspring of crossbred couples making a complete return to one or the other of their natural ascendant types."[127] If Sanson was right, racial normality was both dynamic and permanent. It was always working its way back to the natural ascendant type. That is why, he said, attempts to create new races of domestic animals were theoretically unsound and practically unproductive: "Opinions in circulation about this question are based on illusions of observation. No zootechnical method has the power to create new races. Experimenters merely exercise their skills on physiological aptitudes, which have nothing in common with the characters of race."[128] Physiological qualities might change, but not essential anatomical characters. There were no new cranial shapes, and consequently no new races.

By taking the theory of racial normality to its most radical conclusion, Sanson was building a fortified position that would soon turn into a theoretical trap for Paris-based physical anthropology, and some members of the Society were quick to see this. A member named Mortillet, who had recently joined and who would later become one of the Society's leading evolutionists, made it clear that he himself did not believe species to be fixed. As soon as one took a long-term perspective, he said, "permanence of types and unlimited reproduction through time do not exist."[129] The fossil record seemed to support his claim, but Sanson would not concede as much. So there was no way to advance the debate. This stalemate, it should be noted, was taking place in 1866, at a time when British natural science was at its most ebullient. Four years earlier, Darwin's *Origin of Species* had appeared in French, and at about the same time Clémence Royer, who had herself produced the translation, was admitted as the Society's first woman member. Royer was to prove a well-informed and eloquent spokesperson for Darwinian evolutionary theory, and it was clear that what French anthropologists called "transformism" was gaining ground, especially among the younger members.[130] To maintain a "fixist" notion of race and to insist that hereditary normality must always express itself in unchanging characters was to close off any scientific conversation with evolutionary theory. Ever since the time of Étienne Geoffroy Saint-Hilaire, scientific normality had been conceptually marked by the notion of resistance to change from without and indeed by that of a capacity to impose change around itself, but Sanson's understanding of racial normality was narrower, less flexible, and finally less productive. He had taken the logic of racial heredity through to a final conclusion and made of normality something permanent. Following where Broca and others had seemed to lead but going beyond the positions they had been content to hold, Sanson now found himself claiming that race was material normality carried in perpetuity in the bones of animals.

NOT QUESTIONING THE NOTION OF TYPE WITHIN THE SOCIETY

In the transactions of the Society, hardly any term recurred more often than "type," but the notion of type was not subject to contestation as other key notions were. Speaking of Broca in particular, Claude Blanckaert observes that he "never pointed precisely to" the types of which he spoke, and that he "did not really attempt to define what he meant by the word."[131] That may appear to be unwonted negligence on the part of someone given to close analysis of theory and method, but it actually reflects the view that, whatever the type was exactly, it was not such as to require methodological rigor or reward close conceptual analysis. The term was available through a long tradition of natural history, having been used by Aristotle, by Linnaeus, and by Cuvier, and that tradition was not generally subject to critical interrogation in the work of the Society. As Nancy Stepan observes more generally: "To the typologist, every individual human being belonged in some way or another to an undying essence or type. However disguised or hidden the individual's membership of the type might be, the scientists expected to see behind the individual to the type to which he belonged."[132] To do natural history was by scientific habit to work with a set of assumptions about the type. So we ourselves will not enter into the narrow theoretical business of defining the type for the reason that no one did so at the time. Instead, we will attempt to give an account of the familiarity and the ease of reference that attended the notion, and we will ask at the same time what it meant in practice for such a comfortably established term to hold a place alongside other terms whose signification was being chiseled out in the most exacting manner. The challenge to our history is to understand how the synthesizing, quasi-immediate perception of the type could have cohabited with measurement and calculation in such a way that typology and mensuration came to be effectively, if loosely, conjugated as knowledge practices.

It might have seemed from a statistical point of view that there was no need for any accommodation such as this. It was possible to think of the type in frequentist terms, as the form or set of forms that recurred more often than any other, and was thus able to be determined by calculation. After all, that was what Quetelet had done when he had affirmed that the average man of a population simply was the ideal type in its well-proportioned beauty. While not endorsing Quetelet's rather lyrical view of the average man, Jacques Bertillon rearticulated appositely within the Society his father Louis-Adolphe's definitions of averages. On the one hand there were "indexical averages" in which "the most different values were relentlessly mixed together": that was the case in the much-quoted example of building heights in an unregulated street. On the other were those of the statistically valuable kind, which Bertillon now called "typical averages." These corresponded to Quetelet's true means, being found in measures that were "naturally grouped" and reflected in more or less symmetrical curves of distribution where the mean and the median coincided. That was generally the case, Bertillon pointed out, with the average measurements of French conscripts.[133] Understood thus, he said, an average in the proper sense should be "the expression of a type, or rather the expression of the most usual fact."[134] His statistical reasoning allowed the hitherto imprecise notion of the type to be supplanted by that of the most frequent occurrence. By the same logic, it became possible to declare that wherever a normalized statistical pattern occurred one was dealing with a "natural group," and that the calculated average of that group could be identified accordingly as the type. But it was not generally accepted in the Society that statistics would in fact provide a satisfactory way of redefining the type. Broca and others were unpersuaded that the type of a species or the type of a race would correspond straightforwardly to the "most usual fact." How could anthropologists be sure that the averages produced by their calculations deserved to be accorded such general significance? How could they be sure, in particular, that the overall average of a set of characters was not one of those meaningless numbers produced when "different values were relentlessly mixed together"?

Broca began with the classic natural historian's assumption that the type was defined as a singular set of shared characters. That was his guiding principle, for example, when attempting to study the Celtic race as it was still to be found in France. By focusing on those regions where for historical reasons the Celts were deemed to have been least subject to miscegenation, he hoped to reconstitute their type. But he was obliged to recognize that the task could not be carried out by statistical means. It was perfectly possible to do a distributional study for each character but, to put it in the plainest mathematical terms, those averaged characters did not add up. There was little or nothing to be gained by aggregating the figures to produce a general picture of racial distribution: "When one attempts to assess in each locality the relative proportion of individuals who represent more or less exactly one race or the other, of those that belong more or less equally to each, and of those who belong largely to some foreign race, one is quickly brought to a halt by the impossibility of producing a multiple set of statistics that would bear successively and specifically on each racial character."[135] Translating measured individual characters into racial types at this level was an impossible task. The only thing to do was to "set aside [*faire abstraction de*] the individual varieties and grasp [*saisir*] common characters that make it possible to reconstitute the type of the Celtic race."[136] This might be considered a surprising lapse on the part of the great champion of statistical method, but Broca could see no alternative. In addition to the statistics—in the long run, instead of them—there needed to be some synthesizing grasp of the characters. Unable to build knowledge of the Celtic type in a mathematically incremental way, Broca eventually found himself speaking of something very like the knowing glance. In his view, there was no secure methodological bridge between the accumulation of data and the perception of the type.

Broca could only address the question of type by deviating from his strict line on the necessity of measurement and calculation. Having declared that there were "a very great number of human races," he went on to note that there were "numerous affinities" between certain of them. The very choice of the term "affinities" made it clear that it was not a matter here of mensuration or even of precise definition. Some races simply formed "natural groups," and they did so because "the set of common characters" constituted "the *type* of this group."[137] Topinard put it somewhat differently, but the main point was much the same. The type was recognizable in practice by the presence of a set of characters occurring together. Blue eyes went with blond hair. Flatter faces went with smaller arches over the eyes and flatter noses. "In reality," he said, "it is this conformity [*concordance*] of characters that gives rise to the notion of *type*."[138]

To observe the set of characters and the type was not to observe anything resembling a particular cranium with actual dimensions. It was not even to pick out a typical cranium from the set under examination, thereby retrieving at least one positive artifact that might reveal the type. Broca simply had to concede that the type was "ideal." It could perhaps be "grasped" on sight, but it could not actually be held in the hand. Insofar as the type could be contemplated, it was more akin to a pictorial representation: "All that can be done to represent in the best possible way the type of a race in a drawing or a photograph is to look among the skulls one has gathered for the one that seems by its principal characters to be closest to the type. As for the absolutely typical skull, everyone knows that it exists only in the virtual state."[139] The knowing gaze was being called on to do the job once again. As objects of knowledge, the cranium chosen by sight and the drawing or photograph of the racial type were approximately equivalent. Indeed, the type could not be observed without some degree of approximation. When craniometry was carried out on a large enough series, it might perhaps have seemed as if the ideal type or the virtual cranium were actually present. Craniometry made it possible, said Broca in a moment of methodological enthusiasm, to "constitute that virtual cranium and determine all its dimensions, all its measurable characters with as much certainty as if it were held in one's hand."[140] But the type remained ideal, and the hand that held it could be only metaphorical. It was a mistake, as Broca observed on another occasion, to see these ideal types as really existing.[141] He was obliged to recognize in all modesty that the description of the principal types involved "a methodical, but not a rigorous technique [*procédé*] aimed at facilitating the comparison of human races by forming a certain number of groups."[142] Whatever method was used to know the type, it could not in Broca's view be a "rigorous" one.

In scientific practice, what mattered most about the type was the clarity with which it could be recognized when it did come into view. Many physical anthropologists were wary of placing undue emphasis on the recognition of race via skin or hair color: they believed that there was something superficial about those characters, by contrast with the painstaking observations of craniometry and cephalometry. But those reservations did not seem to apply to the identification of the type. Perceptions of the type in the minds of experts, while not to be thought of as facile, took place with a kind of self-confirming ease. It does indeed seem in this instance that the lack of a strong definition was somehow productive. Analogies brought to light by comparative anatomy, computations of frequency, even perceptions drawn from physiognomy, all seemed to converge, if not exactly to coincide, in the type. To know the type was somehow to know all those things at once, and to know them as approximately the same thing. Reference to the type allowed these forms of anthropological knowledge to be treated as equivalent or convergent while remaining unanalyzed and unresolved.

Lorraine Daston and Peter Galison, in their important book *Objectivity*, provide a broader and longer context for understanding how the type worked in scientific discourse at this time. They do not discuss Broca or Quetelet, nor do they reprise in their joint essay any of Daston's earlier influential work on the history of statistics, but they do give a historical account of synthesizing observation that corresponds well to the manner in which physical anthropologists used the notion of type. Without offering a theory of the type, they explain in effect why there might have been no felt need for one. Linnaeus serves for them as exemplary of a whole tradition in his use of the notion. The task of the naturalist as Linnaeus saw it in the eighteenth century was "to extract the typical from the storehouse of natural particulars."[143] "Extraction" in his sense was not an incremental or algorithmic procedure. Rather, the type was to be rendered in the form of an "idealizing representation."[144] "Atypical variations and extraneous details" needed to be "weeded out," in much the same way as Broca sought prima facie to exclude abnormal individuals from his anthropological series. In the Linnean tradition of natural history, it was not a matter of representing a given specimen exactly.[145] Images drawn by hand were offered as images of the type.[146] The art of drawing, with the idealization or refinement that entailed, was in fact germane to scientific perception, since "the type was truer to nature—and therefore more real—than any actual specimen."[147] Daston and Galison follow their history through by showing how this preferential representation of the typical tended to be displaced toward the end of the nineteenth century by a focus on methods of scientific observation that required subjective judgment to be treated with suspicion.[148] What appears to make the history of physical anthropology distinctive in this regard is that Broca, in particular, was actually a pioneer of objective methods while also making a place in his work for a classical understanding of natural types. In physical anthropology of this kind, objectivity and the perception of types were not successive phases but concurrent modes of inquiry.

CONCLUSION

In mid-nineteenth-century physical anthropology, it was possible to identify hereditary normality as an object of study while taking the notion of race as the main gathering place for anthropological knowledge. But there was considerable uncertainty in this milieu about the very notion of race. Even the most fastidious measurement could not secure an unequivocal outcome as long as there was ongoing uncertainty about the scientific topos. It was certainly possible to engage in measured observation, and that was often done with the greatest of care. Observations could be organized in series and regular distributions produced by the use of methodical protocols that excluded irregular individuals. Indeed, for each character examined, an average could be identified and, quite often, serial regularity displayed. But for a group of individuals found by circumstance it was still not clear how to organize knowledge of them as a properly constituted set. Quetelet had supposed that it would always be possible, at least for a given population, to find the ideal type through statistical calculation, but it was clear to Broca and his colleagues, precisely because they were so attentive to the finer points of mensuration, that the averages of different characters could not simply be added up to produce a single type of person. "Race" was not given in that way. Anthropological science was committed to speaking of racial types but found itself unable to arrive at the type by computation. It did find a concept of the type in natural history, but the type in that sense was not accounted for in terms that made it accessible through quantitive reasoning. A thoroughly modern scientific apparatus was being deployed in the service of an undertheorized traditional notion. Anthropologists were attempting to identify race as a set of inherited characters passed on through heredity. Biological normality, they thought, could thus be studied by the serial observation of shared characters. But their carefully rehearsed methods did not allow them to arrive at a statistically grounded norm. Instead, they found themselves speaking of something that was in effect a typical norm. This mixture of styles of knowing was in fact constitutive, and in a sense productive, for talk about normality. It was only when the average and the type were able to cohabit in loosely managed ways within the same space of inquiry that normality could be functionally recognized and approximately described.

CHAPTER FIVE

The Dangerous Person as a Type: Criminal Anthropology, c. 1880–1900

The term "normal" and its derivates came into use in the course of the nineteenth century by serving specialist purposes in scientific contexts. A research program, that of the Italian school of criminal anthropology, maintained some of those knowledge practices but spoke to a broader public. The leader of the school, Cesare Lombroso, was at one point the Italian author who could boast the greatest number of readers in the world, and the very renown achieved by his *scuola positiva* marked a turning point in the history of our theme.[1] Anthropometry came to be more widely recognized for its social significance, and some key notions associated with normality were now no longer restricted to narrow scientific circles. It would be misleading to see Lombroso and his colleagues as mere vulgarizers. They did not just build fame by simplifying and popularizing physical anthropology, or even by extending its reach while maintaining its methods. They shared certain premises with the anthropologists associated with the Parisian Société that had built physical anthropology as an influential discipline but made sense of those premises in different ways. While the Société d'Anthropologie de Paris had been preoccupied with the study of race as hereditary normality, the Italian *scuola positiva* tended to take the concept of normality for granted. Whether or not that was done purposely, the fact of taking the normal for granted is itself worthy of a place in our history because that became standard practice in the twentieth century. It may well be, in fact, that the normal continues to be most comfortably influential precisely insofar as it remains unanalyzed. It can thus be pleonastic without being perceived as vacuous, as its frequent discursive recurrence serves to indicate that it is indeed conceptually "normal," and thus not requiring scrutiny. That was not so in the work of such nineteenth-century thinkers as Étienne Geoffroy Saint-Hilaire or Paul Broca, who paid close attention to the concept as they made it do scientific work. If in the course of the twentieth century normality settled into a quiet thematic routine, it may well have done so through an easy combination of remembering and forgetting. We are uncertain whether Italian criminal anthropology actually helped to bring that about on a broader scale, but it is the first example of such scientific—and unscientific—thinking that we have found. So criminal anthropology will serve as an emblematic program in which remembering the normal and forgetting normality were productively and seductively articulated. The *scuola positiva* showed in practice how normality could be relied on as a conceptual reference for scientific endeavor without actually being made an object of critical analysis.

THE NORMAL IN LOMBROSO'S WRITING

Lombroso was able to use the term "normal" quite regularly without ever appearing to examine closely what it represented. The focus of nearly all his works, evident in their titles, was on particular kinds of unusual and often dangerous people. The normal or the usual was present in some general conceptual sense, but it was taken as an unanalyzed point of reference. That was the case, for example, in the first (1889) edition of his most influential text, *L'uomo delinquente* (*Criminal Man*). There, Lombroso used the normal implicitly to identify anomalies as such. Anomalies in males and females were declared to be apparent by contrast with *normali*, normal people. The same move was made later in the essay with "savages," whose savage state was not in fact opposed, as one might have expected in texts of that time, to "civilized" people, but identified by contrast with *normali*.[2] At another point in that essay, during a discussion of the wearing of tattoos, Lombroso contrasted the tattoos worn by criminals with those worn by *normali*, meaning at that point mostly noncriminal men and soldiers, who also wore tattoos.[3] It was not at all clear whether nonsavages, noncriminal people with tattoos and other groups notionally used for contrast actually constituted a single set of people who could be considered normal. Was there a class of people simply presenting no anomalies at all, or was normality merely identified circumstantially, so to speak, by the absence of a particular anomaly? And if it was indeed the latter, how was the anomaly identified as such in the first place? *Normali*, for Lombroso, did seem in fact to be a rather opportunistic notion, called on at various points for the purposes of contrastive measurement. Contrast with normal people—or with people taken to be normal—was often a first step in Lombroso's research, but there was hardly any actual measurement of normal individuals considered in their own right. Looking at a set of crime statistics gathered in a particular region, he declared that they could be of scientific value "only if they were confronted with the normal figures of the same area."[4] But he did not adduce those "normal figures" for examination. The point he was making was an entirely orthodox one in principle: it was only possible to speak of anomalies if the normal was available as a yardstick. But numerical data about the normal were almost never available in Lombroso's essays because, by contrast with Broca, he did not collect or measure samples of individuals deemed to be normal. The normal had a key referential function in his scientific discourse, but it did not at any stage become an object of study in itself. As a scientist, Lombroso paid attention to the normal for just so long as he was required to do so by a logic of contrast, referring in circular fashion to "the normal character of the normal honest man."[5] The normal was thus a key signifying element within a pattern of knowledge even though its content was largely unspecified. To put it a little crudely, Lombroso needed the concept and indeed the dimensions of the normal, but did not bother to say what they were.

This opportunistic use of the normal often allowed Lombroso to move freely through lists of things that were all taken to be other than normal. We are obliged to use the circumlocution "other than normal" in our account of Lombroso's work because he himself almost never used the term "abnormal." The central working opposition for him was not the normal-abnormal pair, but an asymmetrical relationship between a declared anomaly and a nondescript normal character or condition. In the following example taken from the 1897 edition of *L'uomo delinquente*, the word *normali* does not actually appear, but a certain idea of the normal is still at work: "Doctor Virgilio studied 266 condemned men, affected however by chronic illness, among whom were 10 insane people and 13 epileptics. He encountered madness in 12 percent of their parents, predominantly here also in the father (8.8 percent). He encountered epilepsy with even greater frequency, 14.1 percent without counting the 0.8 percent of collaterals, and without counting a deaf-mute who was also the father of a rapist, 6 fathers and a mother affected by eccentricity, and a father who was a semi-imbecile."[6] This is an extraordinarily disparate list, bringing together epilepsy and eccentricity, madness and deafness. It might recall for a moment Borges's

Chinese encyclopedia, analyzed by Foucault as an example of categorical unfamiliarity.[7] But the list ceases to be exotic when one perceives that all of these disabilities and disorders are identified by Lombroso as anomalies claimed to be more frequent in criminals than in the undescribed *normali*, who are here barely designated. From the point of view of scientific logic, this was remarkably loose. But there is also something about it that is more modern than the scientific discourses examined in our study up to this point. We saw how Isidore Geoffroy Saint-Hilaire made the identification of anomalies a matter of studious comparative method, balking as he did so at the aggregation of anomalies into some general notion of the abnormal. In *L'uomo delinquente*, Lombroso seemed to be doing something like the opposite. He was treating all variations as anomalies and taking them all as equivalent for the purpose of piling up evidence of typical difference.

These habits of thinking and writing can be better understood if one recalls the institutional context in which Lombroso conducted his research. Whereas Broca, for example, had access to collections of skulls made available by agents of the state during the reallocation of ground formerly devoted to cemeteries, Lombroso conducted anthropometric and other examinations on subjects made available to him by the administrators of prisons or asylums. The general pattern in Lombroso's work was to take as given the institutionalization of these confined people, simply asking why they belonged where he had found them. Their criminal character, their madness, their alcoholism, their epilepsy were facts to be scientifically ordered and accounted for. So normality had no particular place on Lombroso's scientific agenda, either as a condition found in particular persons or indeed as a list of characters. He saw no compelling need to provide a working definition of it or to engage in the study of normal people. He was content to treat normality as if it were more or less a matter of common sense. The true task of criminal anthropology as he saw it was to produce scientific knowledge that justified the confinement of dangerous individuals. When Broca and his colleagues spoke of normal subjects, they did so primarily for reasons of method: that was for them a way of understanding and summing up a properly constituted natural series. But for Lombroso, normal individuals posed no particular scientific problem. That this was a rule in his practice is demonstrated by an exceptional occasion in which a normal individual did command all his attention. In the 1897 edition of *L'uomo delinquente*, Lombroso reported on a prison visit in the company of his students during which he encountered a man serving a life sentence for fraud. Having examined this man in his usual manner, he found to his astonishment (*meraviglia*) that this was "one of the most normal types that had ever fallen into my hands." A series of measurements indicated just what he meant by that. The measures of this man's height, weight, cranial capacity, and cranial index were published in the essay, although no statistical standard was cited in support of the claim that these measures were normal: that was taken to be manifest. The man had "no anomalies on his face, except that his jawbone was a little large." He showed normal sensitivity to pain. On further examination he revealed a set of moral or psychological qualities that confirmed the anthropometric evidence. He was neither religious nor an unbeliever; he was not interested in politics, and never got drunk. Moreover, he had no mad or criminal relatives. His neighbors declared him to be a model of honesty. All these were for Lombroso uncontested qualities of normality. Lombroso was moved to observe to the students accompanying him: "If he had not been condemned to life imprisonment, this person would represent for me the very type of an honest man." At which the person being examined declared that he was indeed honest and that he could offer legal proof of it. Two kinds of proof thus converged, and Lombroso was moved to declare that this was a case of wrongful conviction. Clearly—measurably—the man did not belong in jail.[8] It was indeed a "marvel" for Lombroso to encounter such a normal person while doing his prison rounds.

CRIMINAL ANTHROPOLOGY AND PHYSICAL ANTHROPOLOGY

The great difference in intellectual styles between Paris-based physical anthropology and Italian criminal anthropology becomes clear if we compare this event with an equally rare instance in which Broca found himself involved in the examination of a criminal. In 1867, Broca reported to the Société d'Anthropologie de Paris on an autopsy that he had helped to conduct on the brain of a convicted murderer named Lemaire, following a practice whereby the bodies of those who had just been executed were made available to senior medical figures for dissection. Working with a colleague, Broca had found a series of characters in the body of Lemaire that he now drew to the attention of the Society. Among the most noteworthy were deformed feet, genital organs whose dark color was taken to indicate frequent masturbation, and an asymmetrical skull with a very small frontal region. The face was prognathic, and there was ossification of a suture in the skull bones that could be declared pathological. The brain weighed only 1183 grams, far below the average for "the white race," which was about

1400 grams. These and a number of other observed anomalies could all be gathered under one heading: they showed a defective "organization and development of the body, especially of the skull and above all of the brain, whose frontal lobes recall those of idiots."[9] All these anomalies belonged in the same category, that of anatomy. But that was not the only kind of thing Broca had observed. A number of other anomalies were the consequence of "chronic diffuse meningitis." The lesion caused by meningitis was "superimposed," showing it to be "relatively recent, although nonetheless clearly dating from before the crime was committed."[10] Whereas Lombroso would later take any and all anomalies as aggregate evidence for his theories of criminality, it was important in Broca's view not to confuse the anomalous cranial shape and the pathological lesions. The first was straightforwardly anatomical, and racial norms were an explicit reference where brain size was concerned. But the second was of another order: there was clear anatomo-pathological evidence that this man had suffered from a grave illness. The conclusion to be drawn from the latter was inescapable: "this poor unfortunate, at the time when he conceived and carried out his crime, was suffering from an illness that destroys reason. They thought they were punishing a guilty man, but in fact they guillotined an insane one."[11] There were thus two axes of normality and two kinds of anomaly. At the intersection of the two, according to Broca, stood a sick, intellectually defective person who should have been held doubly irresponsible under law.

In Broca's eyes, there had been in the case of Lemaire a miscarriage of justice for the lack of proper medical expertise, and that needed to serve as a reminder to all: "before giving itself the right to kill a man, society would need to be infallible, and you have just seen that it is not."[12] Society had not succeeded in policing the border between the normal and the disabled or the normal and the pathological in such a way as to ensure that "judicial error" of this kind could be reliably avoided. Lombroso, it should be noted, was quite unpersuaded three decades later by anthropological and medical arguments of this kind. He had no sympathy for what he called "tiresome lamentations about the death penalty."[13] Those who were opposed to the death penalty, he said, had failed to grasp the logic of criminal anthropology in its fully developed form. It had now been established that the proportion of "mad and feeble-minded" individuals was greater among those charged with crimes. Their "absolute responsibility" for their crimes was therefore less than had previously been thought, but that was no reason for reducing sentences. Born criminals were to be feared. They were given to recidivism, and could be properly "neutralized" only when their capacity for harm was taken away by "selection" or seclusion.[14] For Broca, there were too many independent variables in criminality for science to justify the death penalty. For Lombroso, each new anomaly brought to light by science contributed to a cumulative effect of difference between criminals and *normali*. The weight of all this evidence made the execution of born criminals a rational solution.

In 1867, during the discussion of the Lemaire case within the Société d'Anthropologie de Paris, Broca expressed the opinion—shocking to his mind, but proper enough from Lombroso's point of view—that "hundreds of insane people [*aliénés*]" had probably been executed over the years.[15] His report on Lemaire raised a number of issues for the Society. Broca was drawing a strong ethical conclusion against capital punishment, but his younger colleague Eugène Dally refused to follow him. Dally did concede that reliable judgments about responsibility could not be made by the judiciary, but asserted that society had to be defended: "the defense of society is more compromised by insane people than by accidental criminals."[16] "Accidental" criminals were people who had simply taken improper advantage of an opportunity that presented itself, but certain insane people had to be thought of as a continual threat to public order. The intersection of insanity and crime, later to be taken by Lombroso as the essential given at the heart of hereditary criminality, was something of a scientific embarrassment for Parisian physical anthropology. Delasiauve, a doctor at the Salpêtrière Hospital specializing in the care and treatment of the insane, considered that being "in the normal conditions of humanity" meant precisely that people could "weigh up their acts," whereas the acts of sick people "bore the mark of fatality."[17] To be outside the bounds of normality was to be outside the space in which ethics applied. And since that was so, Dally argued, there could be no question of dissuading or deterring insane criminals. It was precisely because of the pathological nature of their behavior that an unbending response was required: "the more a criminal seems to commit his crimes under the influence of a sick impulse, the more severity is called for."[18] While the notion of punishment was no longer pertinent in the absence of moral responsibility, repressive justice could still be seen as a social necessity in the case of the dangerously abnormal.

The key scientific question in 1867 was whether it was possible for the Paris-based physical anthropologists to identify such pathologies as these before they resulted in crimes. Given the Society's collective engagement with craniometry, it might well have been thought that dangerous criminals, when their skulls were measured, would have some anthropological characters that revealed a predisposition to crime. Was that the case with the characters found in Broca's autopsy: the asymmetrical skull, the small frontal region, the anomalous suture? Broca did not say as much, but it was an easy inference to draw. In the discussion that followed, Broca found himself resisting on two fronts. Pruner-Bey considered the possible craniometric evidence in Lemaire's case and cast doubt on its validity. He had observed crania with those characters in people who were not criminals at all. But Broca did not insist that the craniometric observations could provide evidence of that kind. It was the meningitis that proved insanity, he said, not the cranial shape.[19] In Pruner-Bey's question, Broca must have seen the ghost of phrenology, since he made a point of saying in reply that he was not attempting to revive the doctrine of Franz Gall, the most influential exponent of phrenology.[20] It was not a matter here or elsewhere of using cranial data to study individuals, no matter how eccentric or indeed how dangerous they might be. But there was another possible application of craniometry that was, so to speak, closer to home — one that raised more difficult questions for the Society.

Delasiauve, the alienist, was convinced on the basis of his experience that craniometry could be applied to criminals as a group. He had observed in a provincial *colonie pénitentiaire* that, out of two hundred prisoners who had been moved there over time from a high security Parisian prison, "eighty bore on their skulls the imprint of their vicious moral tendencies, which could be perceived in the malformation of the skull and an asymmetrical sloping forehead. In many of these subjects, post mortem examination revealed internal lesions, notably the anatomico-pathological signs of meningitis."[21] Here, in undeveloped form, was the basis for a theoretical reframing of Broca's report. It might be the case, Delasiauve was suggesting, that forty percent of all serious criminals presented the same characters as those revealed in the autopsy of Lemaire: not just asymmetrical skulls and narrow, sloping foreheads, but even the lesions of meningitis. The ostensibly exceptional individual examined by Broca might prove in fact to be typical of a whole class of people. Delasiauve was pointing to a major potential task for craniometers, that of measuring and analyzing criminals as a group. But Broca was disinclined to take up the challenge in the terms in which it was offered. He had said many times in the past that the business of craniometry was to study racial characters passed on through heredity, and it made no sense from his point of view to study criminals as an anthropological group if they were not linked hereditarily. It was patent that criminals were not a separate race with their own hereditary line, so it was in his view theoretically unsound and methodologically inappropriate to proceed as if criminality were somehow biologically akin to race.

But that assumption too was effectively called into question in the course of the debate. Moreau de Tours, another prominent alienist, went against the grain of Broca's work by the simple fact of contending that research was needed into the family background of Lemaire. In cases of suicidal madness, Moreau claimed, a hereditary tendency to such behavior was present ninety-eight times out of a hundred.[22] No tabular evidence of that was offered or even referred to by him. Moreau's ninety-eight percent appeared to be of the same order as Delasiauve's notional forty percent of criminals with cranial anomalies, a numerical rendition of the expert's coup d'œil. Furthermore, said Moreau, as the focus turned to heredity in criminals, it would become clear that craniometry no longer deserved to figure as a privileged mode of inquiry. Against Broca and against Delasiauve's attempt to annex craniometry for the study of criminal insanity, Moreau said: "But one should not rely overmuch on the significance of cranial shapes and encephalic lesions. There is a host of individuals in whom it is impossible to establish a clearly observed link between anatomical or anatomico-pathological facts and their actions. Many people who present the lesions observed in Lemaire are not murderers at all. They are feeble-minded at most."[23] The more urgent scientific task, Moreau was suggesting, was to look for the signs of an inherited tendency to crime, allowing for the fact that those signs might not prove to be cranial ones. If this line of inquiry into hereditary criminality were to be pursued, physical anthropology as it was practiced in the Society would undoubtedly need to be left behind in favor of some unspecified way of identifying and comparing inherited individual qualities. There was no basis here for disciplinary conciliation, and discussion went on for some time to little effect. As if to call the Society back to its principal self-appointed task, Dally moved that the meeting return to its set agenda, observing that these matters were outside the scientific brief of the Society and perhaps outside the general domain of science.[24] The motion was carried. This group of physical anthropologists was drawing a border between the study of race, which it claimed as its own field of expertise, and the problematic, as-yet undefined study of hereditary criminality.

It was in fact possible at the time to raise a broad question about the incidence of criminality without attempting to measure the bodies of criminals. Adolphe Quetelet had pointed the way in an essay on moral statistics published as early as 1846. He spoke there of the tendency to crime (*le penchant au crime*) as something that could be inferred from statistical tables, in this case using data produced in France. His research was simpler in its principle than the undissolved mixture of anatomical typing and statistical calculation usually found in the Société d'Anthropologie de Paris. It was a straightforwardly frequentist analysis that owed nothing to anthropometry. Quetelet simply demonstrated that the occurrence of criminal behavior varied quite regularly according to age. These were not just deviations from the average, but patterns of deviation recurring from year to year whatever kind of crime was considered. Understood thus, the *penchant* for crime was not an individual predisposition detectable by anatomical measurement, but an inference drawn in retrospect from a tabulation of behaviors that had resulted in judicial action.[25] Because of the regularity of statistical patterns, probabilistic calculations could be made about likely future aggregate trends, although that would give no account of how a particular individual was likely to behave. Frequentism thus appeared to offer little in the way of practical application where crime was concerned, as Quetelet himself effectively recognized on another occasion.[26] In sum, a strictly frequentist approach could have nothing to say about any given person, and the school of physical anthropology based in Paris had effectively declared that it was scientifically improper to describe criminals as deviating from an anthropological norm. No one was bidding during the middle decades of the nineteenth century to provide an account of criminality that would bring together a statistical analysis of the frequency of criminal acts with the anthropological description of a putative criminal type.

Before about 1880 indications that it might be possible to construct an anthropology of criminality did not serve to draw the support of a body of researchers. No one in the 1860s rallied around the general suggestions of Delasiauve or Moreau de Tours. That task was redefined and undertaken from about 1880 onward by the Italian school of criminal anthropology, which adopted a less fastidious approach than Broca's, deploying a wide range of knowledge practices around the topic. The success of the Italian school can be fully understood only by analyzing the forces supporting and opposing its rise in Italy. Lombroso's standing within his own country, as Silvano Montaldo notes, made of him "the best known representative of the public engagement of men of science in the age of Positivism, and the point of departure for verifying the influence and the importance that scientists had at the time."[27] Mauro Forno's study of Lombroso's correspondence shows that at the height of his fame he was continually being asked to express an authoritative view on a range of topics extending far beyond his area of scientific expertise.[28] Editors of publications tended in fact to treat him as an all-purpose maestro. An essay by Valeria Babini points to the broad significance of this activity for intellectual life in Italy. She writes: "Cesare Lombroso was not only a scientist but a politician of scientific culture in Italy."[29] Babini points to an intriguing set of questions that have to do not only with the rise of positive science in Italy, but with the manner in which Lombroso played a public role as a scientist. However, we will concentrate on the narrower business of how Lombroso claimed expertise on criminality and how he leaned on a certain notion of normality to support his claim. Yet we cannot fail to recognize at the same time that the "science" of criminal anthropology as he championed it brought together a remarkably diverse, not to say disparate, range of knowledge practices. Where Broca, ever attentive to method, tried to make physical anthropology into a strictly disciplined enterprise, Lombroso made criminal anthropology a subject of wider interest. He wrote—and one might say indeed that he thought—in a more publicly accessible manner.

LOMBROSO'S THEORY OF ATAVISTIC CRIMINALITY

Seen from the point of view that dominated the Société d'Anthropologie de Paris, criminality could not be a proper object of study because criminals were not a race. There could be no direct biological transmission of characters, and it was thus not possible in principle to constitute a statistically ordered series of criminals that might serve to reveal a hereditary norm. But Lombroso made this difficulty disappear. He began with an empirical assertion unsupported at the outset by tabular statistical evidence. An experienced observer could not fail to note, he said, that certain anthropological characters were common to many criminals. The list of physical characters he offered was quite a long one: "the relative lack of hair, the small cranial capacity, the sloping forehead, the greater thickness of cranial bone, the enormous development of the jaw and cheek bones, the prognathism," and so on.[30] This commonality, said Lombroso, could and should be understood as racial, but it was atavistically so. By that he meant that the kind of heredity involved was not serial, since it was not passed on from parent to child, but discontinuous, inherited across a gap in time. It was not reversion to a current norm, but an anomalous relapse into a much older norm that ought to have been left behind in the course of evolution. From somewhere in an archaic past, there reappeared characters that did not properly belong in an evolved, civilized world, although they would formerly have been valuable for warriors or tyrants. In Lombroso's view, these were racial characters in two senses of "racial" made complementary by his theory. When the characters he listed appeared in Europeans, they were to be seen as a throwback to an earlier stage in the evolution of the European race. And by the same token, since Lombroso, unlike Broca and Quatrefages, understood non-European races as each having its place along an evolutionary line leading upward to a single white race, the characters displayed by European criminals showed consistently enough that they belonged on that line at about the same point as certain "savages." In the final edition of *L'uomo delinquente*, Lombroso summed it up thus: "Anyone who looked at the first edition of this essay will have been persuaded . . . that many of the characters present in savage men and colored races often recur in born criminals."[31] Comparing criminals to savages was hardly unprecedented: comparisons of that kind had long been available as hyperbolic metaphor, but Lombroso was seeking to establish the scientific equivalence of the two kinds. In his evolutionary terms, criminals simply *were* savages. By a poorly understood phenomenon of occasional heredity, characters that had been present in the prehistory of the European race and that were still observable in modern "colored savages" could thus reappear in modern Europe. Broca clearly did not consider criminals to be either a natural group or a natural type, but Lombroso was asserting that they were both, with a full set of moral characters corresponding to their physical ones just like the moral characters reputedly found in modern savages, namely "complete moral insensitivity, sloth, the lack of any remorse, impulsiveness," and the like.[32] Insofar as Lombroso was able to treat criminals as savages — however contentious that may have been in the eyes of some of his scientific contemporaries — he was marking out a space in which anthropological methods could be deployed in the study of criminality.

Yet whether these methods could be applied systematically was quite another matter. For while it might have seemed that Lombroso's theory of atavistic criminality committed him and his school to a highly specific, if somewhat enigmatic, view of race, that did not in fact prevent him from using the notion of race with great conceptual freedom, not to say looseness. He was ready to suppose, for example, that the concentration of criminal characters and behaviors in certain places was attributable to race. The Société d'Anthropologie de Paris asked questions about whether a set of skulls found in a given place would prove on examination to belong to the same race, but Lombroso seemed happy to suppose from the outset that race was an anthropological given wherever human settlement was concerned. High crime rates in particular areas, he noted, were "undoubtedly [*certo*] dependent on the race, as is revealed in some cases by history."[33] By that he meant that any history of settlement or invasion by a race with "known" criminal tendencies provided a likely explanation for the prevalence of criminal persons and criminal acts. Broca, for his part, had regularly warned his colleagues against the dangers of relying on history of this kind, saying that it should not be treated as a source of scientific knowledge about race: anthropometric methods were to be systematically preferred.[34] Clearly, Lombroso's view of race was far less constrained and far less disciplined in the scientific sense. He often posited racial history as a key influence without engaging in either anthropometry or statistics to establish whether shared physical characters were actually present. Patterns of behavior, unlike Quetelet's statistically documented "tendency to crime," were often simply reported by Lombroso as matters of notoriety. "We know," he said, "that the majority of thieves in London are children of Irish people or natives of Lancashire."[35] This was racial knowingness without evidence. And it led, by an intellectual shortcut, to explanation in which race was proffered as a cause without any material demonstration of how causality might work in a given circumstance. Observation of a general kind—the knowing look—was sometimes invoked. Referring to certain towns in which crime was more frequent than in the areas surrounding them, Lombroso was "inclined to suspect" that the pattern was a matter of race since in some places he had noticed that the inhabitants were taller.[36] It was a matter of notoriety, indeed, that thuggery was rife "in the notorious [*famosa*] valley of the Conca d'Oro in Sicily, where rapacious Berber and Semitic tribes had settled long ago."[37] Whether provided by an expert gaze, by general notoriety, or by anecdotal ethnic history, evidence of "race" was available on all sides, serving in convergent but unrationalized ways to feed the enterprise of criminal anthropology. It was all grist to the mill. So much was attributed to race that the racial explanation of criminality was rendered ambiguous by the interpretive weight it carried. If Lombroso was right, criminality in the Conca d'Oro might well have been passed on directly from one generation to another, rather than being atavistic. He spoke of it in fact as if it were somehow both serial and atavistic. Crime was an evolutionary aberration, but in some places aberration of that kind constituted an ongoing local rule.

Where Broca had largely abstained from the anthropological study of crime and Quetelet had limited his attention to statistical regularities, Lombroso pursued a range of ambitious goals, not least of which was to find the etiology of criminal behavior. Like so many of the key figures discussed in this book, he was trained as a doctor, and was thoroughly aware of the claims of medical knowledge. Certain characters found in epileptics—a group he saw as very close to criminals—might appear as "abnormal or pathological [*morbosi*]," but the fact of classifying them as such, he suggested, might simply reflect a lack of "embryological and philogenetic understanding." Instead of being pathological, epileptics were probably atavistic.[38] Or rather, he added, "in many cases pathology and atavism stem from a common cause."[39] There were "influences generating an illness that could provoke atavistic morphological regressions." Traces of meningitis, in particular, were sometimes found in atavistic criminals.[40] Here, in principle at least, was a Lombrosian answer to the examination conducted by Broca on the murderer Lemaire: the lesions of meningitis might happen to be characters regularly found in criminals. Whether or not the illness actually came first, those lesions might then be considered a typical character found in criminals in whom meningitis might have actually aroused or triggered atavistic behavior. Understood thus, explanation by atavism and explanation by pathology turned out to have much in common. So while for Broca Lemaire's meningitis was a pathological accident that subsequently exposed him to injustice, for Lombroso the illness was most likely to have been the event that brought to the surface a criminal character somehow already present in him as hereditary potential.

ANTHROPOMETRY

In his later work, Lombroso himself briefly told the story of his involvement with anthropometry as one of disappointment. Reflecting in 1893 on his youthful scientific enthusiasm for the kind of knowledge represented by craniometry in particular, he wrote: "When I began this research, I swore by anthropometry, and especially by craniometry applied to the study of criminals. I saw in them something solid to hold onto as a way of resisting the metaphysical views and supposedly self-evident truths that dominated the study of everything to do with man."[41] But his enthusiasm could not be sustained over decades of activity. There were disappointing instances in which cranial measurement revealed no significant discrepancies between criminals and others. In one particular sample, he noticed no fewer than thirty-three "anomalies in skulls in which measurement did not reveal salient differences."[42] It was not that the anomalies were imperceptible to his practiced eye: it was just that they did not show up in the numbers produced by measurement. He considered in 1893 that persisting so determinedly with anthropometry had in fact diverted his attention from other significant factors: "mensurative research had the unfortunate effect of distracting me from anatomico-pathological research and led me to the overconfident conclusion that there were no anatomico-pathological phenomena to investigate."⁴³ Eventually, "use degenerating into misuse revealed to me the vanity of my hopes, and indeed the cost of having placed too much trust in [anthropometry]."⁴⁴ He was forced in the long run to recognize that "the differences in measurements between the abnormal and the normal [here, for once, he did use the word "abnormal"] are so slight that, without carrying out the most finely grained [*delicatissima*] research, they cannot be brought to light."⁴⁵

Broca would have been the first to agree that craniometry involved very fine distinctions, and it might well be argued in retrospect about Broca's own work that he himself largely failed to establish distinct measurable differences between races, being compelled as a consequence to sustain his scientific activity by developing ever-renewed and more subtle techniques of measurement. Broca could not be sure in the long run that he had found general anthropological truths, but he had certainly invented and refined a set of methods. Lombroso, for his part, clearly did not consider that the point of anthropological science was to perfect its methods while continuing to produce inconclusive results. He was sure that criminals presented abnormal physical characters, and needed to find a method or methods that would allow him to offer positive proof of that. It is not surprising, then, that in the 1897 version of *L'uomo delinquente*, four years after telling this story of failure, Lombroso was still adducing craniometric evidence in support of claims about criminals. Among "free men," he said at one point in the essay, the proportion of individuals with "diminished cranial capacity" was 11 percent, whereas among living criminals it was 18 percent, and could reach 59 percent in dead ones.⁴⁶ Broca, had he still been alive, would doubtless have had a host of questions to ask about how these figures had been obtained. He would have wanted to know, in particular, how the series of criminals and "free men" were constituted. Lombroso gave no indication of how that had been done, but it was clear that, for all his disappointment, he had not simply renounced craniometry. There had been no decisive moment of falsification. In an appendix to the 1897 edition of *L'uomo delinquente*, he reported on some recent Italian research on the lacrimal bone involving series of normal, criminal, and insane people. The study had found the bone to be normally developed in 73 percent of the normal people, 69 percent of the criminals, and 71 percent of the insane. Lombroso might well have commented that these were just the kind of "disappointing" figures so often produced by craniometric studies, but instead he tried to make the best of them: "Therefore in criminals the study found a greater number of progressive anomalies (rudimentality of the lacrimal bone) than in normal or insane people."[47] Differences of only two percent were called on to bear the weight of this generalization, as he endeavored to make them seem significant in ways that no statistician could have condoned.

Lombroso was, so to speak, let down by craniometry, although he remained undefeated in his conviction that significant physical differences existed between criminals and noncriminals. He was let down also, in his own view, by the response of later French physical anthropologists to his craniometric research. A number of them objected strenuously, not only to his methods, but to the very conception of his research program. In Lombroso's eyes, this was disappointingly uncooperative and ungracious. He was bidding to extend the influence of craniometry beyond the study of race as it had been narrowly understood in Paris, and he was promising among other things to give a broadly anthropometric account of innate physical differences between criminals and noncriminals. But the very breadth of this approach provoked critical responses from Léonce Manouvrier and Paul Topinard, who were the most intellectually distinguished and influential of Broca's pupils. As far as they were concerned, there was to be no passing of the baton from France to Italy, no renovation of physical anthropology, and no ongoing elaboration of a shared theory of normality. Any conceptual developments that might take place around Lombroso's work could therefore only do so in an atmosphere of strenuous contestation. Repeatedly, the "rise" of normality and its cognates during the nineteenth century was, as Bertolt Brecht said in another context, thoroughly resistible.

PHYSICAL ANTHROPOLOGY AND CRIMINAL ANTHROPOLOGY

Manouvrier was Lombroso's most direct opponent because he had his own scientific investment in criminal anthropology and in the concept of normality. There existed in France an active group of self-identified criminal anthropologists based in Lyon and gathered around Alexandre Lacassagne. In 1886, Manouvrier had contributed a substantial article to the *Archives de l'anthropologie criminelle et des sciences pénales*, of which Lacassagne was the principal editor. In that article, Manouvrier appeared to be practicing, albeit more cautiously than Lombroso, a form of criminal anthropology. His text laid out some key assumptions. First, it seemed reasonable to suppose that the way in which people behaved was connected to their "organization." He was using "organization" just as it had been used sixty years earlier by Étienne Geoffroy Saint-Hilaire, to mean "anatomical composition."[48] If, he argued, organization was connected to behavior, it would not be absurd to suspect that "particular characters existed in murderers."[49] Scientists needed to determine whether criminal behavior was brought about by factors additional to the obvious one of social environment. We know in general, said Manouvrier, that "the conformation of the body predisposes people to certain acts and makes them averse to others." So it was a matter of finding out whether that was the case with murderers.[50] Research of this kind, he warned, had to avoid the pitfall of phrenology: the murderer's bump on the skull identified by Gall, which supposedly corresponded to a cerebral organ, had been discredited by modern anatomical and psychological analysis.[51] The more careful approach Manouvrier was proposing would attempt to discover whether it could reasonably be claimed that murderers had distinctive "cerebral and even cranial" characters.[52] He then went on to discuss the theory of atavism, which he declared to be a simplistic explanation of criminal behavior.[53] But he did allow that there was a category of "savages through atavism" who belonged among the "abnormal."[54] As noted, Lombroso himself rarely used the term "abnormal," and we shall see what was at stake in that terminological difference.

A key question for Manouvrier was whether people who killed for motives of theft were to be included in the category of those who were "abnormal, pathological, or atavistic." He did not think they ought to be. But those who engaged in murder for its own sake—called in French *assassins* and in Italian *assassini*—might be describable in just those terms. It was useful, then, to distinguish between two kinds of brutality (or brutishness). While those who killed in order to steal often displayed a range of psychological qualities that included "mediocre intelligence," most *assassins* revealed "brutal selfishness." This tendency made of *assassins* "a normal variety."[55] It was not just that murderers of this kind could in Manouvrier's view be classified as abnormal with respect to the population as a whole: as a group, they presented their own kind of normality that lent itself to scientific study. But he needed to specify how that study would proceed. Was the normality in question something that could be analyzed according to the standard procedure of Broca's anthropology, by taking a series of measurements able to be ordered statistically around an average? The key question was whether the supposed normality of *assassins* would show up in anthropometric analysis. Manouvrier wanted "to see whether the [murderers'] skulls present anatomical characters that correspond to their psychological particularities."[56] He had designed a study to answer this question involving eighty crania of decapitated criminals located in French museums. These were compared with "thousands of ordinary [*quelconques*] ones" that he had also studied. The outcome was unimpressive, and he found himself obliged to conclude that those who—like Lombroso, whom he did not name—"found pathological and abnormal characters to be particularly numerous in series of murderers' skulls" were expressing a view that simply revealed their lack of *expérience*—presumably, given the ambiguity of the French term, both their lack of experience and their lack of appropriate experiment.[57]

But Manouvrier was able to salvage something of scientific interest from his craniometric study. There were anomalies visible in some of the murderers' skulls. Some of them did have narrow foreheads and strong jaw lines—both characters included in Lombroso's long list.[58] It might have been, Manouvrier conceded, that these characters corresponded to greater height or greater muscular strength, but they were due in part to "cerebral inferiority correlated with physio-psychological inferiority." The differences were not great. The measured "craniological inferiority" of *assassins* was less than that of imbeciles. It was "not even as great as that of savage races, even the most intelligent of them."[59] Manouvrier had to recognize that no decisive conclusions could be drawn from his research. While narrow foreheads and strong jaw lines tended to be present in murderers, it did not follow, he duly noted, that they occurred only in murderers. It had to be acknowledged that they were present in all classes and in most social categories. In other words, contra Lombroso, these characters did not—or did not yet—define a type. The most that could be said was that narrow foreheads and strong jaw lines were rare in "men of distinguished intelligence and morality."[60] Further conclusions could have been drawn only had there been sufficient crania to put together subsets of decapitated murderers allowing particular kinds to be identified among them, and it was unlikely that the numbers of criminal skulls available would ever be sufficient for that to occur.[61]

Lombroso, who was an indefatigable reader of publications in his field and was in correspondence with Lacassagne, could hardly have failed to read Manouvrier's article, which must have further contributed to his disappointment with the methodological fastidiousness of Broca's disciples and with craniometry in general. In *La donna delinquente* (*Criminal Woman*) of 1893, he described the current French anthropologists as overemphasizing craniometry just as he had earlier done himself. They should have paid more attention to anatomico-pathological characters: "In the same way, Topinard and Manouvrier, when they found that the anthropometric differences in the skulls of murderers and in that of Charlotte Corday were not sufficiently salient, lacked the anatomico-pathological notions needed to notice the very great [*enorme*] anomalies in those skulls and denied their existence."[62] They had not seen what was there to be seen because of the narrowness of their methodological focus. Topinard was in quite fundamental disagreement with Lombroso about the coherence of criminal anthropology as a field of research, but Manouvrier, precisely because of his engagement with the field, continued to be the frontal opponent of the *scuola positiva*.

THE PLACE OF THE NORMAL IN CRIMINAL ANTHROPOLOGY

The conflict between Manouvrier and Lombroso was as public and as resonant as any of the polemical encounters examined in our history. A formal academic confrontation between them occurred during the second international congress on criminal anthropology, which was held in Paris in 1889. Lombroso held a place of honor there, having been allocated the first topic in the congress, "the latest research in anthropology."[63] That was the perfect subject for this most alert and most up-to-date of scientists, and Lombroso duly carried out his task, celebrating recent "victories" and others that would soon come. He gave full metaphorical value to the etymology of "progress," declaring that scientific knowledge was to be achieved step by step. But there was nothing celebratory about the second topic on the agenda, which had been allotted to or claimed by Manouvrier. It was a searching set of questions aimed at some of Lombroso's key working assumptions: "Are there anatomical characters proper to criminals? Do criminals present on average certain particular anatomical characters? How must these characters be interpreted?" The explicit aim of this paper was to "criticize the work that has been done and prompt new research in which the errors committed so far will be avoided."[64] It had to be asked first of all, Manouvrier asserted, what kind of scientific analysis was most suited to the task of identifying specific qualities in criminals. In his view, the task would have been best accomplished by studying "physiological elements," but that was not practicable. So the study of anatomical differences between criminals and "honest people" had to be envisaged as a compromise.[65] That might lead to a clearer understanding of what was "vaguely called in current parlance the tendency to crime" if indeed a correspondence could be established between "well-defined psychological elements" and anatomical ones.[66] But for criminals to be properly examined, said Manouvrier, the first thing to be done was to "distinguish physiologically and anatomically between the normal state and the abnormal state."[67] The favored binary of

French nineteenth-century medical talk was thus brought to bear on the question, and it was soon made clear by Manouvrier that Lombroso's research was failing to respect and apply the key distinction. If the normal and the abnormal were not defined at the outset and their mutual opposition maintained throughout, all that would remain to science would be "a physiological hodgepodge along with an anatomical hodgepodge."[68] Instead of this binary, which Manouvrier regarded as fundamental to the medical sciences, Lombroso had begun with another that Manouvrier considered highly questionable from a social point of view. He had "divided humans into two categories: those arrested by the police and those not arrested."[69] Others who studied criminality might have come to Lombroso's defense at this early stage by asserting that this was a perfectly practical way to identify criminals for study, one that had been followed in much of their own work. But the problem as Manouvrier saw it lay in the fact that Lombroso's school was committed in practice to discovering physical anomalies presumed to be distinctively present in the group of those arrested, making the assumption that they bore visible marks of a natural tendency to crime. No attention was paid to the group of those not arrested, either to their physical characters or to their possible practice of lesser or simply unpunished crimes.

Manouvrier developed this critique by applying the concept of normality, not as a narrowly defined condition, but with the nuances that had accrued around the notion since about 1830 under the influence of statistical thinking. Normality was not to be understood, he said, as opposed in every regard to a supposedly distinctive condition of abnormality. The "healthy, normal man" was not "a man without defects and without any tendency to vice": he was simply located somewhere within the range that deserved to be considered normal. The normal was not the acme, and any man who happened not to be perfect should not be regarded as abnormal or inferior merely for that reason.[70] By the same logic, the science of criminality needed to make allowance for crimes that actually fell within the range of normal behavior. The approximately normal was precisely what made up the normal. In the case of "ordinary crimes," there was "no compulsion to suppose that they are connected with a morbid or abnormal physiological state."[71] That was what made the crimes ordinary. When the normal or healthy state was taken without examination to be a state of perfection, every departure from it, every anomaly could be read in error as a symptom of abnormality. That is what had happened in the work of the *scuola positiva*. Lombroso and his colleagues had failed to take account of what Manouvrier later called in discussion "normal crimes."[72] Determined to confirm that criminals were a natural group, the Italian researchers had attempted to build a body of evidence by looking for anomalies in the individuals they examined, then compiling those anomalies in a list of supposedly typical characters. But it was not proven, in the absence of a proper definition of normality, that these anomalies taken together constituted an abnormal state or defined an abnormal class of human being. Furthermore, this method of investigation provided no check on the proliferation of anomalies. As new signs of a natural tendency to crime were being produced in this manner, a kind of interpretive inflation was occurring: "So the number of characters goes on multiplying daily, with the result there will soon no longer be an honest man who cannot be shown to have half a dozen criminal characters. Out-and-out criminals [*criminels qualifiés*] will doubtless by then have reached a dozen, and a question will need to be asked whether another social category, professional or otherwise, will not have a higher tally than the criminals."[73]

Manouvrier was clearly of the view that the whole project of Lombrosian criminal anthropology was mistaken in its conception, but that was not its only failing in his eyes. Researchers in the Italian school were using undisciplined methods that were out of keeping with the requirements of physical anthropology. Without naming his teacher, Manouvrier was effectively taking Broca's methods as a model for anthropology and censuring those who failed to follow them: "Averages have been presented on the basis of series that are not large enough. Numbers have been collected by more than one observer or gathered using defective techniques, sometimes by novice observers trying out their skills for the first time on criminals. Insignificant differences have been put forward that can in fact be found between two series of ordinary [*quelconques*] men, as well as between ordinary men and criminals."[74] There was a specific answer to the problems created by this lack of methodological discipline. It should not be simply a matter of seeing every supposed anomaly as a mark of the extraordinary, but of assessing the extent and the direction of departure from the averaged norm. If averages were calculated with care, it might be shown that a certain difference had appeared because one character was stronger in criminals than in "virtuous men," whereas other characters might be weaker. And it should never be forgotten, as it regularly seemed to be within the *scuola positiva*, that "the category of 'honest people'" was "riddled with epileptics, idiots, imbeciles, degenerates, brutes, and vicious, inferior people of every kind."[75] Manouvrier was declaring that Lombroso's school had misconceived the whole business of criminal anthropology, and was aggravating its misconception at every point by ongoing failures of method.

Broca's other leading pupil, Paul Topinard, did not confront the Italian school as directly as Manouvrier, but that was because he dismissed the whole enterprise of criminal anthropology out of hand. In 1888, Topinard read to the Société d'Anthropologie de Paris a report on observations done by French colleagues on the skulls of two murderers. Research of that kind was certainly of interest, said Topinard, as were studies of skulls belonging to geniuses or insane people.[76] Those three kinds of people, as he knew very well, were the ones Lombroso had identified for study. But Topinard flatly rejected the conceptual apparatus Lombroso had constructed for the task: "There is no criminal anthropology. The association of those two words is a form of usurpation. There is no anthropological type of the criminal in the sense that we give to those words. The theory that considers criminality to be an atavistic phenomenon, a resemblance to our prehistoric ancestors, is just an idea in someone's head, a complete fantasy of the imagination, a hypothesis."[77] Lombroso was not named here, but his notoriety and that of his school ensured that the polemical address of these remarks was unmistakable. Topinard could hardly have been more trenchant about the whole enterprise and the theory that underlay it, but his opinion was not fully shared by Manouvrier, who intervened immediately after: "Anthropology must have applications," said Manouvrier. "It can render a service to the law, and not just to criminal law." Criminal anthropology, in Manouvrier's view, was "a very good thing."[78] To that Topinard replied that anthropology needed to remain a pure science. It needed to follow the path opened up by Broca and Quatrefages, for whom anthropology was purely and simply the zoology of man.[79]

When the international congress took place in 1889, Topinard could hardly have stepped forward to play a significant role without contradicting the dismissive position he had adopted. He simply wrote a formal letter to the participants, published in the full version of the proceedings, that once again questioned the very idea of a criminal anthropology. He did ask specific questions about Lombroso's work, applying to it the concept of normality as it had been developed in Paris-based physical anthropology. The first thing to be said was that some of the characters identified by Lombroso as making up the type of the "predestined criminal" were in fact "absolutely normal" ones.[80] The Italian school had failed to remember that variations were not the same things as anomalies: "In every type of race there is a central expression around which universal variations oscillate, variations that can be sizeable without deserving to be considered anomalies. Included in this group are the differences in cephalic indices [a craniometric measure developed by Broca], the prominence of the brows, the development of the forehead. We have no reason to say that prominent eyebrows, which are the rule in people from the Auvergne, for example, are a criminal stigma."[81] In other words, Lombroso's implicit conception of the normal was too narrow and his understanding of anthropology too broad. It was indeed the case, said Topinard in a statement that Lombroso might initially have welcomed, that "abnormal or strange characters occur much more frequently in criminals." But this proved only that the individuals concerned had not developed properly in the course of their growth. Apparently distinctive characters found in criminals were simply individual pathologies. The collection of anomalies produced by Lombroso would "never succeed in constituting a hereditary type continuing another type that belonged in the past."[82] Topinard insisted that he was not denying the existence of hereditary characters in criminals, but it was "heredity that did not go far back": it was "family heredity."[83] Criminality was therefore unsuited to the anthropological methods developed for the study of race: "The idea of race, that is, of transmission through the ages, has absolutely nothing to do with criminality."[84]

While Manouvrier confronted Lombroso directly, Topinard was delivering from the sidelines a critique that was even more fundamental. All of that called for a strong reply from Lombroso, and it cannot be said with any certainty that he gave one. He began by making a concession that amounted to a retraction, declaring that "cranial capacity was not a character of criminality," but went on immediately to say that two other characters had recently been discovered. A large number of hernias had been found in criminals, and that was a regressive character, known thereby to be a mark of atavism. Moreover, "the role of ptomains in criminal manifestations" now "appeared certain."[85] Far from disarming Manouvrier, this offer of a two-for-one-exchange could only serve to show that Lombroso's well-known readiness to review and correct his previous work provided no guarantee against the proliferation of anomalies in the course of his research. It must also be said, however, that when the published proceedings of the Paris congress were reviewed in the Italian school's official journal, the *Archivio di psichiatria*, a suggestion was made that the telling replies Lombroso had made to Manouvrier on the spot had been largely omitted from the final version. R. Laschi wrote: "It is to be deplored that the record of the sessions bears only a pale trace of the brilliant, courageous ripostes made by Lombroso in defense of the school, as so many who were present at the congress will recall with great satisfaction."[86] Whether or not Lombroso did reply brilliantly in the heat of battle—and no other comments in the *Archivio* explicitly support Laschi's assertion—it is a challenging task for an intellectual historian to reconstitute the defense that he could or should have mounted. We will briefly sketch out such a defense below and indicate what it signified for the dissemination of ideas about normality.

In place of the kind of binary analysis pressed on him by Manouvrier, which involved working to and fro between the normal and the abnormal, Lombroso tended to move by association from one loosely defined, more or less given category of strange or dangerous person to another. This was often discovery by analogy and extension rather than consolidation by distinction-making and exclusion. Lombroso himself put it simply and disarmingly in *Pazzi ed anomali* (*Mad People and Anomalies*): "I have thought of myself for many years as working to find the differences between insane people and criminals, but I have always found the analogies between them to be ever greater!"[87] It sometimes happened that he encountered subjects who combined all those qualities, thereby seeming to confirm the dynamic of analogy and equivalence: "In these people, the criminal type and that of imbecility are inextricably mixed together, and are actually exaggerated... here, imbecility is truly added to moral insanity, which, like epilepsy, affects the extremes of the intellectual scale that goes from genius to idiocy."[88] This very accumulation of qualities appeared to Lombroso as a revealing fact. The dynamic of discovery by analogy found its confirmation, not in any large statistical sample, but in a relatively small number of extreme cases where the approximate categorical equivalence of these disorders was confirmed by their supposedly typical coexistence. There was thus a practical and intellectual dynamic in Lombroso's work ensuring that, despite an initial determination to mark differences between criminals, insane people, and epileptics, scientific discovery would usually take the form of more clearly perceived analogies. The progress of discovery led, not to the elimination of all difference between categories, but to the proliferation and enrichment of analogies between them. The more Lombroso examined criminals and mad people, the more he found similarities and forms of equivalence. And the more he examined epileptics or reflected about hidden forms of the disorder, the more he perceived analogies with the other two categories. That was, in some sense, the overriding general fact that he saw as emerging from his research. There was a family resemblance between criminals, the insane, and epileptics.[89] The study of normal people must have appeared to him, then, as an unrewarding distraction from this dynamic of discovery.

Historian Renzo Villa makes some perceptive remarks about this aspect of Lombroso's work. He comments, as Manouvrier had done in his own way, on the absence in Lombroso's writing of a definition of the normal, but he does not suppose with Manouvrier that that was an oversight or a failure on Lombroso's part.[90] Villa identifies in Lombroso a "negative anthropology": "criminal anthropology is a kind of inverted, negative anthropology in which the protagonist is not a general subject who is tautologically defined. It is the anthropology of the negative, of the edges and the confines within which the 'normal,' like a kind of besieged species, locks itself in and defends itself."[91] Villa is interested in the politics of Lombroso's scientific endeavors, suggesting that the he was preoccupied with various kinds of "deviants" who were a threat to normal people, but did not ever make normality an object for scientific examination. We will not follow Villa in his use of the term "deviants" because that statistically-derived concept did not appear until thirty years or so after Lombroso's death. However, Villa does help us to understand that "the study of madness constitutes for Lombroso a possible means of better understanding normal man, but his interest leads him rather to establish connections between madness and other aspects of human behavior such as crime and genius."[92] In putting forward the notion of "deviant" as a way of referring to criminals, the insane, and *tutti quanti*, Villa is attempting to summarize Lombroso's thinking in a manner that Lombroso himself could not have directly accepted. He made repeated attempts to associate the born criminal, the epileptic, the mad person, the degenerate, and the atavistic person, but did not actually name a general category that could have included all of them.[93] In a sense, the perception of analogies had its full value only because these kinds of people were given to observation as different. He certainly did not call them all abnormal, as his French critics might have expected, not to say required. To find one common name for all would have been to enter into the binary logic that Lombroso preferred to avoid. It would have made one capacious category of abnormal people, identified by some measurable net difference between them and a supposedly recognizable standard of normal people. What Lombroso sought and found instead is what Villa elegantly calls "the solidarity of the stigmata"[94] that mark all his subjects as different, without ever claiming that they were different in the same way or to the same degree, and without ever noting specifically or even generally to what unblemished norm their difference was referred.

It may seem surprising, then, that in 1893 Lombroso began *La donna delinquente* with a section on the normal. Was Lombroso changing his approach in response to the criticisms of Manouvrier and Topinard made a few years earlier? He certainly quoted some of their criticisms later in the 1893 essay, but it would be hasty to suppose that the self-conscious change of approach was a response to their strictures.[95] It seems most likely that this was for Lombroso a different intellectual circumstance, requiring by its nature that he depart from the approach he had previously favored.[96] He certainly affirmed the need to establish "the normal profile" of woman in order to contrast it with that of criminal women, but he had after all sometimes stated that principle in other works without allowing it to constrain his actual practice. Here, he went further, to the point where it seemed to him that he was exposing himself to criticism by attending at length to normal woman: "If anyone accuses us of wasting time on the honest woman, may we remind them that none of the phenomena in criminal woman can be explained if one does not have to hand the normal profile?"[97] To our knowledge, no one had accused him in the past of spending too much time on normal people. Quite the contrary. But when he did set about the task, there were none of those specific measures or tables of statistics that Manouvrier had asked of him, although some anthropological research carried out by others was considered in passing. The discipline of primary reference in this first section of the essay was in fact zoology: "it is not possible," Lombroso said, "to undertake the study of criminal woman without first analyzing normal woman, or rather the female on the animal scale."[98] He then went on to make a series of observations about zoological regularities. The female is larger than the male in lower species, but smaller in higher ones. Inferior beings develop earlier than superior ones, have lesser cranial capacity and smaller brains. All of those differences were found to be present as biological gender differences in humans, and were read by Lombroso as signs that normal woman was, scientifically speaking, physically and intellectually inferior to normal man, being less evolved than he was.[99] Topinard and Manouvrier had both pointed out that woman's smaller brain size was proportionate to her lesser stature, and Lombroso did not fail to record their argument, but he concluded nonetheless: "that does not put an end to her true inferiority."[100] Invidious zoological comparison was given a racial dimension when Lombroso pointed out, without quoting any studies that might have supported his claim, that in the lower races women's faces were more virile. In other words, the physical development deemed to occur through evolution made white, civilized woman more feminine, thereby increasing the gap between her and the normal man of her own race. The inferior races he took here as examples were the usual suspects of late nineteenth-century anthropology: Hottentots, Kaffirs, and Bushmen.[101] Because white woman tended to remain in a state of arrested development while white man went on evolving, white woman had more in common with inferior blacks than white men did with black men. In that racial evolutionary sense, the normality of normal woman extended across a range of races as it did across evolutionary time. Here was a specific reason to begin a study of criminal woman with an analysis of the normal: the characters of normal woman were more widely shared across time and space than those of normal man. In that distributive sense, normal woman was far more normal than normal man.

This rather lengthy examination of "normal woman" in zoological and evolutionary anthropological terms enabled Lombroso to produce a series of propositions about normal woman that could serve specifically as references when he came to elaborate his views about *la donna delinquente*. He claimed to have shown, for example, that woman was more like a child (*infantile*) than man was.[102] Comparison with animals, female and male, allowed him to make a general statement about sexuality: "The greater frigidity and passivity of women in coitus is in fact common to females in all animals."[103] Indeed, he declared, woman was "naturally and organically monogamous and frigid."[104] The two adverbs in this last sentence showed how normality was constituted in principle: it was a natural order inscribed in the animal organs of women. Lombroso's own physiological experiments had also demonstrated to his satisfaction that normal women were on average less sensitive to pain than men. This was in his view necessary to the species so that women could bear the pain of childbirth. He had to concede that women displayed signs of pain more than normal men, but this was not because women were more sensitive: it was simply because they were more irritable in the physiological sense.[105] And precisely because of this natural combination of lesser sensitivity and greater inclination to react visibly, women were better able than men to feign emotion for seductive and deceptive purposes. Evidence of that supposed anthropological character was provided, not by experimentally based statistics or indeed by citing the findings of other scientists, but by a series of proverbs.[106] In Lombroso's eyes, no further proof was required. "To demonstrate how mendacity is habitual and almost physiological in woman would be superfluous given that this is found even in popular tales."[107] For Lombroso, proverbs were by definition about normal woman, and the anthropological characters of normal woman were proverbial in the literal sense.[108]

SCIENCE AND PROVERBIAL KNOWLEDGE

It is clear from all this that Lombroso was developing a form of knowledge that moved freely between science and common sense. Or, to put it in a manner that is more challenging to thinkers of our own time, he had found a way to include proverbial common sense within what he called science. Mary Gibson, in one of the most notable studies of Lombroso's work available in English, suggests in effect that this was no more nor less than bad science. She says rightly enough that "his books were characterized by haste, slipshod logic, and a tendency to ignore data that did not support his theories."[109] From there she goes on to formulate a twentieth-century critique of Lombroso's research method: "More damaging to the credibility of Lombroso's research was that most of his studies lacked control groups against which to compare his data on criminals. In the successive editions of *Criminal Man*, for example, his tables enumerating rates of various anomalies in the prison population only occasionally offer comparative data on normal men."[110] This seems to pursue Manouvrier's criticism of 1889, but it does so in rather anachronistic terms: no one, including Manouvrier, spoke of "control groups" in Lombroso's own time. The question that interests us most here as intellectual historians remains: how and why did it happen that the author of those hasty, slipshod books should have been so widely read and so broadly influential in the last decades of the nineteenth century?

Gibson pursues her own criticism by observing that "Lombroso, as a self-proclaimed positivist, mixed various types of 'soft' qualitative evidence with his statistical data."[111] What Gibson calls a "strange" practice is likely to appear to modern readers as a noteworthy failure to meet the methodological requirements of modern science. That is certainly how it seems to her: "Most surprisingly, *Criminal Man* and other writings of Lombroso are packed with quotations from literature and folklore, a characteristic that displayed his wide reading in a number of fields but casts further doubt on his claim to be scientific."[112] How could Lombroso have mixed folklore and statistics in this way? This certainly appears to Gibson to be a strange form of eclecticism. But the opposition between hard and soft evidence—methodological routine for the social sciences of our own time—appears not to have been as stable or as widely shared in the 1880s as it is today. Some insight into this question may be gained by returning once again to the Manouvrier-Lombroso exchange of 1889. To Manouvrier's attack, with its list of errors of method and logical inconsistencies, Lombroso replied in the first instance with a proverb. He had begun his own talk earlier in the congress with a proverbial reference to the Greek philosopher who proved movement by walking, and he was continuing in the same mode. It is not far, he observed elliptically in response to Manouvrier, from the Capitoline hill to the Tarpeian rock.[113] What exactly was the point of this classical metaphor in the circumstances? Did Lombroso mean that one can occupy a high position and then be quickly cast down, referring to his possible fate as leader of the Italian school? Or did he mean that there is no great distance between the honest man and the criminal? The former could be considered the standard meaning, but the latter might nonetheless apply here, since Lombroso went on to talk about the differences between criminal types. Whatever he might have meant exactly, there was a noteworthy generic or discursive discrepancy between Manouvrier's critique and Lombroso's reply.

The use of proverbial language should not be taken here as a lapse of conversational decorum in a scientific milieu. It was effectively the assertion of a different view of what science ought to be, and how one might talk about it. In Manouvrier's presentation, Lombroso identified the characteristic language of logic, which for him was bound to lead to the inhibition of scientific endeavor by philosophy. Manouvrier was trading in "syllogisms," the kind of talk that Lombroso considered "the biggest enemy of great truths."[114] Logic was not in his view the friend of practically minded science. By dint of abstract or hidebound reasoning, it could readily find itself disjunct from positive observation. That was undoubtedly the discursive stake for him in answering a tightly logical argument with a proverb. And as if to insist, Lombroso added another piece of classical wisdom: "You can see that by dint of logic we find ourselves, like the father, the son, and the donkey in the fable, in a situation where we cannot make a choice or take a step."[115] This was presumably an allusion to Buridan's fable about the donkey that starves because it could not decide whether to drink first or to eat. Strict binary choices of any kind were likely to lead to inaction. That was why Lombroso did not want to enter into discussion in Manouvrier's narrowly syllogistic terms. He did not think in that manner himself, and it was not how he expected the science of criminal anthropology to move forward. Narrow methodological argument might mean that scientific inquiry would never be able to reach a worthwhile destination. Logical disputation was likely to provoke only irresolution and immobility.

Lombroso knew, however, that he could not simply hold to common sense and claim it as science. In *Pazzi e anomali*, published in 1886, he had discussed just how his school could manage to square those two kinds of knowledge. He began by acknowledging that "scientific truths can rarely be demonstrated with a few easy-to-read pages and in terms that anyone can understand."[116] The imperative for scientists was to follow the difficult path of discovery, then render the results of their work as comprehensible as possible for a general public.[117] By the same token, it did have to be recognized, in his view, that some of the knowledge that criminal anthropology wished to organize and present was already abroad in the form of popular wisdom. That was one of the great advantages the *scuola positiva* could claim over its "metaphysical" and philosophically minded adversaries. When it came to considering "crime in popular consciousness," it was clear that the classical school was "at odds with popular convictions."[118] That was weakness on their part, a symptom that they could not see what was plainly visible to everyone else. So should they or anyone else contend that there was no such thing as a criminal type, Lombroso and his colleagues could appeal directly to common sense: "everyone admits that they recognize a scoundrel by his external appearance, and popular proverbs are proof of that."[119] This was not the expert coup d'œil of the experienced scientist, but the routine wisdom of the popular gaze. Proverbs carried its truth, and Lombroso proceeded to list quite a few of them at that point.[120] Proverbs provided *evidenza*, to make the Italian pun that was also possible in French: evidence in the forensic sense, but also "obviousness." This kind of evidence was—like so many other things—grist for the mill for the positive school. It revealed the general availability to knowledge of the objects studied by criminal anthropology. Proverbs, fables, and the like might at times be vehicles for prejudice, but they often provided opportunities for what Lombroso and his colleagues described as scientific "confirmation."[121]

That did not simply remove the difficulty of communicating with the general public referred to by Lombroso at the beginning of his book. The existence of proverbial wisdom did not ensure that the general public would accept a restatement of common sense truths in scientific terms: "Just try to tell them that in [the recognizability of criminals] one finds the criminal type. They will completely refuse to believe you."[122] Even when the truth was evident to all, widespread habits of thinking made it difficult to accept scientific formulations. Science still had to "overcome prejudices and preconceptions that are no more than a way of judging things according to one's own habits and those of one's ancestors."[123] So a kind of routine paradox obtained. The general public was likely to be resistant to the ideas of the school because the public was always inclined to resist new ideas, but the positive observations of the school were so thoroughly true that they had already found their way into popular consciousness: "And yet where many affirmations of our school are concerned, the evidence [*evidenza*] is so strong and so wide that it even finds itself reflected in public consciousness, limited as that consciousness may be. The trace of it is found in proverbs, in popular songs and in the poetic lines of authors which have, as in a clear mirror, captured the reflection of common people's ideas."[124] To appeal to proverbial knowledge was to trouble the locus of scientific authority as it was understood by the philosophically minded, and doing so exposed the school to criticism. But to keep science as close as possible to common sense was to assert that the things being studied by the school were positively observable to "all who had eyes to see," as Lombroso's colleague Virgilio Rossi had put it.[125]

CONCLUSION

It was possible for Italian criminal anthropology to refer regularly to normality while hardly ever giving close scientific attention to people considered normal. In the *scuola positiva*, the normal was taken as known and supposed data about normal subjects referred to in passing without actually being gathered or tabulated. To Broca's successors in France, that seemed a culpable omission. How, they asked, could Lombroso escape the binary logic that opposed abnormal kinds of people to normal ones, and how could he escape the obligation to study the two in parallel? As it happened, Lombroso was generally untroubled by such criticism. He referred to the normal with a mixture of familiarity and inattention, and that unresolved combination became itself a characteristic figure of knowledge, one that would recur in twentieth-century usage. In the writings of Étienne Geoffroy Saint-Hilaire, the normal had been introduced and refined as a narrowly scientific concept. Geoffroy could never have taken the term for granted, but Lombroso was able to do just that, and doing so was in a sense the mark of his modernity. The normal found a loosely drawn but conceptually fundamental place in criminal anthropology precisely by being taken for granted. It seemed to Lombroso, after all, that ordinary people could themselves identify the normal with no particular help from science. Since the characters of the nonnormal were visible to all, it was clear where the urgent scientific task lay: not in the routine description of normal individuals, but in the study of those who posed a threat to society. Lombroso and his school helped to develop a genre of writing in which the authority of science was called on to support a range of claims about humans that were not actually grounded in experimental or frequentist evidence. Statistics were used in opportunistic ways to confirm what was already believed; epistemic status was given to the established verities of common sense; the natural and the organic were invoked in order to privilege certain physical and psychological characteristics over others; and strongly constraining generalizations were produced about "natural" human sexuality. That generic and epistemic mix became a broad set of discursive habits in talk about normality. Lombroso may not have been the very first to produce this knowledge formation, but he was certainly one of the earliest and most influential to do so.

CHAPTER SIX

Anthropometrics and the Normal in Francis Galton's Anthropological, Statistical, and Eugenic Research, c. 1870–1910

In the late 1870s, the anthropologist and statistician Francis Galton began a series of experiments in a photographic procedure he termed "composite portraiture."[1] Composite portraits were produced by layering a number of individual portraits onto a single photographic plate, through the partial exposure of each image.[2] The result was a sort of photographic palimpsest: the ghostly impression of a composite face that retained the hazy outline of each of its originals (see figure 1). The purpose of these portraits, Galton explained, was not to represent "any man in particular," but rather to "portray an imaginary figure possessing the average features of any group of men." "These ideal faces have a surprising air of reality," Galton wrote. "Nobody who glanced at them for the first time would doubt its being the likeness of any living person, yet . . . it is no such thing; it is the portrait of a type and not an individual."[3] Galton seemed here to be following the concerns of earlier statisticians like Quetelet and Louis-Adolphe Bertillon. He was continuing to speak of averages while using composite images to produce an "ideal" image that stood for a type. Composite portraits provided a technical solution to the difficulty of knowing and representing the type, which had beset anthropology for much of the nineteenth century, evident in the debates held at the Société d'Anthropologie de Paris. Galton argued that the combination of photography and statistics, and more particularly the use of photography to undertake statistical analyses, represented a significant advance on previous anthropometric methods. In the first place, photography provided the means to represent complex data in a more compelling and persuasive way than numerical tables: "the reasonableness of the results will become more apparent when they are displayed in pictorial form," he claimed.[4] This made composite portraits accessible to a much wider audience than anthropometric studies. The portraits became suitable for display in public lectures, popular journals, and international exhibitions.

But composite portraits were in no way intended as gimmicks for the public or even as visual translations of numerical data. They were themselves properly scientific documents, Galton claimed, and were best understood as a new form of "pictorial statistics,"[5] designed to provide "generic pictures of man, such as Quetelet obtained in outline by the ordinary numerical methods of statistics."[6] What had been "ordinary" about numerical methods was, in Galton's eyes, an undue reliance on the calculation of averages: "in respect to the distribution of any human quality or faculty, a knowledge of mere averages tells but little."[7] More important in his view was "to learn how the quality is distributed among the various members of the Fraternity or of the Population, and to express what we know in so compact a form that it can be easily grasped and dealt with."[8] Quetelet, for all his emphasis on the "average man," was given credit by Galton for the fact that he too had worked on distribution. For Galton, however, it was clear that the emphasis had to be shifted. Instead of talking about the average man as the ideal type, it was now important to focus on "the mathematical law of deviation."[9] And the great advantage of composite portraits was to display distribution while at the same time indicating the existence of a type. They produced in outline an averaged representation of a group while at the same time including in the very blurring of their edges "the features of every individual of whom they are composed."[10]

THE WORD "NORMAL" FINDS A PLACE IN STATISTICAL DISCOURSE

By questioning the centrality of averages in the practice of statistics, Galton was effectively making way for a properly statistical understanding of the normal. Quetelet had spoken some decades earlier of "the normal state" in a manner that borrowed informally from already established medical discourse. And when one tracks the use of the word "normal" in the discourse of other statisticians of the mid-nineteenth century, it becomes clear that under-theorized usage of the term extended well beyond Quetelet's work. Even Louis-Adolphe Bertillon, who generally sought to provide his colleagues with an example of rigor, felt free to use the adverbial phrases *en moyenne* (on average) and *normalement* (normally) as if they were equivalent.[11] The expression "normal state" was also used in a conversation about statistics when, in 1862, Alfred Legoyt, director of the Division of General Statistics in France, was attempting to defend the official use of a statistical formula that had been described by Bertillon as unsound. Legoyt conceded that the formula was "only applicable to populations in which the stationary state has become, as one might say, the normal state."[12] Used thus, the expression was effectively a figure of speech, but the fact that precisely this figure of speech should have been used by a government official in the Société de Statistique de Paris serves as evidence that the concept of the normal state was no longer confined to medicine. Toussaint Loua, another public statistician, when speaking of the "statistical laws of marriage" in the style of Quetelet, referred to "the normal conditions [*conditions*] without which marriage would be unproductive and purposeless."[13] Clearly, the term "normal" could be used by statisticians from about 1860 onward, even if only as a short-term loan from medical discourse.

A more thorough terminological change began to occur in the late 1870s. Even as Galton was seeking to limit the importance of averages for statistical thinking, a German statistician introduced the term "normal" in order to deal with a persistent difficulty in the use of averages to study populations. At the Exposition Universelle held in Paris in 1878, statistician and economist Wilhelm Lexis presented a paper entitled "Sur la durée normale de la vie humaine et sur la théorie de la stabilité des rapports statistiques" ("On the Normal Duration of Human Life and on the Theory of Stability of Statistical Relations"). Lexis began by recognizing Quetelet's "remarkable research," but went on to express regret that the benefits of Quetelet's work had not been felt beyond the domain of anthropometry. Quetelet's approach, said Lexis, could be applied just as well to mortality, and he went on to provide a demonstration of that.[14] The first thing to which he drew attention was that the "average age" of all the members of a population alive at one time was nothing more than an arithmetical average—that is, to recall Quetelet's examples, the kind of average found in building heights, rather than the one found in Scottish soldiers. The difficulty of providing a proper account of the age of a given population was well known to statisticians,[15] but it could be overcome, said Lexis, if children were eliminated from the set of people being measured, just as they needed to be eliminated from the set used to calculate the average height of the population: "Children who are taken from us in the preliminary phases of life do not contribute in any way to the formation of the normal type of human life, any more than they contribute to the normal type of human stature."[16] This coupling of the word "normal" and the word "type" was new, and Lexis made it do statistical work. If pressed to identify a type among children, he would undoubtedly have had to concede that children could be taken as a group and studied in their own right, just like any other set of living creatures. But he wanted now to constrain the work to be done by averages, and he did so by talking about the normal type. To do that was to turn away from the established medical term "normal state" in order to establish a properly statistical concept. This called for a new rule of calculation. When the normal type was taken as a reference for the population as a whole, it served to narrow the group to be studied, since children were declared for this specific purpose not to be normal. And once a rule of normality had been applied, a regular distribution could be expected to appear, producing something that had the shape of a "natural"—and not just an arithmetical—average. The proper way to allow this pattern to emerge, said Lexis, was simply to calculate the average age of adults.[17] That gave rise to an average with a regular distribution in which the mean and the median more or less coincided—what Quetelet himself would have called a real average. This was what Lexis proposed to define as the "normal life" of a population: the average age of those who, as adults, belonged within the range of the "normal type." By excluding a whole group who were declared not to belong to the normal population, he had succeeded in transforming an arithmetical, statistically pointless average, into an ostensibly significant one. That move was to be repeated many times in the subsequent practice of statistics, as classes of individuals were excluded on technical grounds, thereby producing orderly data and "typical" outcomes. By the practice of such normalizing exclusions, statisticians would often be able to transform what would otherwise have been insignificant averages into something that could be said to represent a "normal type."[18]

The set of moves made by Lexis found a place in subsequent statistical work in which the normal served as a reference and as a kind of algorithm: first, one excluded irregulars (in this case, children) who would otherwise have rendered the average artificial and made the distribution asymmetrical by displacing the mean from a median position; second, one proceeded to study those remaining, establishing them as a group to which Quetelet's "natural" law of distribution applied. Galton himself did something comparable when he "transmuted" a set of anthropometric data so as to make allowance for unequal measurements resulting from net general differences between males and females: "The artifice is never to deal with female measures as they are observed, but always to employ their male equivalents in place of them. I transmute all the observations of females before taking them in hand, and thenceforward am able to deal with them on equal terms with the observed male values."[19] This was an entirely convenient way to do such calculations, explicitly recognized by Galton as an "artifice." For Lexis, children had to be left out when the population age was considered; for Galton, women needed to be "equalized" by recalibration of their measurements. Both procedures allowed regularities of type and distribution to be identified. Mathematically speaking, such normalizing did no more than allow data to be organized and interpreted. But as statistical routines of this kind became established, they were bound to produce equally routine effects of oversight and discrimination.[20]

Galton's composite photography, as we have seen, addressed the difficulty of including the full range of subjects while also making the average salient, but it did so only insofar as a particular group could be affirmed and confirmed as natural. Karl Pearson, in summarizing Galton's view, spoke of adults and children, but also of races: "If, however, we take any one of the principal races of man and confine our portraiture to adult males, or adult females, or to children whose ages lie between moderate limits, we ought to produce a good generic representation."[21] To consider adults as a group for statistical analysis was, for quite technical reasons, to exclude children; to make males the object of attention was to exclude females unless some artifice was found to transpose and include their data; and to consider one of the "principal races" was to exclude all others. Only when moves of this kind were made would composite photography provide "a good generic representation."

Galton himself first used the word "normal" in a talk he gave to the Royal Institution in 1877. He was describing a quite complex device of his own fabrication that distributed pellets poured in at the top in such a way that patterns appeared in regular form at several levels below. The basic geometrical component of this apparatus was the quincunx, which dispersed the pellets in the same manner regardless of their number.[22] As a result, there appeared in the lowest compartment "a heap of which the scale of deviation is much more contracted than that of the heap from which it is derived." That "heap," said Galton, was "perfectly normal in shape."[23] This particular sentence is listed by the OED as the first known use of the term "normal" in statistics in English, although Galton's rather casual use of it may not fully justify that attribution. He may simply have meant that the pattern of distribution was regular or predictable. He certainly did not repeat the word in his talk, and used it only once, in exactly the same sentence, in a set of three articles based on the talk that were published the same year in *Nature*.[24] At any rate, Galton displayed none of the ceremonious innovation evident in Lexis's 1878 paper, or indeed in Jacques Bertillon's enthusiastic reprise of Lexis's work in the Société d'Anthropologie de Paris in 1879.[25]

But Galton did indeed come in his later work to use the word "normal" in a carefully articulated manner that marks a decisive turn in our history. To understand the context in which this change took place, we need to recall that Galton's work found its broad scientific purpose in engagement with the evolutionary theory of his cousin Charles Darwin.[26] Indeed, the pellet-distributing device we have referred to was constructed as a geometrical way of asking and answering a question provoked by Darwin's work. In the preceding decades, Quetelet had applied statistics to anthropometry and the study of populations, and Broca had used statistical methods in his analysis of hereditary anthropological characters. But Galton was now using statistical thinking to understand the distributive patterns produced by heredity.[27] The problem was that, in principle, living organisms might be expected to become more and more different from each other over time through the workings of natural selection. The process hypothesized by Darwin might be expected to result in limitless dispersion as emergent new characters allowed certain individuals to become successful at random. But when species were observed in nature as a group, they displayed an orderly distribution of characters. How did it happen, despite the chance effects of the struggle for life, that a distribution corresponding to what Galton called the "mathematical law of deviation" could reappear in each generation of a given species considered according to the stronger or lesser presence of a given character?[28] Galton wanted to understand "what the laws of heredity must . . . be [in order] to enable successive generations to maintain statistical identity."[29] The device showed how, after the pellets had followed a wide range of paths as they passed through the ordered set of quincunxes on their way down through the upper section, a regular curve of distribution would appear at an intermediate level, then, after the same pellets had been allowed to fall through the quincunxes in the lower section, reappear at the bottom in a modified, but equally regular form. Each iteration of the dispersing operation produced a new distribution corresponding to the known law of deviation. For Galton, the conclusion was inescapable: "the processes of heredity must work harmoniously with the law of deviation, and be themselves in some sense conformable to it."[30]

In his essay *Natural Inheritance*, first published in 1889, twelve years after the lecture and the articles we have just been discussing, Galton incorporated "normal" into his statistical vocabulary, all the while pursuing his interest in the theory of heredity. One of his chapter titles in that essay, "Normal Variability," coupled two key terms.[31] This was not Lexis's corrective adjustment whereby "average life" became "normal life." It was an application of the concept of normality to an issue Galton considered to be of greater importance, that of distribution. The curves based on mathematical tables presented in his essay might "with propriety," he said, "be described as 'Normal.'"[32] Without fanfare, he was claiming no more than the terminological propriety of a capitalized adjective. But the word was proper in his view because it affirmed a single mathematical law of distribution applicable to any salient character across a group of subjects having the same type. What deserved to count as "normal" was presumably both the regular shape of the distributive pattern and the "universality," as he described it elsewhere.[33] So Galton was not just affirming the law-bound nature of Quetelet's "binomial" curve: he was laying down the basis for a mathematical study of distributive patterns. He was not concerned merely with the general facts of regular variation, but with "the measurement of variability."[34] By contrast with Quetelet, he was making deviation itself into an object of mathematical knowledge.[35]

The word "normal," which has served as an index of our intellectual history, was here being called to a more prominent role, one that subsumed a variety of significations that had been in play since the early nineteenth century. We have seen that "normal" was used earlier in the nineteenth century to signify geometrical and/or anatomical regularity. And we have also seen that, at much the same time, the notion of physiological regularity had underpinned a medical usage opposing the normal and the pathological. None of that history was foregrounded in Galton's writing, but it was broadly supposed by his theorizing of evolutionary heredity. His key move, summed up in his use of terms like "normal variability" and "normal curve," was to speak in the same phrase of the geometrical and the frequentive — of the geometry of nature conceived since about 1820 by medical thinkers like Étienne Geoffroy Saint-Hilaire and of the natural recurrence of phenomena that had since the late eighteenth century held the attention of mathematicians such as Laplace. In that double sense, the apparently modest use of these rather new noun phrases signified a natural/normal order describable by biology and a natural/normal order of recurrence calculable by statistics. Taken together, they allowed Galton to present a properly statistical theory of biological heredity.

The curve of which Galton spoke was recognizable first by the properties of its middle area. Where a group of men had been measured to determine their strength, it would always be possible to identify one who was "Middlemost." Such a man, said Galton, "is of mediocre strength." He went on to add: "the accepted term to express the value that occupies the Middlemost position is 'Median,' which may be used either as an adjective or as a substantive."[36] There was no celebration here of the middle position as such, simply a proliferation of words beginning with "m": the middlemost and the median were grouped together in the alliterative neighborhood of mediocrity. This was a far cry from Quetelet's definition of the average man as the "ideal type" of a population, where "ideal" referred to a mathematical abstraction, but also to a kind of perfection. That, from Galton's point of view, could only have amounted to an overvaluing of the average as such. He himself was content to identify the middling position elliptically and algebraically: "it will usually be replaced in this work by the abbreviated form M."[37] The normal curve could thus be defined in practice by the fact that all of these m-words signified much the same thing: "whenever the Scheme is symmetrically disposed on either side of M . . . then M is identical with the ordinary arithmetic Mean or Average. . . . The reader may look on the Median and on the Mean as being practically the same things, throughout this book."[38] So here was the normal curve, but it was not normality as a property of particular persons or creatures. For Galton, it was the curve itself that was normal, not those who stood close to M. "Normal" was in no sense an epithet to be transferred to particular members of the population being measured. Those in the middle were simply clustered in mediocrity.

For Galton, the pellet machine showed in a nonverbal way that there were necessarily two compensatory processes at work in heredity. In addition to natural selection, there had to exist a general tendency toward what he called reversion or regression. It was patent that heredity occurred within families, as characters were passed on from parents to children. But if that were the only process at work, natural selection would result in endless diversification as new generations encountered new challenges to survival and adapted to them: "If family variability had been the only process in simple descent, the dispersion of the race would indefinitely increase with the number of the generations."[39] For normal patterns to reappear in each generation like the "heaps" that reappeared at different levels in Galton's machine, the general tendency to dispersion had to be countered by a general tendency to reversion: "Reversion is the tendency of that ideal mean type to depart from the parent type, 'reverting' towards what may be roughly and perhaps fairly described as the average ancestral type."[40]

By conceiving of generational heredity and ancestral heredity as mutually compensatory tendencies, Galton was finding a way to conjugate evolutionary biology with an anthropology of type. Reversion, as he saw it, served to maintain racial types by effectively working against natural selection. It acted to bring any increase in the rate of dispersion—not dispersion itself, simply its rate of increase—to a "standstill."[41] "The same process continues in successive generations," said Galton, "until the step-by-step progress of dispersion has been overtaken and exactly checked by the growing antagonism of reversion."[42]

By identifying the relative newness of Galton's view of normal distribution, we can hope to recover the specificity of the cultural networks and configurations in which the statistical concept of the normal first developed, and to draw attention to the role and significance of anthropometry and anthropological theories of the type that underpinned this increasingly quantitative idea of the normal. Despite the fact that Lombroso, as we saw, was able to use the word "normal" in a casual way, taking for granted that its meaning would be generally understood, as late as 1909, Galton felt required to introduce this word formally to the readers of *Essays in Eugenics*. In this text he flagged normal as a "new word," and defined it as a specialist statistical term: an organizing principle of mathematical regularity that was used in the new phrases the "normal curve" and "normal distribution."[43] Outside of this specifically statistical context, Galton did not use the word, referring instead to the average, the typical, or the generic. He had entirely turned away from the midcentury use of the medical term "normal state."

This part of Galton's work did much to establish the statistical idea of normal distribution as an organizing principle for large data sets in a wide range of turn-of-the century institutions and discourses. Moreover, he introduced this concept to the general public through his tireless schedule of public lectures, publication of articles in popular magazines, and participation in international expositions. While Lombroso, too, certainly attempted to reach a broader public, most of the work analyzed in our genealogy so far was produced behind the closed doors of learned societies or in the pages of academic journals. Composite portraiture, although an important part of Galton's statistical research, represented a significant shift in the critical genealogy of the normal we have traced in the preceding chapters, in that it helped to move discussion about the normal into the public sphere. This shift was further enabled by the development of new visual and recording technologies, like the photographic camera. From the end of the nineteenth century, the normal would no longer be the subject of professional debate alone but would also begin to take its place as an object of popular display in the new visual and spectacular cultures that were emerging at the time. As late as the start of the twentieth century, however, the "normal" remained a term with which not even an educated reader could be presumed familiar. The statistical concept of the normal remained fragile and precarious; a "blurred, nervous configuration," as Allan Sekula has memorably described Galton's composite portraits.[44] We will read this blurriness here not simply as evidence of conceptual uncertainties in Galton's own methodologies or theorizations, but as a means of better understanding the changing technological and epistemological configurations in which the statistical concept of the normal emerged at the end of the nineteenth century.

VISUAL ANTHROPOMETRICS: COMPOSITE PHOTOGRAPHS AND THE STUDY OF THE TYPE

Whereas the pellet machine had served at one point to represent hereditary processes, composite photography remained for Galton an enduring preoccupation, serving as it did to build an anthropology while usually confirming and occasionally disconfirming the existence of types. Galton continued in fact to produce composite images until his death in 1911. His earliest experiments were undertaken using donated photographs of family portraits and professional men, such as railway engineers and physicians. From family portraits he carefully removed each of the heads, and combined the individual images to produce composites of the family group. He used the portraits on ancient coins to produce composite profiles of Roman emperors. He also worked in collaboration with a range of public institutions—including prisons, hospitals, and schools—to access or commission large photographic collections, from which he produced portraits of such types as thieves, phthisis patients, and Jewish boys. For Galton, each of these groups, like the members of one family or professional organization, could in principle be identified and categorized according to shared characteristics that followed identifiable patterns of biological inheritance. Like Lombroso, Galton was interested in the mental as well as the physical characteristics of each group, and believed there was a correspondence between the two that anthropometry could measure. Composite portraits were designed to demonstrate this correlation. In this, as Galton's protégé Karl Pearson acknowledged in his intellectual biography *The Life, Letters and Labours of Francis Galton*, he had, "like most men of his generation and probably like most of us today, consciously or unconsciously given weight to physiognomy."[45] Galton's belief that mental characteristics and capabilities were biologically inherited, and visible in facial features, may have been founded on the false science of physiognomy, Pearson argued, but he brought to his study mechanical and mathematical ingenuity that made a significant impact on nineteenth-century science: Galton "was really attempting to make a true science out of the study of physiognomy."[46]

Galton's first large experiment with composite portraiture was enabled by the donation of an extensive collection of identity photographs by the Director-General of Prisons, Edmund Du Cane. Like Lombroso, and influenced by his work in criminal anthropology, Galton was interested in examining the correspondence between cephalic indices and an inherited tendency toward different forms of criminality. In a paper delivered in 1878, Galton described himself as having embarked on composite photography "while endeavouring to elicit the principal criminal types by methods of optical superimposition of the portraits."[47] Du Cane's donation of these photographs demonstrates that an economy of scientific research was continuing to function in late nineteenth-century England as it did in France and Italy, where anthropologists like Broca and Lombroso had been offered sets of skulls from institutional sources. At the same time, it also marks an epochal and technological transformation of this history, in that it represented a shift in the objects of scientific collection and study, from physical specimens to photographic representations, from skulls to images of living heads. In addition to Galton's interest in criminal types, Galton could not fail to value Du Cane's collection of prison photographs for a number of practical and technical reasons. First, they provided him with a large collection of photographs at a time when such collections remained difficult to source. Galton would complain all his life about the difficulty of obtaining sufficient quantities of photographs, frequently noting that a lack of materials was hindering his progress: "I am sure that the method of composite portraiture opens a fertile field of research to ethnologists," he wrote in 1883 at the start of *Inquiries into Human Faculty*, "but I find it very difficult to do much single-handed, on account of the difficulty of obtaining the necessary materials."[48] He often called for amateur photographers to send him duplicate or unwanted prints. Several years before his death, he was still requesting that members of the general public send him their "waste photographic portraits," for the "research on resemblance" he was currently undertaking.[49] Second, the prison photographs were particularly useful because, as institutionally commissioned portraits, they were highly standardized in size and composition. This was crucial for the production of a clear composite: if the original images were not uniform in proportion and perspective, the composite would be both visually blurred and unhelpful to the business of statistics.

Galton undertook two further large institutional studies, in both cases commissioning the photographs himself. The first was a pictorial study of 400 patients undergoing treatment for phthisis, primarily at Guy's Hospital, but also at the Brompton and Victoria Park hospitals, with photographer F. A. Mahomed. The second was a smaller study of schoolboys at the Jews' Free School in London, with the photographer Joseph Jacobs. Galton's intention in producing these composites was to identify the key characteristics of each of these types—the criminal, the consumptive, and the Jew—and to demonstrate that inherited mental traits and predispositions could be correlated with particular physical characteristics. His composite portraits in this respect sought to combine photography with statistics. As Pearson succinctly described this part of Galton's work: "he accordingly sought to isolate types and to measure deviations from facial type in order to determine whether facial variations were correlated with mental variations."[50] Over the course of the nineteenth century, anthropological studies of types had focused increasingly on the measurement of bodies, and particularly on that of skulls and brains, advancing their scientific research through the analysis of such anthropometric data; however, it was Galton who moved a step beyond Quetelet by making statistical analysis the key interpretive framework by which types were to be identified, and who, in so doing, embedded normal distribution as the organizing principle within anthropometrics. A clear and distinct composite portrait meant that the anthropometric data of the individual subjects fell within the range of normal distribution: composite portraits were thus "a perfect test of truth in all statistical conclusions."[51]

The function of normal distribution was pivotal to Galton's conceptualization and study of the type, and composite portraits were valued because they provided an excellent test of whether a type had been correctly identified: "when the portraits in the series make a good and clear composite it shows that medium values are much more frequent than extreme values, and therefore that the series may be considered a generic one; otherwise it is certainly not generic."[52] While Galton was here using the term "generic" in a way that might have been drawn from the natural history tradition, he made it mean something quite specific by anchoring it in statistical practice, just as Quetelet had done with the notion of the type. Praising Quetelet as "the first to give [the word "typical"] a rigorous interpretation, and the first whose idea of a type lies at the basis of his statistical views,"[53] Galton explained that statistics could be used equally to detect the existence of a coherent type. As he wrote in *Generic Images*, "The word generic presupposes a genus, that is to say, a collection of individuals who have much in common, and among whom medium characteristics are very much more frequent than extreme ones. . . . No statistician dreams of combining objects into the same generic group that do not cluster towards a common centre, no more can we compose generic portraits out of heterogeneous elements, for if the attempt be made to do so the result is monstrous or meaningless."[54] Here again we see a concern that averages or composites might turn out to be meaningless computations, like building heights in an unregulated street. A fuzzy portrait represented a sort of statistical "monstrosity," in which no system of order or regularity could be detected. Only if the portrait had clear and boldly defined features could the image be properly understood as "generic" or "typical"; if it was too hazy for those features to be clearly seen, then its component elements could not be said to constitute an identifiable genus or type. Statistical analysis found its true purpose when a type was made to appear; likewise the appearance of a distinct portrait guaranteed that a type had been properly identified.

Galton's theory of normal distribution was thus developed in the context of his anthropological research on the type. We have seen that the concept and category of the type, having largely been taken for granted in the first part of the nineteenth century, had increasingly become a subject of debate in the final decades of that century: physical anthropologists disputed Lombroso's identification of the criminal as a type, for instance, by arguing that criminals did not constitute a race, because criminality was not a biologically inherited characteristic, and so could not be properly subjected to anthropometric study. By the end of the nineteenth century, as Nancy Stepan observes, the concept of the type was under further pressure as a result of the theories of evolution developed by Darwin. Where the type in physical and criminal anthropology was understood as the expression of "fixed and unchanging essences . . . whose differences were taken to be deep and socially significant," she argues, evolutionary theory saw biological inheritance as subject to "processes of continual change, in principle deeply opposed to typological or essentialist thinking."[55] As we have seen, Galton found a way through this theoretical difficulty by his double theory of dispersion and regression. Unlike French and Italian anthropologists of the preceding decades, Galton was attempting to understand the type in the context of evolutionary theory. Broca and his colleagues had resisted evolutionary theory for decades, preoccupied as they were with fixity. They were determined to study "race," and race was for them the bone-and-flesh incarnation of stability. Lombroso, for all his insistence on atavistic criminal types, was more or less committed to a theory of dangerous individuals as the product of aberrations in the evolutionary process itself. Galton's work, by contrast, served to shift the theoretical basis of anthropology toward evolutionary biology, leading to a form of anthropometrics that Pearson would later term biometrics. Biometrics took little or no interest in the desiccated mensuration of craniometry. Its practices of mensuration were increasingly brought to bear on living bodies and their faculties.

Galton's concept of the type, as developed through his practice of composite portraiture, reflected the increasingly important function that change over time played in the concept of the type, and the role that his developing theories of normal distribution played in this process. The new technology of photography was integral to this, just as the development of new mensurative technologies had been pivotal to the work of Galton's predecessors in anthropometrics. Galton's early adoption of photography as a new tool for the anthropological study of the type was far from an isolated event: throughout the 1870s and 1880, anthropological associations in the UK, United States, and Europe undertook large projects in which they sought to compile photographic collections of various types.[56] The Anthropometric and Racial Committee of the British Association for the Advancement of Science began a study in the 1870s designed "to compile data on the physical characteristics of human beings in the British Empire, and to publish photographs of the typical races to be found there," as Anne Maxwell notes.[57] However, the anticipated photographs proved as difficult for the committee to acquire as they did for Galton over the same period. As Maxwell observes, while "the committee's reports show that within five years some 24,000 physical observations ... had been collected," only 400 photographs were acquired during the same period, many of which "were purchased from professional photographers and were commercial in style."[58] While Galton's interest in photography was a part of this larger anthropological uptake of the new visual technology, his photographic project was of a markedly different order: composite portraits were designed to produce exemplary images that revealed typical faces, rather than individual portraits of faces offered as exemplary of known types. That is, his aim was to use photography in the service of anthropometrics, as "pictorial averages," rather than simply to record visual images of individuals who might have been perceived at a glance to be anthropologically typical.

Galton's use of photography is better understood if it is contextualized not simply within late nineteenth-century anthropological research but within both the developing visual cultures enabled by the mechanical production of images and the statistical and scientific research of his day. Because photographs were widely understood as unmediated by human hand or eye, both by scientists undertaking early experiments with the medium and by the general public, they were considered to provide objective and accurate images of the objects they depicted.[59] As David Green notes, photography was taken to be "independent of the prejudices and interests of the observer, and thus uncontaminated by the potential subjectivism of theory."[60] As such, photography not only fascinated the general public, it was also widely taken up as a new method of scientific research, because it satisfied an increasing demand for new "modes of empirical observation and documentation, and techniques of quantitative measurement."[61] Certainly, Galton understood photography to function in this way, claiming that his composite portraits were "independent of the fancy of the operator, just as numerical averages are."[62] It is precisely this quality that made photography so characteristic of, and such an important catalyst for, the emergence of objectivity as a principle and practice of scientific research in the final decades of the nineteenth century, as Daston and Galison argue in their history of scientific observation. Photographs, which "let the specimen appear without that distortion characteristic of the observer's personal tastes, commitments, or ambitions," served at that time to transform the nature of scientific illustration and its role in the production of scientific knowledge.[63]

Until the late nineteenth century, as Daston and Galison show, the purpose of scientific illustration was to represent "not the actual individual specimen . . . but an idealized, perfected, or at least characteristic exemplar of a species or other natural kind."[64] The production of these idealized typical images required the careful discernment and expertise of the trained professional, who was able to see past the appearance of individual variations to record images of the type itself in all its perfection. During the second half of the nineteenth century, however, this form of illustration slowly gave way to a new preference for mechanically produced and highly individualized images.[65] This represented a new "orientation away from the interpretive, intervening author-artist," which in turn shifted "attention to the reproduction of individual items—rather than types or ideals."[66] Composite portraiture represented the convergence and cohabitation of both of these traditions: on the one hand, the production of typical faces through the fabrication of "imaginary portraits"; on the other, images produced through the "mechanical precision" of the camera. As Galton writes: "A composite portrait represents the picture that would appear before the mind's eye of an individual who had the gift of pictorial imagination in an exalted degree. But the imaginative power of even the highest artists is far from precise, and is so apt to be biased by special cases that may have struck their fancies, that no two artists agree in any of their typical forms. The merit of the photographic composite is its mechanical precision, being subject to no errors beyond those incidental to all photographic reproductions."[67]

Galton thus used the new technology of photography in the service of an older way of seeing: his composite portraits were mechanically produced images that had been manipulated to create idealized portraits of a given type. Even in a period in which photography was still so new that all uses of it were effectively experimental, conventions for photographic portraits not yet being firmly established, Galton's composite portraits seemed strange and anachronistic to his contemporaries.[68] As Galton himself acknowledged, they were not popular with their subjects: "I have made several family portraits, which to my eye seem great successes, but must candidly own that the persons whose portraits are blended together seldom seem to care much for the result, except as a curiosity," he wrote. "We are all inclined to assert our individuality, and to stand on our own basis, and to object to being mixed up indiscriminately with others."[69] Photographic portraiture, in popular culture as in anthropological research and institutional administration, was already focused on portraits of individuals. Galton's project to produce composites of types was at odds with the developing photographic conventions of the period.

Like other experimental and innovative photographers on the time, Galton was not primarily focused on producing objective and accurate images of "the thing itself," but rather sought to use the camera as a scientific tool that allowed the researcher to see things not otherwise visible to the human eye. In this respect, composite portraits were like the photographs of his contemporaries Jules-Étienne Marey and Eadweard Muybridge, which revealed for the first time the frozen movement of a galloping horse's legs and bird wings in flight. Marey's photographs, which captured such movement as the motion of a man swinging an axe within a single frame, have many visual similarities to Galton's composites: their compression of a sequence of motions into one image produces the same blurred quality that is so striking in Galton's photographs. Marey, like Muybridge, is often understood to have represented the passage of time in new ways through photography, and in this respect again both have resonances with Galton's attempt to capture the inheritance of biological traits in his composite portraits.[70] What all three showed was that photography could be used not simply to record visual images in "mechanically precise" ways, but also as a new instrument for the measurement of bodies. We see this in Galton's insistence that composite portraits were statistical as well as visual documents: "the trustworthiness of the final result must be estimated on the same principle as if we had been dealing with numerical averages," he argued of his composites of phthisis patients.[71] Like photography and with it, statistics was seen to provide an objective and unmediated access to facts. Composite portraiture satisfied the need for empirical observation along with the need for quantitative measurement, and it is in this respect that it constituted a visual anthropometrics.

EUGENICS: "THE ACTUARIAL SIDE OF HEREDITY"

Where the physical and criminal anthropologists had concentrated their research on a limited range of racial or degenerate types, and an even more restricted range of physical and mental traits, Galton sought to extend dramatically the range of both what could be measured—widening the field of anthropometrics to incorporate the study of what he referred to as psychometrics, and Pearson would later influentially term biometrics—and of who should be measured. Galton's practices of mensuration were potentially applicable to all subjects, rather than just the racially exotic or socially problematic. He worked closely, for example, with schools to introduce testing for such faculties as sight and hearing. He also ran public "Anthropometric Laboratories," and encouraged members of the public to record family data in two workbooks published in 1884, the *Life History Album* (1884) and *Record of Family Faculties* (1884), even offering prizes for the most complete set of family records sent back to his laboratory. Such activities demonstrated that Galton intended anthropometric systems of measurement and management to be applied to all subjects. In this way, his work was designed to increase the application of anthropometrics exponentially, and Galton's ambitions in this regard far outstripped those of his predecessors. Anthropometrics, he argued, could and should be used not simply to measure and classify bodies, but as the basis of a new social policy designed to address "the problem of the future betterment of the human race."[72] Where previous anthropometricians had been content to measure and assess bodies, Galton sought to use the data he was gathering in a program of social engineering, which he termed eugenics. Pearson praised the boldness of Galton's vision in devising and proposing the application of eugenic policy, writing: "If it is given to few men to name a new branch of science and lay down the broad lines of its development, it is the lot of fewer still to forecast its future as a creed of social conduct."[73] Where earlier physical and criminal anthropologists argued that anthropometric study was of great potential value for diagnostic and/or classificatory purposes, Galton was the first to argue that anthropometrics could form the basis of an applied social policy. In this respect, Pearson argued, he was the first "social Darwinist."[74] Because "the whole course of modern social evolution has served to suspend the action of natural selection," it was imperative for science to intervene to ensure the health and future of the race, Pearson argued.[75] The mechanism by which Galton sought to determine the best means of correcting the effects of modern urban life, which tended to suspend the effects of natural selection, was to identify the normal distribution for each race or type through the practice of composite portraiture. It remained to be seen just how that statistical notion could find expression in social policy.

Galton coined the term eugenics in *Inquiries into Human Faculty and Its Development* (1883), noting that its purpose was to examine "the practicability of supplanting inefficient human stock by better strains, and to consider whether it might not be our duty to do so."[76] He provided a more detailed account in *Essays in Eugenics*, in which he explained: "The creed of eugenics is founded upon the idea of evolution; not on a passive form of it, but on one that can to some extent direct its own course. Purely passive, or what may be styled mechanical evolution, displays the awe-inspiring spectacle of a vast eddy of organic turmoil. . . . But it is moulded by blind and wasteful processes, namely, by an extravagant production of raw material and the ruthless rejection of all that is superfluous."[77] Where evolution and natural selection were tumultuous and unruly processes, characterized by extravagant waste, the benevolent and improving hands of science could bring order and efficiency to racial development, ensuring its progress and betterment: "Eugenics co-operates with the workings of Nature by securing that humanity shall be represented by the fittest races," Galton argued. "What Nature does blindly, slowly, and ruthlessly, man may do providently, quickly, and kindly."[78] Nature could be aided and ameliorated if racial characters were clearly identified and organized through the practice of anthropometrical statistics: "the statistical effects [of the laws of biological inheritance] are no longer vague, for they are measured and expressed in formulae," Galton argued in *Essays in Eugenics*.[79] Quetelet and others, including more recently Lexis, had deployed statistics in the tabulation and analysis of such things as death rates, and Galton's own mathematical work on heredity was adduced to support the claim that heredity "now admits of exact definition and of being treated mathematically, like birth and death rates, and other topics with which actuaries are concerned."[80] Basing eugenics in quantified forms of knowledge appeared to ensure both scientific precision and political neutrality: "Eugenics seeks quantitative results. It is not contented with such vague words as 'much' or 'little,' but endeavors to determine 'how much' or 'how little' in precise and trustworthy figures."[81] Much as P. C. A. Louis had done when defending numerical method seventy years earlier in the French Académie de Médecine, Galton was seeking to proscribe from scientific language mundane adverbs of degree. Statistics, Galton was claiming, would support the scientific rationale of eugenics, just as the mechanical precision of the camera had supported and guaranteed the objectivity of his composite portraits.

So eugenics should be thought of as a consequence — if not exactly a strict logical outcome — of Galton's attempts to apply statistical analysis to the study of biological inheritance. Central to this theorization of eugenics, and underpinning its claims to scientific objectivity and accuracy, was the law of normal distribution, again understood to be made visible through composite portraiture:

> It is the essential notion of a race that there should be some ideal typical form from which the individuals may deviate in all directions, but about which they chiefly cluster, and towards which their descendants will continue to cluster. The easiest direction in which a race can be improved is towards that central type, because nothing new has to be sought out. It is only necessary to encourage as far as practicable the breed of those who conform most nearly to the central type, and to restrain as far as may be the breed of those who deviate widely from it. Now there can hardly be a more appropriate method of discovering the central physiognomical type of any race or group than that of composite portraiture.[82]

Here we see how the law of normal distribution was used by Galton to form the statistical basis of eugenic theory. The normal curve and the normal law of frequency enabled him to calculate the patterns of biological inheritance as well as the hereditary basis of various traits. The law of normal distribution, which decreed that variance would inevitably regress toward the mean, could thus be used to predict directions of biological evolution: the "filial center falls back further towards mediocrity in a constant proportion to the distance to which the parental center has deviated from it, whether the direction of the deviation be in excess or in deficiency."[83] In this way, Galton used a mathematical law as the basis of a proposed social law: mathematics, like evolution, was understood as a readily available intellectual practice that the benevolent and rational hands of science could use to improve the race. Eugenics as it was initially conceived by Galton could thus be conceived in quite narrowly mathematical terms as a policy of normalization. Its concern, in principle at least, was not with disciplining dangerous individuals, but with maintaining normal patterns of distribution across a population even while seeking to inflect them and ameliorate their effects.

However, Galton's own account of this process, in which normal distribution played a central role, problematized a number of his own key assertions. The function of normal distribution, as Galton himself acknowledged, meant that the "unfit" would not in fact continue to become "increasingly inferior." Considered statistically, unfitness and inferiority were to some degree self-correcting: "The law of 'regression towards mediocrity' insures that their offspring, as a whole, will be superior to themselves."[84] Galton's repeated identification of this law as one of "regression towards mediocrity" is a further reminder of how slowly and how late "normal distribution" took its place as the preferred term for such a law. Furthermore, it reveals a complication at the heart of his statistically based theory of heredity. For Galton, the eugenical improvement of a race or type was both a naturally occurring dynamic, inscribed in the operation of "passive or mechanical evolution," and one that could and should be harnessed by scientists to increase the rate of progress and racial improvement with greater efficiency. "The most merciful form of what I ventured to call 'eugenics,'" wrote Galton, "would consist in watching for the indications of superior strains or races, and in so favoring them that their progeny shall outnumber and gradually replace that of the old one."[85] This was like Darwin's account of artificial selection applied to the cultivation of domestic animals. It was in fact a cultivation of race. And it would later become known as positive eugenics. Galton equivocated in his account of how eugenicist processes could be expected to work. "The aim of eugenics," he wrote, "is to represent each class or sect by its best specimens; that done, to leave them to work out their common civilisation in their own way."[86] But what exactly was the relation between Darwinian survival of the fittest and the notion of a common civilization? The gap between the two could be closed only by action of a kind Galton was disinclined to consider. As Pearson noted, Galton's own ambitions were largely restricted to the application of positive eugenics: "Galton would have been content to grade physically and mentally mankind, and to have urged that marriage within your own grade was a religious duty for those of high grades and castes."[87]

Galton was not only the first person to develop a properly statistical theory of the normal, then, but also the first to suggest that it be applied as a practice of social and biological normalization—as distinct from biological normativity, which, as we have seen, was understood as a natural function of biological development. That this occurs only at the start of the twentieth century makes the history of normalization itself much shorter than is often assumed to be the case. Foucault identified normalization as the key dynamic of the disciplinary society, and dated its emergence to the end of the eighteenth century. However, while standardization and quantification had become increasingly important cultural dynamics in institutional and governmental practice over the course of the nineteenth century, it is only here, in the context of Galton's theorization of eugenics, that the idea of normalization itself—that is, the application of the idea of the normal as a cultural practice—emerged. Such a practice was needed because the conditions of modern urban life had disturbed natural evolutionary processes to the extent that the British race was now degenerating, and the unfittest of its classes were now proliferating at a dangerous rate, Galton argued. In a letter to the *Times* designed to promote his ideas and the recent publication of *Essays in Eugenics,* Galton asserted that his anthropometric studies had scientifically established that "the bulk of the community is deteriorating, judging from results of inquiries into the teeth, hearing, eyesight, and malformations of children in Board schools, and from the apparently continuous increase of insanity and feeble-mindedness."[88] In *Essays in Eugenics*, he proposed a number of solutions for this. The first was that it would be in the best interests of the nation if the criminal and other undesirable classes "were resolutely segregated under merciful surveillance and peremptorily denied opportunities for producing offspring. It would abolish a source of suffering and misery to a future generation, and would cause no unwarrantable hardship."[89] Such a proposal should not be confused with that of biological extermination, he claimed: "There exists a sentiment, for the most part quite unreasonable, against the gradual extinction of an inferior race," he wrote. "It rests on some confusion between the race and the individual, as if the destruction of a race was equivalent to the destruction of a large number of men. It is nothing of the kind when the process of extinction works silently and slowly through the earlier marriage of members of the superior race."[90]

Despite Galton's assertions, eugenics was nonetheless immediately associated with the extinction of particular, living individuals and the destruction of large numbers of people. This is most evident in the work of Karl Pearson, who had founded the journal *Biometrika* with Galton and with the evolutionary biologist Walter Frank Weldon in 1901, and was appointed first professor of eugenics at University College London in 1911, a position funded by Galton. Even before the publication of *Essays in Eugenics*, Pearson had made his own position quite clear on this front: "My view — and I think it may be called the scientific view of a nation — is that of an organized whole, kept up to a high pitch of internal efficiency by insuring that its numbers are substantially recruited from the better stocks, and kept up to a high pitch of external efficiency by contest, chiefly by way of war with inferior races. . . . History shows me one way, and one way only, in which a high state of civilization has been produced, namely, the struggle of race with race, and the survival of the physically and mentally fitter race. If you want to know whether the lower races of man can evolve a higher type, I fear the only course is to leave them to fight it out among themselves."[91] Statements like this, which chillingly foreshadow the later application of eugenics as a policy in Germany under Nazism, have had a significant influence on the way this pivotal moment in the history of the normal has been understood. They foreshadow the violent applications to which normalization has been put.[92]

Whereas Gobineau's militant racism, for instance, was sharply criticized and kept at a distance by the Société d'Anthropologie de Paris under Broca's leadership, Galton's eugenics cannot be so easily quarantined from Pearson's politics of aggressive racial conflict. Pearson, Galton's greatest advocate, was the most vocal proponent of eugenics in Britain at the turn of the century, and his work would have a great influence on the American Charles Davenport, the founder of the Eugenics Record Office in Cold Spring Harbor, New York, in 1910. Davenport cultivated the American eugenicist movement that in turn fostered German eugenics in the 1920s and '30s. Given this history, it is easy to interpret Galton's understanding of racial normality in the light of subsequent practices of oppression. Galton did indeed advocate normalization through eugenics, and his theory of biological normality lent authority to that view. At this point, an explicit theory of normality came for the first time to be conjugated with an idea of social normalization. Many scholars in queer, race, and disability studies might say that this is — at last — where our long genealogy allows the normal to be revealed in its true colors, as the ground for a politics of dominance. But our history shows that the colors of normality are in fact rather mobile and sometimes quite blurred. For this reason, it is important to recognize the specific way in which eugenics was received in Britain at the turn of the century, and to acknowledge what happened then as a distinct moment in the history of eugenics. By localizing this moment, we can better understand its place within the long history of the normal. As Daniel Kevles notes in *In the Name of Eugenics: Genetics and the Uses of Human Heredity*, eugenics was far from persuasive in Britain in the early 1900s: the members of the Eugenics Education Society (renamed the Eugenics Society in 1925) numbered only in the hundreds, it did not enjoy broad institutional support, and few of its policies were implemented.[93] Nancy Stepan observes in turn that the public and professional reception in Britain was highly critical: "From the beginning, its critics labeled its claims as at best naive, at worst exaggerated and dangerous."[94] Moreover, the repeated emphasis on "efficiency" and "racial betterment" found in both Galton's and Pearson's early accounts of eugenics signaled its increasing divergence from the law of normal distribution itself. What was at stake in eugenics was not normalization but optimization, in its British and American incarnation, and purification, in its German incarnation. In turn-of-the century Britain, Galton's ideas had very little traction, and that itself has much to tell us about the precarious and still-formative concept of the normal emergent at this time.

THE RECEPTION OF EUGENICS

For Galton, eugenics was so demonstrably reasonable and beneficial that the general public would inevitably come to recognize its rightness and embrace its application: "when the subject of Eugenics shall be well understood," he asserted, "and when its lofty objects shall have become generally appreciated, they will meet with some recognition both from the religious sense of the people and from its laws."[95] The subsequent history of eugenics—from its application as a program of biological normalization in the 1920s United States to that of biological extinction in Nazi Germany in the 1930s—suggests that there was in eugenicist thinking a logic that led almost inevitably to the emergence of a disciplinary technology, one that would, so to speak, impose a governmental practice of normality in the first decades of the twentieth century. And yet, from the beginning of the century, proposals to apply eugenics were met with widespread skepticism. We can see this in the transcript of Galton's 1904 address to the Sociological Society, which includes over a dozen formal responses made by prominent scientists, public intellectuals, and writers such as H. G. Wells and George Bernard Shaw, and thus provides a detailed record of the reception of eugenics in Galton's own day, much in the manner of the nineteenth-century debates we analyzed earlier. Shaw revealed himself to be a keen supporter of Galton's views: "Nothing but a eugenic religion can save our civilization," Shaw proclaimed. "What we must fight for is freedom to breed the race without being hampered by the mass of irrelevant conditions."[96] But Shaw was in the minority among Galton's contemporaries, in that he foresaw no significant resistance to Galton's plans for marriage to be restricted on eugenic grounds: "men and women are amazingly indiscriminate and promiscuous in their attachments," he asserted, and would prove not to be "particular as to whom they marry, provided they do not lose caste by the alliance."[97]

Most respondents to Galton's address, however, ranged between the cautiously reserved and the robustly critical. One biologist dismissed the central claim of Galton's eugenic theory—that both moral and mental attributes were biologically inherited—as highly tendentious. Even Weldon celebrated Galton's biological theories only for the "triumph of statistics" they represented, sidestepping the question of eugenics altogether.[98] H. G. Wells was more openly critical. He questioned Galton's assumption that the criminal classes lacked mental acuity; not all criminals were slow-witted or "feeble-minded," Wells argued. On the contrary, "many eminent criminals appear to me to be persons superior in many respects—intelligence, initiative, originality—to the average judge."[99] Such statements seemed designed to infuriate Galton, not least because the reference to judges seemed to belittle his own choice of judges as a standard of intellectual ability.[100] Galton's closing remarks were curt and impatient: he declared himself "extremely unhappy" with the quality of the debate,[101] and concluded that the Society would have to make a much better effort if it was to contribute to the public acceptance and social application of eugenics.[102] When he included the transcript of his lecture in *Essays in Eugenics* five years later, Galton excised the entire discussion that had followed.

The view of these early critics foreshadowed that of the general public in Britain, which showed very little interest in eugenics. Membership of the British Eugenics Society over the entire course of its operation "never exceeded seventeen hundred," notes Kevles,[103] and was mostly limited to upper-middle class progressives and intellectuals.[104] In his own history of eugenics, Galton acknowledged that scientific interest had been slow to build: "The direct pursuit of studies in Eugenics . . . did not at first attract investigators. The idea of effecting an improvement in that direction was too much in advance of the march of popular imagination, so I had to wait."[105] Yet eugenics did not die a quiet death, even in Britain. In the twilight of Galton's career, it seemed suddenly to reignite as a subject of institutional interest and public debate. As Galton told it, he had "kindled the feeble flame that struggled doubtfully for a time until it caught hold of adjacent stores of suitable material, and became a brisk fire, burning freely by itself."[106] In 1901, the launch of *Biometrika* under Pearson's leadership appeared to mark a new collaborative phase with, Galton and Weldon as co-editors.[107] By 1904, Galton felt that eugenics was an idea whose "time [was] ripe."[108] He endowed a Research Fellowship in Eugenics at the University of London, so that eugenics could gain "the recognition of its importance by the University of London and a home for its study in University College."[109] By all appearances, it continued to gain academic and institutional standing, as Karl Pearson was invited to deliver the Boyle Lecture at Oxford University in 1907, and chose to give it on eugenics; two weeks later Galton followed suit in his Herbert Spencer Lecture at the University College of London. In the same year he released a new edition of *Inquiries into Human Faculty and Its Development*, and the Eugenics Education Society was founded. In 1909, he was knighted. When his collection *Essays in Eugenics* was published shortly afterward, the author's name was preceded by his new title.

While Galton certainly did attain a level of prominence and respectability for his work, then, eugenics itself did not become a mainstream part of the science of his day, nor were there any systematic attempts to apply eugenicist policies in any institutional or governmental context in the first decade of the twentieth century. The history of eugenics rather traces its increasing divergence from that of the normal and normalization: American eugenicists almost never referred to the normal, and neither did the later German eugenicists. Nonetheless, it is at this historical moment that the possibility, and then the application, of normalization entered into cultural life and institutional practice. In order to recover the specificity of this moment, and the processes by which this took place, it is necessary to recognize that neither the idea of the normal nor practices of normalization were dominant or persuasive at their point of emergence, but rather often failed in the applications for which Galton intended them. This process was exemplified by the fate of his composite portraits.

COMPOSITE PORTRAITS: THE PRODUCTIVE FAILURE OF A SCIENTIFIC EXPERIMENT

Although composite portraiture was integral to the development of Galton's theory of the normal as a statistical concept, and although he extolled its benefits for scientific and eugenic research in a series of public lectures and articles written for both professional and popular audiences, his enthusiasm was not widely shared. Composite portraiture found very little uptake or practical application in the scientific communities, public institutions, or popular cultures of Galton's time. The American Henry Pickering Bowditch, a contemporary of Galton and dean of the Harvard Medical School, made a number of composite portraits of American soldiers and of professional gentlemen. He introduced the subject to American audiences in an article "Are Composite Photographs Typical Pictures?" published in the popular magazine *McClure's* in 1894. His composite portraits were exhibited after his death at the Second International Congress of Eugenics in New York in 1921. The British anthropologist Flinders Petrie and the Italian criminologist Lombroso also regarded them as scientifically valuable.[110] There were occasional attempts to apply the form to other cultural fields, such as Walter Rogers Furness's attempt to apply composite portraiture to etchings and busts of Shakespeare.[111] However, none of the large institutions with which Galton collaborated to collect source photographs adopted composite portraiture as part of their own administrative practices. Some portraits found further application in physical and criminal anthropological studies of the type, but in a very limited and largely unsuccessful way. Thus, although cultural historians such as Josh Ellenbogen have recently claimed that composite portraiture was "making inroads into France" in the mid 1880s,[112] and the art theorist Allan Sekula asserts that its use "proliferated widely over the following three decades," enjoying "a wide prestige until about 1915,"[113] there is little evidence that composite portraits had a significant effect on statistical research or the wider scientific or popular cultures of this time. Given the absence of any significant application or uptake of composite portraits, we might question their status as scientific documents, and ask whether they deserve to be taken in the way we are proposing as a kind of epistemic emblem for the emergence of the normal in Galton's work and a gauge of how that work was disseminated in the wider scientific circles and public spheres of which it was a part. As we have seen, however, changing concepts of the normal often took shape within the context of scientific debate and experimental failure or dead ends. Composite portraits, too, precisely because they remained so marginal to the culture in which they were produced, have much to tell us about the messy genealogies and the precarious cultural formations in which the statistical idea of the normal emerged at this time. If the project of composite portraiture and the idea of the typical face as determined by normal distribution in which it was based proved to be somewhat out of step with contemporaneous developments in the scientific cultures, institutional practices, and photographic conventions of its time, this itself indicates that the statistical concept was not borne along by a great sweep of historical change.

This is evident in Galton's first series of composites, which focused on criminal types. Galton's study of criminal types had something in common with that of Lombroso, and specifically Lombroso's theory that criminality was the result of innate biological characteristics evident in anatomical and physiognomical traits. In criminal anthropology, measurements of cranial asymmetry and irregular cephalic indices were taken as indicative of intellectual, physiological, and moral atavism. But Galton's understanding of criminal heredity deserves to be considered more nuanced and more consistent than Lombroso's. Galton was indeed committed to the notion of a criminal type, and was seeking to use composite photography to make that type visible, but he spoke of heredity in rather different terms. Reflecting on his composite portraits of criminals, he wrote: "it is unhappily a fact that fairly distinct types of criminals breeding true to their kind have become established and are one of the saddest disfigurements of modern civilization."[114] For Galton, criminals were in fact "breeding true to their kind," whereas for Lombroso that was not necessarily the case. Lombroso did speak at times of criminality as endemic in certain places, certain races, certain families. At others, he spoke of it as a remarkable throwback to a distant past, an aberration in the process of evolution. Galton saw the difficulty of speaking at the same time in sociological and biological terms. He talked about "the criminal class" in a way that was foreign to Lombroso, who tended to resist sociological accounts of criminality. But at the same time Galton did speak of "a type of humanity" in a manner that Lombroso could only have found congenial:

> The perpetuation of the criminal class by heredity is a question difficult to grapple with on many accounts. Their vagrant habits, their illegitimate unions, and extreme untruthfulness, are among the difficulties of the investigation. It is, however, easy to show that the criminal nature tends to be inherited. . . . The true state of the case appears to be that the criminal population receives steady accessions from those who, without having strongly marked criminal natures, do nevertheless belong to a type of humanity that is exceedingly ill suited to play a respectable part in our modern civilization, although it is well suited to flourish under half-savage conditions, being naturally both healthy and prolific.[115]

It was clear enough in any case—to both Galton and Lombroso—that composite portraits had something to offer anthropologists who were seeking to identify the criminal type.

Galton's composite portraits of criminals were among his first experiments with this form, and criminal anthropology was one of the few fields in which composite portraiture found any subsequent uptake. Havelock Ellis's *The Criminal* (1890) included several composite portraits provided by the physician of the Elmira Reformatory, Dr. Hamilton Wey (see figure 2). *The Criminal* was designed, Ellis wrote, to introduce English audiences to "a science now commonly called criminal anthropology,"[116] whose key texts—including those of Lombroso—were not yet available in English. For Ellis, as for the criminal anthropologists examined above, craniometry was the most reliable means by which to identify and study criminal types, and this was the basis for his interest in composite portraiture. Despite his evident admiration for Galton's work, however, Ellis was forced to acknowledge that the composite portraits he had examined had not provided any useful information, had proven no correspondence between craniometry and criminality, and had, if anything, undermined the theoretical basis of criminal anthropology: "the average size of criminals' heads is about the same size as ordinary people's heads," Ellis conceded.[117] The composite portraits, along with the anthropometric data, suggested no particular statistical regularity among the individuals included in the composite: "Nothing very definite can be said of the cephalic indices."[118] Once this had been noted, however, the remainder of his book proceeded as though these cautionary remarks had never been made, continuing to assume a relationship between head shape and atavistic tendencies toward crime. We see a similarly inconsequential acknowledgment of the failure of composite portraits to produce their expected results in Galton's (favorable) review of *The Criminal*: "Although numerous dissections and measurements have led to no well-established important fact, they have, however, narrowed the field within which speculation may legitimately ramble."[119] Galton did not explain the curious logic by which the failure of craniometry to produce useful empirical results in criminal anthropology constituted a definition of the proper field of scientific inquiry into criminality. He similarly acknowledged that in the composite portraits included in *The Criminal* "the outlines of the heads are very hazy, testifying to large and various differences in the component portraits" that together "show no prevalence for any special deformity in head or features,"[120] without reflecting on how this might problematize his own belief in a measurable correspondence between head shape and criminal type. Just as Ellis described Franz Gall, the great champion of phrenology, as a man "popularly known chiefly for his mistakes,"[121] the history of composite portraiture is in many ways the history of its failure to find the scientific application Galton intended for it.

One of the most significant of these was Galton's attempt to fabricate images of types so exemplary they did not exist in actuality, which left him profoundly at odds with the broad historical shift in scientific illustration toward individual images "untouched by human hands" noted by Daston and Galison. We can see how that contributed to the failure of composite portraiture to find an application as a tool for scientific research at the time if we compare Galton's experiments with composite portraits to the anthropometric photography of one of his contemporaries: Alphonse Bertillon, the son of Louis-Adolphe and the brother of Jacques, both of whom figure in our history. Alphonse Bertillon was the head of the Service d'Identité Judiciaire at the Paris Prefecture of Police and was responsible for the management and organization of the vast quantities of data being accumulated by the prison system in the late nineteenth century. He created the identity files for anyone arrested in the greater Paris area and developed the first filing system for the rapidly expanding archive of police records. As Josh Ellenbogen recognizes, there were many points of correspondence between Bertillon's and Galton's anthropometric projects, including the key role that photography played in both:

> Galton's work in photography took place during almost exactly the same span as Bertillon's. His first experiments in the medium date from 1877, just two years before Bertillon began work at the prefecture, and occupied him until his death in 1911, three years before Bertillon died. Although Bertillon was French and Galton English, the two operated in the same photographic milieu, deploying the technology in relation to identity, statistics, and criminology . . . Galton's first published references to Bertillon date to 1884, at just the moment Galtonian photography began making inroads into France. On a lark, Galton posed for Bertillon when he visited him in 1893, receiving a mock criminal identity card as a souvenir.[122]

Despite this collegiality, the systems of anthropometrics developed by the two men were in many ways in competition, and Galton wrote a rather acerbic review of Bertillon's *Identification anthropométrique: Instructions signalétiques* when it was published in English translation.[123] As Allan Sekula recognized, "the projects of Bertillon and Galton constitute two methodological poles of the positivist attempts to define and regulate social deviance. Bertillon sought to individuate. His aims were practical and operational, a response to the demands of urban police work . . . Galton sought to visualize the generic evidence of hereditarian laws. His aims were theoretical, the result of eclectic but ultimately single-minded curiosities of one of the last Victorian gentlemen-amateur scientists."[124] Where Galton was a "compulsive quantifier," Bertillon was a "compulsive systematizer."[125] Where Galton attempted to reduce the archive to "a single potent image,"[126] Bertillon "invented a classifying scheme that was based less on a taxonomic categorization of types than upon an ordering of individual cases with a segmented aggregate."[127] What we see in Galton and Bertillon are two different kinds of anthropometric vision, two different ways of collecting and analyzing anthropometric data: one focused on the type, the other on the differentiated individual. Of these two approaches, Bertillon's system of individuation was incomparably the more influential. His system of anthropometrics would be used as part of the training of all incoming members of the French police force in his day.[128] Galton's photographic studies of types, on the other hand, would leave little trace in the institutional culture or scientific practice of his time.[129] Images of "ideal" and "imaginary" typical faces were of little use to large public institutions focused on the identification and management of individuals.

To judge the success or failure of composite portraiture by the extent of its institutional uptake is, however, to judge it by a standard to which Galton was neither working nor accountable. Unlike Bertillon, Galton worked in private research, free from the practical concerns and daily administration of a large institution like the prison. But here again composite portraits served rather to problematize than to prove the theories about physiognomy and heredity Galton intended them to demonstrate. According to Galton's increasingly eugenics-influenced theories of the type, some types were inherently and essentially superior to others. Criminal types, for instance, were "perhaps incapable of improvement."[130] Galton expected that, represented in composite form, the visual evidence of this criminality and its correlation to particular physiognomies would come into focus. Once again, however, the results of Galton's composite portraits ran counter to those he had anticipated: "It will be observed that the features of the composites are much better looking than those of the components. The special villainous irregularities in the latter have disappeared and the common humanity that underlies them has prevailed. . . . All the composites are better looking than their components because the averaged portrait of many persons is free from the irregularities that variously blemish the looks of each of them."[131] Galton had similar results with the composite portraits of Jewish boys photographed at the Jews' Free School: "They were children of poor parents, dirty little fellows individually, but wonderfully beautiful, I think, in these composites."[132] In concluding that composite portraits proved that the exemplary form of any type would appear the most beautiful, or perfect, of its kind, Galton found himself reiterating a principle that had been affirmed in the Société d'Anthropologie de Paris in discussions of race, one that could be traced back at least to Cuvier in the early nineteenth century, and to Quetelet. The type was beautiful precisely because it was uncluttered by accidents of variation.

Galton thus found himself conceding that his composite portraits of criminals were "interesting negatively rather than positively" because "they reproduce faces of a mean description, with no villainy written on them. The individual faces are villainous enough, but they are villainous in different ways, and when they are combined, the individual peculiarities disappear."[133] The purpose of these composite portraits, it should be recalled, was precisely the opposite: to make visible the most characteristic faces of criminal types, and to show how their physiognomy could be used diagnostically as a principle. The disappearance of all signs of villainy in the composite portrait suggested, according to Galton's own theories, that "villainy" was not a shared physiognomic characteristic of the series of individual portraits from which the composite had been compiled. Yet Galton never questioned whether villainy was a physical characteristic of the criminal type, even though, as we saw above, one of the key functions of composite portraiture was precisely to serve as an empirical test of whether a type had been correctly identified. If Galton consistently disregarded the visual evidence provided by his composite portraits it was, ironically, because he remained so convinced that their "mechanical precision" guaranteed their accuracy and objectivity.

We can see this further in Galton's study of phthisis patients in London hospitals. The findings of this research, based on the extensive collection of patient photographs commissioned by Galton, were published in an essay entitled "An Inquiry into the Physiognomy of Phthisis by the Method of 'Composite Portraiture.'" Galton began by addressing the widespread assumption that a predisposition to disease was a biologically inherited trait: "the belief that certain physical conformations indicate predispositions to certain diseases has always held so prominent a place in medicine from the earliest ages that it is unnecessary to dwell upon its history or its present position at any length."[134] Where earlier physicians may have been wrong to attribute such a predisposition to "the utterly false and erroneous doctrine of 'humeurs,'" Galton noted, their observations of a correlation between physiognomy and susceptibility to particular diseases were "probably correct enough."[135] The stated purpose of this study was to use composite portraiture to test the belief in physiognomy that still underpinned much medical research, Galton argued, and "to ascertain whether there are any facial characteristics in common to any large proportion of cases of phthisis."[136] Galton clearly believed that there were, and proved highly resistant in practice to evidence that contradicted the view he held at the outset. He did acknowledge that when he assessed the individual portraits of phthisis patients, he was "first struck by their diversity" and "the absence of those characteristic faces which we expected to find among them."[137] This level of variation should have suggested to Galton that the individuals in his collection might not constitute a unified type, and he recognized that this diversity did indeed seem to suggest "a distinct answer in the negative to the question, Is there a tubercular diathesis?"[138] However, he noted: "after much sorting and arranging into groups . . . certain results began to unfold themselves."[139] Although those results were not especially evident in the composite portraits, as even Galton's strongest supporters were forced to admit, nonetheless Galton felt confident to conclude that the composite images did demonstrate "a strong hereditary tendency to the disease."[140] According to Galton, the composites showed that phthisis patients could be divided into two distinct types. These corresponded to the types physicians referred to as "stumous" and "tubercular."[141] That these results held only when he selectively removed faces that were physically dissimilar did not trouble Galton. Where in principle Galton saw the empirical testing of hypothetical types as one of the most important functions of composite portraiture, then, in practice he often simply disregarded that evidence, concluding that a particular composite portrait had been unsuccessful or "imperfectly realized."[142]

THE LEGACY OF COMPOSITE PORTRAITURE: FROM THE DISCIPLINARY TO THE DATA SOCIETY

Despite the many failures of Galton's theories to find an institutional application or widespread support, his composite portraits, and the theory of normal distribution he elaborated with their support, did have a significant cultural impact. As we have seen in our study of Lombroso's criminal anthropology, they were failures (in the sense that the initiator did not achieve the application intended for them, and they disproved the theories they were designed to demonstrate) that were nonetheless highly productive. To understand this, we need to turn away from the scientific or intellectual context in which Galton's work was researched and written in order to consider the wider context in which it was produced and circulated. Composite portraiture was again central to this process. As we have seen, sourcing sufficient quantities of appropriate photographs remained difficult in the 1870s and 1880s, and Galton often complained that the difficulty of finding sufficient source images was impeding his research. While Galton was an ingenious inventor of new technologies himself, developing a number of apparatuses for the production of composite images (some of which predated his eventual preferred use of the photographic camera), he evidently had little interest in taking original photographs. As a result, he remained dependent on donations of photographs from both individual and institutional photographers throughout his life. In consequence, he was obliged to work in collaboration with large institutions such as prisons, hospitals, and schools, as well as soliciting photographic contributions to his research from members of the general public. To overcome the material difficulty of acquiring sufficient photographs, Galton needed to persuade institutions and individuals alike of the value of his research, and to encourage their voluntary involvement.

Galton used three key methods to encourage public participation in his various photographic composite projects. First, he produced the two practical manuals designed to encourage and guide the collection of family anthropometrics mentioned above: the *Life History Album* and *Record of Family Faculties*. The purpose of publishing these books, Galton explained, was to "further the accumulation of materials for life histories in the form of adequate photographs, anthropometric measurements, and medical facts."[143] The first was intended, as Anne Maxwell notes, to provide "both a guide for those wishing to create a photographic and written record of their family's health, and a template to be filled by those who ordered one from him."[144] The second was launched, as Frans Lundgren writes: "with a public competition featuring large cash prizes for the most complete copies submitted, in an effort to stimulate 'a custom of keeping family records.'"[145] Lundgren notes that despite their similarities, the two workbooks had a slightly different purpose: where the *Life History Album* was designed to provide families with a "detailed biographic register," the *Record of Family Faculties* was "a tool to accumulate more abstract accounts of entire families."[146] Second, Galton wrote proposals calling for institutions such as schools to undertake large-scale collections of anthropometric data. His "Proposal to Apply for Anthropological Statistics from Schools," published in 1874, outlined a plan for the mass collection of data—primarily regarding height and weight—among what he saw as the fairly homogenous populations of schools. Galton was successful in soliciting the support of a number of schools, including "country" schools, such as Marlborough, Clifton, Wellington, and Eton, as well as "town" schools, such as City of London School, Christ's Hospital, King Edward's School in Birmingham, and Liverpool College.[147] "I do not see why it should be either difficult or costly to schools of the upper and middle classes," Galton claimed, "to institute periodical measurements even of a somewhat elaborate character under skillful itinerant supervision, and to register them in a methodical and uniform manner. It should, I think, become a recognized part of the school discipline to have this done."[148] Third, Galton encouraged members of the public to see the collection of their own anthropometric data and its statistical analysis and interpretation as a social responsibility as well as a contribution to scientific research: "No doubt it would be contrary to the inclinations of most people to take much trouble of the kind about themselves," he acknowledges, "but I would urge them do so for their children so far as they have opportunities, and to establish a family register for the purpose."[149]

Along with the *Life History Album* and *Record of Family Faculties*, which guided members of the public through the process of data collection, Galton advocated the establishment of public anthropometrics laboratories. In "Why Do We Measure Mankind?" he wondered: "When shall we have anthropometric laboratories, where a man may from time to time get himself and his children weighed, measured and rightly photographed, and have each of their bodily faculties tested, by the best methods known to modern science?"[150] Galton sought to demonstrate the potential benefits of such a laboratory by establishing a temporary version during the International Health Exhibition in London in 1884. This laboratory was a great success, with over 9,000 attendees queuing for hours to have a detailed chart of their biometrics measured and recorded.[151] Galton claimed that the door to the exhibition room was "thronged by applicants waiting patiently for their turn, or after a while turning away seeing it was almost a hopeless task to wait."[152] Galton's strategies in promoting his anthropometric laboratory and generating such a strong level of interest were not simply those of a dispassionate scientist but borrowed a little from those of P. T. Barnum: he charged members of the public for the privilege of participating. For their fee (threepence), they had a detailed set of anthropometrics recorded and were sent home with a souvenir copy of their report. Lundgren argues that such practices encouraged the public to see self-assessment as a means by which to contribute to scientific research and progress, in which these reports served as "a token of the visitor's contribution to science as well as proof of having stood up and been counted, or being represented in an imperative new statistical aggregate."[153] In this way, Galton fostered a culture of participatory anthropometrics, which he attempted to further by opening an anthropometric laboratory intended to be permanent at the South Kensington Museum—although this laboratory did not enjoy the popularity of the one that was part of the International Health Exhibition.

No doubt there were many visitors at the International Health Exhibition who saw these reports as novelties rather than solemn scientific records, but it is certainly evident that Galton's anthropometric vision caught the popular cultural imagination. By attaching a sense of moral edification to the practice of self-measurement, Galton made this practice seem appealingly worthy. He encouraged the public, writes Lundgren, to see "self-observation, self-assessment and evaluative comparison as a natural part of responsible life."[154] In cultivating public interest and voluntary participation in this work, he reinforced the growing status of "quantification as a universal cultural value that would set the standard for future public discourse."[155] Galton was very successful in encouraging members of the public to recognize the importance of anthropometrics and to embrace self-measurement as a new form of social citizenship. Through these means, he played an important role in moving anthropometrics into the public sphere. Thus, while Galton's experiments with composite portraiture were not an influential part of his scientific or statistical research, the research it was used to support nonetheless had a discernible cultural impact. In particular, his anthropometric projects encouraged a widespread popular uptake of new practices of self-assessment and a widespread acceptance of the social and scientific significance attributed to them. This stands in marked contrast to the popular and scientific reception of eugenics at the end of the nineteenth century in Britain and to the repeated failures of composite portraiture to find application in the large public institutions with which Galton collaborated. The prison, the hospital, and the school—those key institutions of the disciplinary society identified so influentially by Foucault—were not particularly interested in Galton's composite portraits, nor did they see much scope for application of the images of types he produced. Other parts of his research, such as his development of fingerprint identification, would find uptake in disciplinary contexts, but these, like Bertillon's prison records and photographs, focused on the identification of individual subjects, rather than their aggregation as composites.

Galton's methods and technologies for collecting anthropometrics would also prove influential. As anthropometricians like Galton and Alphonse Bertillon exponentially increased the amount and variety of data collected in the final decades of the nineteenth century, how to store, retrieve and cross-reference such data emerged as one of the most pressing concerns of administrative institutions and practices. Statistical methods of analysis such as the law of normal distribution enabled Galton to sort and analyze increasingly large data sets, but the storage and retrieval of the data presented an additional problem. Pearson notes that Galton's technical ingenuity was also exercised in this area, through the development of a "mechanical selector" or early punch card machine.[156] Although these technologies and the practices of anthropometrics they supported were not distinct from the operation of the disciplinary institutions such as the prison, army or school in which Foucault identifies the emergence of the "law of the normal" and practices of normalization, the repeated failures of Galton's composite portraits to find traction in the disciplinary institutions of his day are also significant. Galton's research and technologies found their greatest uptake not in the century-old disciplinary society but in the emergent data society, which they helped to establish. For Galton developed new technologies of mensuration, and, through their use, collected data not simply on physical characteristics but on faculties and inherited traits such as genius, on the one hand, and criminality, on the other. In this way, Galton shifted anthropometrics increasingly toward the study of biological characteristics, that is, toward the discipline he referred to as psychometrics.

MEASURING MENTAL ABILITY: THE EMERGENCE OF "INTELLIGENCE" AS AN OBJECT OF ANTHROPOLOGICAL STUDY

Adolphe Quetelet had laid out a program for anthropometry that remained incomplete at the time of his death. There were, he said, three sets of faculties that could be studied in parallel: the physical, the moral, and the intellectual. In a major work published in 1870, *Anthropométrie, ou Mesure des différentes facultés de l'homme* (*Anthropometry, or the Measurement of the Different Faculties of Man*), he focused on the mensuration and tabulation of human physical features while announcing two volumes to follow, one devoted to moral qualities and the other to intellectual ones.[157] For each of these categories, Quetelet affirmed, the intensity and range of qualities would be comparable. As it happened, he died four years after the first volume was published, and the other two never took shape, but there is more to this pattern of research than the contingencies of one man's working life. Most nineteenth-century anthropometry ordered its priorities in the same manner, focusing on the physical and deferring the study of the moral and the intellectual. Physical characters, after all, were not just easier to observe: they were more easily defined as objects of scientific inquiry. Quetelet could not be said to have neglected moral statistics. He played a leading role in their interpretation, as we saw. But he did no "moral" measurements of his own, simply tabulating data produced by governmental agencies. And where intellectual characters were concerned, it must be said that he contributed very little. This focus on the physical and the particular neglect of the intellectual it entailed were characteristic of the generation of anthropologists that came to prominence during the middle decades of the century. Broca, who was undoubtedly the most clear thinking of that generation, made a broad claim to resolve the problem without really addressing it in the classificatory terms in which it had been posed by Quetelet. The anthropology Broca championed was simply "physical" in both theory and practice, effectively taking skull size as a proxy for brain size and brain size as a proxy for intellect. This was methodologically efficient, but it hardly offered a rich account of a set of hereditary intellectual qualities that might characterize a race—supposing, as Broca wished to do, that heredity functioned in the same way for characters of all kinds. At the end of an inquiry into physical features, Broca could do no more than hold out the rather vague hope of "finding some data about the intellectual value of the different human races."[158]

Galton shared many theoretical assumptions with anthropologists elsewhere in Europe, but he parted company with them on the importance and the scientific feasibility of studying intelligence as a set of hereditary qualities. That did not amount to a theoretical revolution, but it did constitute a significant shift in research priorities. To set about the task, Galton first needed to defend the very idea of studying "genius"—or, to use the term that tended to displace it in the decades that followed, "intelligence"—as a hereditary phenomenon. He began his 1869 essay *Hereditary Genius* by conceding that the notion had hitherto been derided by many and advocated as an object of study by only a few.[159] Skepticism of that kind needed to be answered scientifically, and he now proposed to do just that by the application of statistical method: "I may claim to be the first to treat the subject in a statistical manner, to arrive at numerical results, and to introduce the 'law of deviation from an average' into discussions on heredity."[160] The pertinence of this law to any study of heredity, he noted, had already been demonstrated through tabulations of "form and physical features," which were clearly "derived by inheritance."[161] It had been shown most impressively in Quetelet's study of the physical measurements of Scottish soldiers, which had by then become a standard of anthropometry.[162] With the physical program already well under way, Galton now moved to extend the field of inquiry by a kind of material inference. It was only to be expected, he said, that the same law of heredity would apply to the full set of human features, including "circumference of head, size of brain, weight of grey matter, number of brain fibres, etc.; and thence, by a step on which no physiologist will hesitate, . . . mental capacity."[163] The way was open to study hereditary intelligence on the assumption that the physical features he had named would not be just inherited characters, but also the physiological vehicles, the literal incarnation of intellectual ability. If appropriate data were to hand, an inquiry based in statistics would then make it possible to ascertain "the laws of heredity with respect to genius."[164] In his 1883 essay *Inquiries into Human Faculty and Its Development* Galton maintained the same order of scientific concerns: "It is with the innate moral and intellectual faculties that the book is chiefly concerned, but they are so closely bound up with the physical ones that these must be considered as well."[165] The physical qualities to which he was referring did not have the compelling primacy they appeared to have for Broca. It was just that for the purposes of an inquiry into hereditary intelligence it was "convenient to take them first."[166] That was because, pace those anthropologists like Quetelet and Broca who had spent most of their lives studying physical qualities, physical anthropometry presented no particular challenges to science: "the differences in the bodily qualities that are the usual subjects of anthropometry," said Galton, "are easily dealt with."[167]

The most convenient way to set about studying hereditary genius, Galton asserted in his 1869 essay, was to study families. He had begun with questions of race, a more general concern of anthropology at the time, but had come to the view that he did not need to range far afield to find exemplary patterns of hereditary intellectual ability: "The idea of investigating the subject of hereditary genius occurred to me during the course of a purely ethnological inquiry, into the mental peculiarities of different races; when the fact that characteristics cling to families, was so frequently forced on my notice as to induce me to pay especial attention to that branch of the subject."[168] It was only when Galton returned in the same essay to claims about racially inherited intelligence that his statements took on disquieting political overtones. He rehearsed the standard anthropological proposition of the time about degrees of civilization found in the various races, but then went on to make a narrower claim about differential intellectual ability within his own national group: "To conclude, the range of mental power between — I will not say the highest Caucasian and the lowest savage — but between the greatest and least of English intellects, is enormous. There is a continuity of natural ability reaching from one knows not what height, and descending to one can hardly say what depth."[169] It mattered little in Galton's political terms whether a notional English race was superior to others on the anthropometric scales generally favored by anthropologists. What mattered far more to him were perceived hereditary disparities in natural ability within his own nation. The politics implicit in eugenics were potentially more acute and more local than those of a global racism. In that figurative sense, eugenics was for Galton a family affair.

Engaging in a statistically based study of hereditary intelligence within groups meant identifying an average that defined mediocrity in Galton's etymological sense while establishing degrees of intelligence ranged about it. Broca's craniometry had established a rigorous set of procedures, but no one had done comparable work for the measurement of intelligence. There was no agreed mensurative technique, and no agreed scale. Without those things, there could be no data for statistical analysis, and no standardization of knowledge. A method of testing was urgently required. As Nancy Stepan points out, eugenics helped to shape scientific inquiry at this point. It established intelligence as a topic for psychology, and in so doing helped create a requirement for intelligence testing in Britain: "It was of course 'genius,' 'talent,' or 'intelligence' (as it came to be called) that was of first importance to the eugenists. It was they who placed the subject of intelligence at the centre of psychological science and made the word so potent in the English language and in English education. Intelligence tests underwent explosive growth between 1900 and 1930, exactly the period that eugenics enjoyed its greatest popularity."[170] The pioneers of intelligence testing were not British, however, and they were not eugenicists. During the early years of the twentieth century, Alfred Binet and Théodore Simon, working mainly in France, developed a series of tests and a corresponding scale for use on children between the ages of three and thirteen.[171] Their research did indeed answer a requirement of eugenicist thinking and was widely taken up in England and elsewhere, but its history ought to be read with care, especially regarding its place in the genealogy of normality. Binet and Simon's immediate institutional purpose had nothing to do with matters of racial superiority or inferiority, or indeed with the transmission of abilities through heredity. They simply made a technical breakthrough that could be turned to a variety of political purposes, one of which happened to be that of Galton's eugenics.[172] Looking back about fifteen years later, Charles Spearman, an eminent psychologist involved in the English eugenicist movement who actually had strong reservations about the theoretical assumptions made by Binet and his colleague, was compelled nonetheless to refer to their work as "epoch-making."[173]

Binet and Simon began their study of the development of intelligence in children by observing that there had been controversy in recent years about whether intelligence could effectively be measured at all. Their approach in the face of such controversy was not to enter into "theoretical discussions" of the matter but to attempt to resolve it as a "factual problem."[174] The key difficulty as they saw it from the outset was that the formal analogy between physical anthropometry and their own anthropometric interest was inadequate and misleading. They made a specific contrast with the business of measuring children's height. It was perfectly possible to establish the height of each child in a given group and arrange the children in separate tables according to age. One could then calculate an average and measure distance from the mean in the case of each individual. This was in fact how Quetelet or Louis-Adolphe Bertillon would have gone about the task, although Binet and Simon did not invoke the authority of their predecessors. They simply characterized the procedure for measuring heights as *très peu artificiel*—involving hardly any artifice.[175] The same procedure could not be used for intelligence, however, because intelligence was not a singular variable: "It is quite a different matter with the measurement of intelligence. If one attempts to apply the same system of comparison between the intelligence of a child and the average intelligence of children at different ages, one is pulled up short by this difficulty: a child is likely to be behind in certain tests for his age, and ahead in others."[176] This difficulty complicated the task, but the development of intelligence tests could go ahead on condition that a convention of some sort was put in place. Height could be measured and compared without recourse to any convention, but intelligence could not be measured unless a way was found to deal with the fact that the object and the means to measure it were not immediately to hand. In that sense, the measurement of intelligence would always have "an artificial character."[177]

Once Binet and Simon's set of tests had been developed, they were available to serve the purposes and policies of eugenics, even though they appear to have done so casually rather than systematically. But there is more to be said about the tests here because the demand to which they responded marks a phase in the history of normality. In an essay published in 1907, *Les Enfants anormaux* (*Abnormal Children*), Binet made it clear just what was at stake for the practice of intelligence testing in France. A quarter of a century after the Third Republic had introduced compulsory free primary schooling, it had become clear that not all children were receiving the hoped-for universal benefit. In the first decade of the twentieth century, the children deemed to be missing out were referred to in official generic terms as "abnormal." "Abnormal" was a catch-all category. Binet himself called it a "heterogeneous" group, offering this indicative list: "deaf-mute children, blind ones, epileptics, idiots, imbeciles, feeble-minded children, unstable ones, etc."[178] Every kind of "abnormality" that might have a systematic effect on a child's success at school seemed to be included. A ministerial commission set up in 1904 to address the issue actually used Quetelet's three categories, referring to "physically, intellectually and morally abnormal children," but those terms were just hopefully inclusive rather than helpfully analytical.[179] There were too many kinds of children in the category of "abnormal" for a single set of policies to respond to their needs.

So what confronted Binet and Simon was a disparate set of pupils identified by teachers as needing expert attention. The teachers were asking, in fact, whether ostensibly abnormal children really were abnormal in some psychological sense yet to be defined with authority.[180] In order to answer that question systematically, Binet first had to sort out the cases and types that had been lumped together. The first critical move he made was to question the applicability here of the term "abnormal," precisely because the term owed so much to medical discourse. "Medical language applies the term 'abnormal' to any individual who is clearly at a sufficient distance from the average to constitute a pathological anomaly," said Binet, pointing out as he did so that there might well be anomalies that ought not to be considered pathological insofar as they posed no threat to the individual's health or that of their descendants.[181] In any case, the category of "abnormal children" needed to be broken down into better defined subgroups. Given the terms in which the question was being put by teachers and school principals, it was possible to exclude some kinds of children from the present problematic. Deaf-mutes and blind children had long been recognized as needing special kinds of teaching: they did not require expertise about intelligence to explain in principle why they might be having difficulties with their schooling.[182] In much the same way, "idiots" could be excluded from the group identified for analytical attention because they were known to require continuous medical treatment.[183] These methodical exclusions left a residual group of pupils who were behind with their schooling for reasons that were still to be explained.

The problem, put trenchantly, was that abnormality was taken to be a manifest phenomenon, while the normal condition that might have defined it contrastively was not: "everyone is ignorant of how much intelligence a child needs in order to be normal."[184] So there were in fact two possible standards and two kinds of normalization possible within the school system. The obvious difficulties that had so far been taken to define abnormality had to do with pupils being behind in their schooling. That allowed Binet and Simon to produce a working definition of the abnormal child defined by performance: "we have defined abnormal children at school purely and simply by the weakness of their schooling faculty [*faculté scolaire*]. Any child who is three years behind in their studies should be considered abnormal unless the delay can be excused by a lack of time at school."[185] But that did not establish of itself whether such pupils were behind with their schooling because of inferior intelligence. Binet and Simon's working assumption, and something their tests tended to confirm, was that many children who were behaviorally "abnormal" were in fact of normal intelligence.[186]

They went about this task by looking for a "law of intellectual development in children," meaning by "law" something very much like what Quetelet and Galton meant by the term: a statistical regularity that offered a range of values disposed about an average.[187] Children were grouped according to age, and for each age group tests were developed that would produce a regular distribution. The key thing was to be able to compare—and thus to normalize—the spread at each age. For example, "five years is the age at which one can draw a square; not until seven can one draw a diamond shape, and even then a fifth of the children fail to do so; at six, half of them fail."[188] It was only possible for Binet and Simon to offer a statistical account of individual variations by measuring age in whole years, although they did seek to mitigate the effects of such standardization by introducing a margin of tolerance into their calculations when dealing with a given child.[189] That did not prevent the introduction into their tests of another kind of normalization that went unrecognized and unquestioned: the normalization of gender. A question about gender identification was taken to provide a key marker of intelligence at a certain age: "Three-year-old children can make a mistake, but a normal child of four always answers correctly when asked

254 CHAPTER SIX

about his or her sex."[190] "Normal" here appears in retrospect to be an unfortunate pun. Testing a child for intelligence in order to manage difficulties in schooling required a technical definition of the normal, but that definition might well be attended by a range of assumptions that were helping to generalize normality, among other things, as a set of sexual behaviors. The irony of history was that systematic, professional attempts to give an account of sexual behaviors were by then coming to serve as a favored terrain for the development of sharper—and more sharply discriminatory—ways of thinking about normality.

CONCLUSION

Francis Galton considerably expanded the scientific reach of statistics and gave it a key role in thinking about the normal. He managed to do that by shifting the emphasis away from the calculation and tabulation of averages toward a set of mathematical and geometrical practices in which "normal" became a properly statistical term. Normality as Galton understood it was not a quality attached to a particular state, or indeed to individual subjects: it named a particular kind of distribution. In the work of Quetelet and Louis-Adolphe Bertillon, statistics had been primarily concerned with averages, but Galton focused attention on what he called the mathematical law of deviation. In his view, the challenge for statistical knowledge was to represent all the members of a given "fraternity or population" in a manner that revealed clustering around the midpoint wherever it occurred while displaying at the same time the full range across which the group or class of persons was spread. That was the scientific value of the newly named "normal curve": the curve itself was normal, not those who stood close to the middle. Similarly with composite portraits, the type could be shown as a property of the whole group, revealed when it was present by neatness of overlap and clarity of definition.

Like Broca, Topinard, and Lombroso, Galton was preoccupied with the study of race, although his perspective on race was complicated by his adhesion to Darwin's theory of evolution. The urgent theoretical question for him was how to reconcile the concept of natural selection, which was likely to produce ongoing dispersion through the effects of individual adaptation to random phenomena, with the notion of race understood as the inheritance of characters belonging to an ancestral type. In response to this difficulty, Galton produced a geometrical model that represented a properly statistical theory of biological heredity, incorporating both a tendency to dispersion and a compensating tendency to reversion. A complex and rather contradictory uptake of Darwinian biology led him to advocate eugenics as a concerted, socially based intervention into the processes of evolution. His stated goal was to improve the national race by attending to those near the middle of the range, fostering their improvement through the appropriate choice of partners for procreation. That was, he said, how the type itself could be improved. A mathematical model of heredity had thus led him to imagine a policy of normalization whereby the health of the race would be managed by government. As it happened, he was quite unsuccessful in winning institutional support—or more generally, public support—for a eugenic program.

Galton also extended the range of characters that were subject to anthropometric study. To physical measurement, so central to the work of Broca, he added a precise interest in mental faculties or internal states. That is how he came to support, both intellectually and materially, the development of the new sciences of biometrics and psychometrics. While failing to persuade institutions to engage in governmental normalization, he certainly contributed to practices of individuation through the dissemination of self-managed anthropometrical practices.

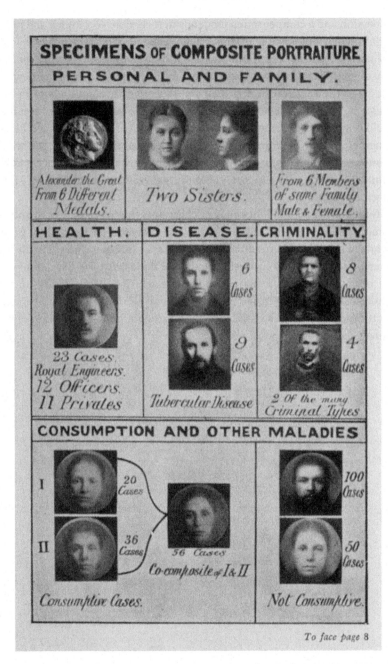

FIGURE 1. Frontispiece of Galton's *Inquiries into Human Faculty*, 1883. Courtesy of the Wellcome Library.

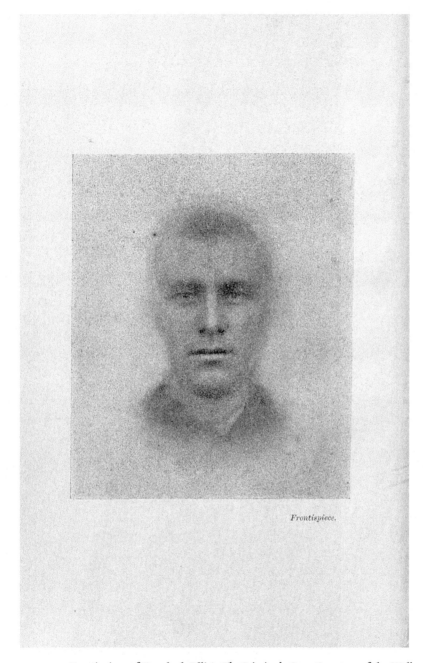

FIGURE 2. Frontispiece of Havelock Ellis's *The Criminal*, 1890. Courtesy of the Wellcome Library.

PART II

The Dissemination of the Normal in Twentieth-Century Culture

CHAPTER SEVEN

Sex and the Normal Person: Sexology, Psychoanalysis, and Sexual Hygiene Literature, 1870–1930

The nineteenth century was a nervous age. Across Western Europe and the Anglophone world, new hospitals were opened dedicated to the treatment of nervous men, women, and children. Alongside flourished a market in private clinics that offered a range of experimental therapies for nervous disorders. In the 1880s, the American doctor George Beard diagnosed a national epidemic of nervous exhaustion in the United States. He termed this condition neurasthenia, and popularized the news of its pernicious influence in two influential volumes: *A Practical Treatise on Nervous Exhaustion (Neurasthenia): Its Symptoms, Nature, Sequences, Treatment* (1880), and its supplement *American Nervousness: Its Causes and Consequences* (1881).[1] In the first half of the nineteenth century, such conditions had been understood through the framework of an earlier epidemic: that of spermatorrhoea, or seminal incontinence.[2] In the second half of the nineteenth century, neurasthenia took the place of spermatorrhoea, and nervousness came to be accepted by medical doctors and general public alike as posing a significant and widespread social problem. The principal cause to which this affliction was usually attributed was the artificiality and hyperstimulation of contemporary life. Nervousness was understood to be endemic to a culture in the midst of rapid urbanization and industrialization, cultural dynamics whose sudden impact was placing too much stress on the nervous system. City dwellers spent their days inside apartment houses and offices, with artificial light and stale air, running on too little sleep on beds that were too soft. This both hyperstimulated and fatigued the body. Nervous disorders were thus framed as a failure of adaptation, or a sign of an unsustainable imbalance in the relationship between individuals and their environment. Although fears of a widespread nervousness flourished at the same time as fears of racial degeneration were shaping the work of Lombroso and Galton, nervousness provided a very different lens through which to view the relationship between individual and group, or subjectivity and biology. For Lombroso and Galton, degeneration was a natural but undesirable process of racial regression caused by the biological inheritance of inferior "characters." That is, it was produced by biological factors that were fairly fixed and physical features that were largely stable. Nervousness, on the other hand, suggested a volatile subject, one who was able to change and be changed, and indeed for whom adaptation was a vital quality.

When Lombroso and Galton wrote of degeneration or deviation, as we have seen, they rarely used the words "abnormal" or "normal" in a quantitative sense, even as they established practices of measurement and data analysis that would allow the statistical concept of the normal to be formulated so productively at the start of the twentieth century. In the final decades of the nineteenth century, the word "normal" did not yet have a strongly statistical meaning: although we have dated the earliest references to a statistical normal to the 1870s, it did not become fully elaborated as a statistical term until the turn of the century. Accordingly, when the large late nineteenth-century anthropometric projects in physical and criminal anthropology examined numerical regularity or cohesion, they did so through the concepts of the average, the typical, or the generic, rather than the normal. At the end of the nineteenth century, then, the normal continued to be much as it had been since the 1820s: a specialist term used primarily in medical discourses, referring either to the normal state, understood as a condition of overall health and biological functionality, or to normal anatomy, the more or less stable and healthy composition of organs. It was an idea debated in the pages of scholarly journals or behind the closed doors of learned societies; it had begun to appear in Anglophone and European dictionaries over the second half of the century; occasionally it could be found in the pages of literary texts, especially those influenced by medical and scientific writing. But there was no popular concept of the normal, nor any general category of "normality." The word was rarely heard in everyday speech, and it had not yet acquired the cultural authority or familiarity it would enjoy in the second half of the twentieth century.

At the end of the nineteenth century, however, the word "normal" began to be taken up and used with increasing frequency and conceptual centrality in a new body of literature, in whose development specialists in nervous diseases played a prominent role. This work was commonly characterized by its authors as the scientific study of human sexuality. It was produced in the context of a diverse range of fields, most of which were broadly medical in approach, including psychology, psychiatry, sexology, psychoanalysis, public health, and marital advice literature. Together, the books published in this field in the last decades of the nineteenth century and first decades of the twentieth constituted an important site at which the idea of the normal began to be elaborated in more detail and to occupy a more central conceptual place than it had in any of the fields we examined previously. The normal here, emerges in relation to the nervous, rather than the degenerate or deviant, and is described in much greater detail than we have seen in criminal and physical anthropology, which focused on skull measurements, or eugenics, which turned its attention to intelligence and other mental faculties. Evaluation of normality in the context of writing on nervous disorders and on human sexuality required assessment of a vastly larger range of considerations, which were not so easily quantifiable. Normality in this context required examination of the subject as a whole. It was no longer simply individual measurements or characteristics that could be identified as normal, but the whole person.

An early account of the normal person was given in Warren Lloyd's *Psychology, Normal and Abnormal: A Study of the Processes of Nature from the Inner Aspect*, published in 1908. For Lloyd, a physician, the normal person was a paragon of equanimity and moderation: "In the normal condition . . . if, for any reason, the feeling becomes more intense with reference to another person or thing, the intensity seldom rises beyond safe and exhilarating enthusiasm. . . . Although accustomed to climaxing his emotions, he descends gracefully to the dead level of humanity in a proper and well-balanced manner. A normal man does even unusual things in a normal way. His acts coordinate. Their causes and effects are a perfect fit and easily discerned. He would be a genius without eccentricities. His judgment as to when concentration upon any subject should stop would be superb."[3] The key characteristics of normality were balance and moderation, argued Lloyd, maintained by avoiding extreme states or excesses of behavior. In describing the normal in this way, Lloyd might be thought to have simply applied to general psychology a concept of normality derived from medicine, whereby the normal was understood as a state of psychological balance or harmony. But this was more than an incremental change. It represented a significant shift in the meaning and application of the word "normal." Lloyd described the normal as an attribute of the whole person. In so doing, he also made it possible to think about normality as a general condition.

Much of the writing on the normal person produced at the turn of the twentieth century focused on sexual psychology and behavior. Writers like Krafft-Ebing believed that sexuality was "the most powerful factor in individual and social existence."[4] As a result, it was particularly in books on sexual health and behavior that the word "normal" gradually came into more regular use in professional discourses and then moved into the public sphere. The extent to which the normal became a more central concept in writing in this field during the early decades of the twentieth century was demonstrated by the proliferation of titles in which it featured, such as William Robie's *Rational Sex Ethics: A Physiological and Psychological Study of the Sex Lives of Normal Men and Women, with Suggestions for a Rational Sex Hygiene with Reference to Actual Case Histories* (1916) and Isabel Davenport's *Salvaging of American Girlhood: A Substitution of Normal Psychology for Superstition and Mysticism in the Education of Girls* (1924).[5] In these texts, the idea of the normal was the key term of reference for sexual behavior, giving it a much more central place than it had occupied in anthropology and eugenics.

Writing on human sexuality from a medical and scientific perspective produced over the period 1870 to 1930 thus constituted a pivotal moment in the genealogy of the normal. In the first instance, it broadened the conceptual parameters of the term, and in the second it provided a key conduit by which the normal finally moved from its almost exclusive use in professional discourses into the popular sphere. Guides such as Robie's and Davenport's played a crucial role in popularizing the idea of the normal at this time in part because they found such a large and receptive public audience. When Sylvanus Stall, an American Lutheran pastor and campaigner for sexual hygiene, published his guide *What a Young Boy Ought to Know* in 1897, it proved so popular that it led to a series of follow-up texts, including *What a Young Husband Ought to Know* (1899), *What a Man of Forty-Five Ought to Know* (1901), and *What a Young Man Ought to Know* (1904).[6] Stall's publisher commissioned Mary Wood-Allen, the American purity movement campaigner, to produce matching guides for women, with *What a Young Woman Ought to Know* appearing in 1897 and *What a Young Girl Ought to Know* in 1904.[7]

At the same time, the distribution of such texts to a general readership did not go unchecked. When Krafft-Ebing published what is often identified as the first medical study of human sexuality, *Psychopathia Sexualis*, in 1886, he sought to restrict the readership to medical and legal specialists by writing it in a technical language designed to exclude lay readers: "In order that unqualified persons should not become readers, the author saw himself compelled to choose a title understood only by the learned, and also, where possible, to express himself in *terminis technicis*," he explained in the preface to the first edition.[8] He took the further precaution of presenting the most sexually explicit parts of his text in Latin rather than German. Despite the professional success and popularity of *Psychopathia Sexualis*, and the evident demand for guides on sexual health and behavior to which that success attested, the circulation of such material remained subject to legal and professional restrictions until well into the twentieth century. In Britain, Havelock Ellis's *Sexual Inversion* (1897), often identified as the first medical textbook on same-sex sexuality, was banned in 1898.[9] The cover page of Robie's *Rational Sex Ethics* carried an instruction it was to be "sold only to members of the recognized professions."[10] As late as 1929, the publisher of the American criminologist and sex researcher Katherine Davis deliberately limited the print run of her study *Factors in the Sex Life of Twenty-Two Hundred Women*, refusing requests for translation rights in order to restrict its circulation to a small circle of medical and public health workers.[11]

Despite such measures, the market for popular guides on sexual health and behavior flourished and had become well established by the start of the twentieth century. As Stall noted in *What a Young Man Ought to Know* (1904), the time for pulling a discreet veil of Latin over discussion of sex, as Krafft-Ebing had done, had now passed: when "Rev, Dr. John Todd [was prompted], in 1854, to write the 'Student's Manual' . . . public sentiment was such that in his treatment of that portion of his subject which related to personal and social purity he felt it necessary to print in Latin the wise counsel which every student should have been permitted to translate from plainest English into daily living."[12] Yet furnishing that counsel in "plainest English" continued to remain contentious until well into the twentieth century. The preface to Marie Stopes's *Married Love: A New Contribution to the Solution of Sex Difficulties*, written by Dr. Jessie Murray and published in 1918, began by acknowledging continued concerns about the propriety of public discussion of sex. *Married Love* dealt "with subjects which are generally regarded as too sacred for an entirely frank treatment," Murray recognized, and "some earnest and delicate minds may feel apprehensive that such frankness in details is 'dangerous.'"[13] Despite such reservations and initial difficulties finding a publisher, Stopes's book caused a sensation, selling thousands of copies a week and making its author "the nation's first acknowledged, if informal, adviser on sexual and emotional problems."[14] Books such as Stopes's were an important locus for the development of a new idea of the normal at the turn of the century and its gradual extension from professional discourse into the public sphere. In tracing the development of this field, we are interested in the elaboration of a newer and more general concept of the normal to which it gives rise. This new idea of the normal emerged in the period immediately prior to its popularization as a statistical or quantitative term. It envisaged the normal for the first time as a site of negotiation and adaptation through which a new form of subjectivity—the normal person—would take shape in the first decades of the twentieth century.

"WHEN NOT IN MASOCHISTIC STATE I AM A PERFECTLY NORMAL MAN": SEXOLOGY AND NORMALITY

Krafft-Ebing's *Psychopathia Sexualis* is often identified as the first study "to describe sexual behavior within a framework of disease categories."[15] The purpose of this text, wrote Krafft-Ebing, was to provide "a description of the pathological manifestations of the sexual life" that would serve as the basis for a new field of medico-legal research in the area.[16] Such research was desperately and urgently needed, Krafft-Ebing argued, because the entire field of sexual pathologies had been neglected as an area of study: "Only he who, as a medico-legal expert, has been in a position where he has been compelled to pass judgment upon his fellow-men, where life, freedom, and honor were at stake, and realized painfully the incompleteness of our knowledge concerning the pathology of the sexual life, can fully understand the significance of an attempt to gain definite views concerning it. Even at the present time, in the domain of sexual criminality, the most erroneous opinions are expressed and the most unjust sentences pronounced, influencing laws and public opinion."[17]

One reason why the study of sexual pathologies remained so neglected at this time, as Sander Gilman reminds us in his study of nineteenth-century German psychiatry and psychoanalysis, is that both the treatment of sexual health and the practice of psychiatry were considered lowly and disreputable areas of medical specialization in late nineteenth-century Germany as elsewhere. Treatment of sexual diseases was considered morally questionable, while psychiatry as a profession was not considered a properly medical field since it was concerned "as much with the administration of the asylum as the treatment of the insane."[18] Asylums were state institutions and part of the legal system; therefore they were not considered purely medical establishments. The focus on sexual pathologies in *Psychopathia Sexualis* was itself a reflection of this: Krafft-Ebing was the superintendent at the Feldhof asylum in Graz during his tenure as professor of psychiatry at the University of Leipzig in the 1870s and 1880s. The period during which he built a reputation as the foremost German specialist in forensic psychiatry and sexual pathology was one in which legislation proscribing an increasingly detailed range of nonreproductive sexual acts was being introduced and enforced across Europe and the Anglophone world, as historians of sexuality have widely documented.[19] He intended *Psychopathia Sexualis* to serve an instrumental purpose in the administration of disciplinary institutions, as Bland and Doan note, by furnishing "a handbook to assist the courts in understanding sexual crime."[20]

In the process of developing a new field of forensic psychiatry that could examine and, if possible, treat sexual pathologies, Krafft-Ebing did much to change the status of both psychiatry and sexual disorders in late nineteenth-century Germany, establishing them as legitimate and important fields of medical research. Krafft-Ebing's own primary training was in medicine, and he had briefly run a private nerve clinic in Baden-Baden in 1869, specializing in the use of electrotherapy. During his time at the University of Leipzig, he lobbied for the establishment of university wards where medical students and researchers could gain clinical experience with nonchronic patients. A prolific author, he published the three-volume *Lehrbuch der Psychiatrie auf klinischer Grundlage* (*Textbook of Psychiatry Based on Clinical Observations*) (1879–1880). This publication "established his reputation as a leader in clinical psychiatry," as Harry Oosterhuis notes, and "became a standard textbook in the field."[21] By the time he published *Psychopathia Sexualis* in 1886, Krafft-Ebing was an eminent figure, recognized as the country's foremost authority on forensic psychiatry. *Psychopathia Sexualis* was by far the most popular of his books, and despite its subject matter its publication only increased his professional standing and public renown: in 1889 he accepted a position at the University of Vienna, the most prestigious university in the country.

In the absence of existing medical research, Krafft-Ebing's study of sexual pathologies drew on a wide range of texts and sources of knowledge from fields as diverse as anthropology, poetry, philosophy, and literature. He derived the terms for which he is perhaps best remembered—sadism and masochism—from the names of the novelists Sade and Sacher-Masoch. He adopted biological theories, especially that of degeneration, from anthropology, in part because these provided the somatic basis necessary to establish sexual pathology as a field of medical research.[22] His development of forensic psychiatry drew particularly on criminal anthropology, and Krafft-Ebing frequently cited Lombroso as an authority, occasionally including in his own case studies Lombroso's anthropometric data on cephalic indices as providing evidence of degeneration. Lombroso was supportive of Krafft-Ebing's work in turn, writing the introduction to the 1889 Italian translation of *Psychopathia Sexualis*. However, Krafft-Ebing's work differed significantly from Lombroso's anthropometric approach and from that of forensic doctors, who measured and examined fixed physical characteristics of those accused by the legal system: Krafft-Ebing focused rather on dynamic psychological states. His primary concern was with the pathological rather than the degenerate, with mental states rather than inherited biological conditions. And his approach to the study of this subject was taxonomic and encyclopedic: drawing on clinical observation and case studies, his own and those of others, Krafft-Ebing sought to identify and classify all the varieties of sexual pathology. Beginning with broad categories such as inversion, fetishism, sadism, and masochism, he proceeded to break them down into an increasing number of subsets, which included, in the case of sadism, such specific practices as "sadistic acts with animals," "whipping of boys," "defilement of women," "sadistic sex-murder," and "violation of corpses." As he collected more case studies, Krafft-Ebing continually expanded the text of *Psychopathia Sexualis*. Between 1886 and 1903, he produced fourteen new editions, during which time the initial 110 pages expanded to 617, and the original 45 case studies ballooned to 238. Many of these case studies were contributed by readers moved to send their own stories to Krafft-Ebing for analysis and inclusion, providing further evidence of the resonance the book had found with a general audience.

From its very first pages, *Psychopathia Sexualis* described the various manifestations of sexual pathology in relation to the normal and the abnormal, giving these terms a conceptual centrality and prominence that they had not previously enjoyed in medical or scientific writing. However, despite the importance of the normal and the abnormal within the new framework he was proposing for understanding and treating sexual pathologies, Krafft-Ebing did not in fact offer definitions of these two key terms. Instead he simply invoked them relationally, in a way that suggested he felt that their meaning could be taken for granted. That is how the terms "normal" and "abnormal" entered medical writing on sex through *Psychopathia Sexualis*: in a way that was both emphatic and self-effacing. They were introduced and used with such insistence that it seemed immediately as if they had always been there, allowing their meaning to be taken for granted. Normal sexual desire, Krafft-Ebing announced at the start of the book, was the kind experienced and acted on by members of opposite sexes for the purposes of potential reproduction: "love . . . is only conceivable in a normal way as existing between individuals of opposite sex and capable of sexual intercourse."[23] This is Krafft-Ebing's first reference to the normal, in the opening pages of *Psychopathia Sexualis*, and it immediately aligned the term with what we (but not Krafft-Ebing) now think of as heterosexuality. This view seemed to restrict normal sexuality to heterosexuality, naturalizing its existing cultural privilege. Here we see a shift from the descriptive to the prescriptive that appears to characterize the idea of the normal from the outset: it refers not to those sexual practices that most people engage in (determined by their average incidence) but those that are most privileged by the established culture and aligned with its dominant values. Normal sexual function, as it was described in *Psychopathia Sexualis*, thus involved more than the physical ability to have intercourse, encompassing in fact a whole series of psychological states and dynamic processes.

This was a decidedly more multifaceted and more encompassing concept of normality than that which we have seen previously, although it was thoroughly restrictive in a normative way. Love was normal only when it was experienced and manifested passively in women and aggressively in men, Krafft-Ebing argued: "If [a woman] is normally developed mentally, and well bred, her sexual desire is small. It is certain that the man that avoids women and the woman that seeks men are abnormal."[24] Female passivity and male aggression were not merely an effect of "good breeding" but of biological imperatives, he claimed.[25] Although Krafft-Ebing did not define the terms "normal" or "abnormal," then, he positioned them at the heart of his discussion of sexual pathologies and made them conceptually central. The abnormal, as it is described in his work, was whatever diverged from the normal: desires that were too aggressive or passive, ejaculation that was too quick or too slow or did not occur, practices that did not focus on reproductive sex. Sexual desire could be abnormally manifested in many ways: it could be experienced at the wrong time (paradoxia), in the wrong amount (anesthesia and hyperesthesia), or take the wrong aim or object (paresthesia).

Although it may at first seem that *Psychopathia Sexualis* conceptualized the terms "normal" and "abnormal" as binary opposites and as synonyms for culturally valued or denigrated forms of sexual behavior, closer analysis reveals a much more complex relationship between the two terms in Krafft-

Ebing's thinking. Sexual abnormalities, according to Krafft-Ebing, were not opposed to but rather intensifications of normal forms of sexuality. "Sexual bondage," for instance, was merely an exaggerated form of "the normal standard established by law and custom," which similarly had its basis in practices of commitment and devotion.[26] Sadism, likewise, was an abnormal expression of normal (male) sexual aggression; it was "an excessive and monstrous pathological intensification of phenomena—possible, too, in normal conditions in rudimentary forms—which accompany the psychical *vita sexualis*, particularly in males."[27] Fetishism, in a similar way, "does not really extend beyond the limits of things which normally stimulate the sexual instinct."[28] This is a point Krafft-Ebing reiterated throughout *Psychopathia Sexualis*: abnormal sexualities varied in degree but not in kind from their normal manifestations. The normal and abnormal were not squarely opposed one to the other but overlapping regions situated within a dynamic field of possibilities. This conception of the abnormal allowed Krafft-Ebing to make new distinctions between pathology and immorality, between perversion and perversity. As Oosterhuis notes, Krafft-Ebing argued that whereas perversion was abnormal, perversity was simply immoral: "Perversion was considered as a permanent constitutional disorder—be it inborn or acquired—that affected the whole personality, whereas perversity was just passing immoral conduct of normal persons. According to Krafft-Ebing, sexual behavior could be abnormal without being perverse."[29] Abnormality thus provided a new conceptual category with which to understand and categorize sexual pathologies. It mapped out a new approach to sexuality in which the normal and the abnormal could not be categorically differentiated from one another but were rather located on a spectrum: "perversions did not form a wholly distinct class, an isolated group of monstrous phenomena, but tended to be considered merely as variations within a wide range of natural possibilities."[30] Examined as possible variations rather than deviations, and as interlinked possibilities rather than binary opposites, the normal and the abnormal emerge together, not as a fixed standard but as a dynamic and rather volatile state. Once sexuality was "placed under the rule of the normal and the pathological,"[31] as Foucault argues, it came to be seen "as an extremely unstable pathological field: a surface of repercussion for other ailments, but also the focus of a specific nosography, that of instincts, tendencies, images, pleasure, and conduct."[32] *Psychopathia Sexualis* revealed the multiplicity of ways in which sexuality could find an abnormal manifestation or expression. This understanding of normal sexuality as a fragile state, easily unbalanced, would be central to sexological discourse well into the twentieth century. Magnus Hirschfeld, for instance, argued in the 1930s that: "any disturbance of the normal course of these phases [of development] causes the sexuality to differ from the normal."[33]

In his discussion of the normal and the abnormal, Krafft-Ebing introduced a further innovation: the category of abnormality as a generalizing rubric. Krafft-Ebing repeatedly referred collectively to different forms of sexual pathology, such as sadism, masochism, and fetishism, as diverse manifestations of a single condition: the "abnormality of the sexual functions."[34] We saw that when the term "normal" took a central place in teratology during the 1840s, Isidore Geoffroy Saint-Hilaire specifically resisted the grouping of anomalies under the general heading of the abnormal, and we have seen since that neither Lombroso nor Galton used the term "abnormality" in this way. Anthropological and eugenic writing had been broadly consistent in that regard across half a century of writing. For instance, the eminent nineteenth-century British specialist on "mental diseases" Thomas Clouston proposed extending "the practices in which criminals have been weighed and measured, observed and described," to the much wider classes of "the idle, the vagrant, the pauper, the prostitute, the drunkard, the imbecile, the epileptic, and the insane,"[35] but did not group these individual types together as manifestations of a single, underlying category of abnormality. As late as the 1930s, the American eugenicist Harry Laughlin catalogued a list of "socially inadequate" types that specified the "feebleminded, insane, criminalistic, epileptic, inebriate, diseased, blind, deaf; deformed; and dependent" along with "orphans, ne'er-do-wells, tramps, the homeless and paupers,"[36] but did not identify these kinds of people as belonging to a larger category of "abnormality."[37] In *Psychopathia Sexualis*, however, abnormality was being elaborated, perhaps for the first time, as a general category.

At the same time, a general idea of normality emerged alongside it, a complex and changeable state that could be evaluated only by taking a wide range of factors into account and assessing them in an overall way that allowed for a great degree of variation. The normal and the abnormal, as Krafft-Ebing described them, could not be determined through consideration of a single data set, such as skull measurements, or even intelligence. Rather, assessment required examination of something more complex, the "whole person." Psychiatry needed "to investigate the whole personality of the individual and the original impulse,"[38] Krafft-Ebing argued, because only in this way could the complexity of the individual's psychology be properly understood, legally evaluated, and medically treated. The overall state of the "whole personality" could not be simply measured and quantified by examining the fixed physical proportions of the skeleton but called instead for expert evaluation of the complexities of the living mind. Even while using the word "normal" in a way that took its meaning for granted, *Psychopathia Sexualis* nonetheless played a pivotal role in the development of a much more elaborate concept of the normal than the one previously encountered in our genealogy, and one with much broader applications. It is in Krafft-Ebing's work that the normal was first conceptualized as a state of overall balance combining a variety of mental and physical factors that were not fixed and did not conform to a single standard. This was what psychiatry had to offer modern science and medicine, Krafft-Ebing told the audience of his 1902 valedictory lecture at the University of Vienna: it is "only in psychiatry you have the opportunity to learn about the whole man . . . whereas any other clinical field deals only with a part of man."[39] The study of normality, as a condition of overall health and harmony, encompassed every aspect of the subject's physicality and mentality, with all the variability that might entail. This aspect of Krafft-Ebing's work radically expanded the range of subjects in which psychiatry took an interest.

Krafft-Ebing's method for examining the whole man was through the collection of case studies. Where police, hospital, and school records contained increasingly detailed biometric and biographical data, generating administrative procedures and technologies that allowed for the storage and cross-referencing of increasingly large quantities of data, the case study went further. It aimed to examine individuals in their entirety, as Foucault argues: "the examination, surrounded by all its documentary techniques, makes each individual a 'case': a case which at one and the same time constitutes an object for a branch of knowledge and a hold for a branch of power."[40] At the same time, and like the other normalizing technologies Foucault examines, the case study also represented "the individual as he may be described, judged, measured, compared with others, in his very individuality; and it is also the individual who has to be trained or corrected, classified, normalized, excluded, etc."[41] Emerging out of the earlier practices of scientific and institutional examination, Foucault demonstrates, the case study was the locus through which different regimes of power and knowledge converged, allowing medical, legal, and, increasingly, psychological records to be combined, thereby providing a much more complete, multifaceted record of the individual. That was what made the case study "a primary instrument" by which individuals were produced in the nineteenth and twentieth centuries as "normative social units," as Lauren Berlant argues persuasively.[42] The power of the case study, she points out, derives from the manner in which "the case hovers about the singular, the general, and the normative."[43] The shift between the individual and the social, the singular and the general is, as we have seen, the key characteristic of normalization as Foucault defines it. Foucault, too, viewed normalization and individuation as closely interrelated processes, both involving a constant movement between the singular and the aggregate, differentiation and homogenization. The case study, in a similar way, made possible the production of new forms of subjectivity even as it gave rise to new fields of knowledge, as Joy Damousi, Birgit Lang, and Katie Sutton have shown: "the emergence of the clinical sciences from the late eighteenth century enabled the entry of the individual into the field of knowledge, even as individual cases provided an anchor for new forms of disciplinary authority."[44]

Yet while the case study may have made the individual a subject of disciplinary knowledge, it also enabled new forms of individuation and sexual subjectivity. Katie Sutton shows how cross-dressing individuals in 1920s Germany appropriated the language of the sexological case study to produce new kinds of "transvestite" subjectivities and narratives.[45] A similar dynamic is evident in reader responses to *Psychopathia Sexualis*. Krafft-Ebing received thousands of letters from men keen to appraise their own sexual normality after reading his text. These letters, many of which Krafft-Ebing published, demonstrate that readers almost immediately adopted his terminology of the normal and the abnormal to characterize and evaluate their own sexualities. And while this correspondence can be seen to exemplify what Ian Hacking has described as the "looping effects" of medical discourse, in which subjects internalize and identify with their medical and scientific classifications,[46] Krafft-Ebing's readers were not always so docile. A number of Krafft-Ebing's readers disagreed with aspects of his book. One correspondent detailed the various ways in which his life and behavior could be identified as normal (happily married family man with respectable employment) and abnormal (enjoyed masochistic sex with prostitutes). He concluded: "When not in masochistic state, as far as feeling and action are concerned, I am a perfectly normal man."[47] Krafft-Ebing himself could be brought around to such readers' point of view, and sometimes adjusted subsequent editions accordingly. In one well-known example, his interviews with and treatment of patients diagnosed with sexual inversion led him away from his earlier belief in degeneration as a major contributing cause of inversion. As Havelock Ellis recognized: "Krafft-Ebing was inclined to regard inversion as being not so much a degeneration as a variation, a simple anomaly, and acknowledged that his opinion thus approximated to that which had long been held by inverts themselves."[48]

While Krafft-Ebing, like Lombroso, Galton, and Ellis, did believe that biological factors played a role in the development and nosology of sexual pathology, his focus was on psychiatric and medical conditions rather than biological inheritance. He saw sexual pathologies as diseases, potentially responsive to medical intervention and therapeutics, not (necessarily) as manifestations of fixed biological states or evidence. That is, they were subject to change, and could thus be cultivated and improved. What examination of the whole person revealed, in Krafft-Ebing's study, was that no one could ever be perfectly or consistently normal. The relationship between the normal and the abnormal was dynamic and unstable. This meant that all people, not just those suffering pathological conditions, were potential targets of medical surveillance and therapeutic intervention. As Renate Hauser argues, Krafft-Ebing's work shifted the medical study of sexual psychology "from an exercise in pathology to a psychological project,"[49] one focused not only on disorders but more generally on "subjective experiences."[50] It also placed change, and the need to negotiate and manage processes of change, at the heart of this idea of the normal, in relation to which all subjects were increasingly encouraged to evaluate themselves. For Krafft-Ebing, the normal was not a fixed standard or type of body against which one could be measured and assessed. Rather, it was a state of overall functionality and conventionality, which encompassed an ever more detailed range of the subject's physicality and mentality.

"WE HAVE ENORMOUSLY INCREASED THE NUMBER OF PERSONS WHO CAN BE ADDED TO THE PERVERTS": FREUD AND THE PERVERSION OF NORMALITY

The idea of the normal elaborated in Krafft-Ebing's study of sexuality—in which it was seen as entwined with the abnormal, so that the two categories always needed to be examined relationally—played an even more central role in Freudian psychoanalysis. Like Krafft-Ebing, Freud did not define the word "normal," although he used it with such frequency and insistence, Jonathan Ned Katz argues, that his writing constituted a veritable "incantation of the normal."[51] Like Krafft-Ebing, Freud saw the normal and the abnormal as connected states, different in degree rather than kind: "In no normal person does the normal sexual aim lack some designable perverse element, and this universality suffices in itself to prove the inexpediency of an opprobrious application of the name perversion," Freud argued in *Three Contributions to the Sexual Theory*, first published in 1905 and translated into English in 1910.[52] We have seen similar statements in Krafft-Ebing's work, although that argument was made in reverse. Krafft-Ebing examined how the fetishist, the sadist, and the masochist were driven by desires that were inherently normal—merely experienced with an intensity that made their practice extreme—so that sadism was simply an excessive expression of a normal aggression associated with male sexuality. Freud made a similar argument about the neurotics and hysterics that made up the bulk of his patients: their behaviors could be understood as variations of degree from normal sex, which Freud, like Krafft-Ebing, saw as a precarious state: "every step on this long road of [sexual] development may become a point of fixation and every joint in this complicated structure may afford opportunity for a dissociation of the sexual impulse," he warned.[53] As in Krafft-Ebing's view, this meant for Freud that no subject was always or entirely normal. However, Freud was interested not just in the way the neurotic or the hysteric were driven by inherently normal desires that were abnormally expressed, but also in the way apparently normal people—like Krafft-Ebing's respectable masochist—were revealed to be inherently perverted. As Freud acknowledged, the result was that psychoanalysis had "enormously increased the number of persons who can be added to the perverts. This is not only because neurotics represent a very large proportion of humanity, but we must consider also that the neuroses in all their gradations run in an uninterrupted series to the normal state."[54] Where Krafft-Ebing saw the normal in sexual pathology, then, Freud saw perversion and perversity in the normal person. In this way, Freud exponentially increased the range and number of people who were the potential subjects of psychoanalytic treatment by decreasing the number of those completely and consistently normal to, effectively, zero. If no one was always or entirely normal, then everyone could benefit from medical and therapeutic interventions, and psychoanalysis had applications across the full spectrum of the population. After all, Freud argued, "we are all somewhat hysterical."[55] Just as Galton had wanted to measure and assess all mankind, so did Freud want to bring even apparently normal subjects under the gaze of the psychoanalyst.

Like Krafft-Ebing, Freud had received his primary training in medicine. He graduated with his medical degree from the University of Vienna in 1881, several years before Krafft-Ebing joined the faculty there. Having specialized in neuropathology, he began to develop psychoanalysis after a period of study with Jean-Martin Charcot at the Pitié-Salpêtrière Hospital in Paris just as Krafft-Ebing had spent time working in the Illenau asylum after his doctorate, studying under Christian Roller, a leading figure in German psychiatry at the time.[56] Despite these similarities in their professional training, Freud's position and prospects within nineteenth-century German medicine were very different from Krafft-Ebing's, as Sander Gilman has shown. Where Krafft-Ebing came from an academic family, Freud's position as a Jew meant that his options to enter the medical profession were largely limited to psychopathology and sexuality.[57] Freud's early focus on neurology was an attempt to elevate the status of these fields within medicine by restructuring them as objects of neurological research. This marks an important distinction between the origins of psychoanalysis and that of psychiatry or sexology: "Psychoanalysis originated not in the psychiatric clinic but in the laboratories of neurology in Vienna and Paris," Gilman argues forcefully. "Its point of origin was not nineteenth-century psychiatry but rather nineteenth-century neurology."[58]

There were other significant differences between Krafft-Ebing's and Freud's work, which informed how they understood both sexuality and normality. As Rita Felski notes in her introduction to Lucy Bland and Laura Doan's *Sexology in Culture: Labelling Bodies and Desires*, Freud and Krafft-Ebing are often seen to fall on either side of a historical divide. Freud's work, notes Felski, is often taken to "erect a seemingly impenetrable barrier between the modern view of sexuality as an enigmatic and often labile psychic field rooted in unconscious desires, and the work of nineteenth-century sexologists . . . with its emphasis on the physiological and congenital roots of human erotic preferences."[59] We have seen that the move away from congenital theories of sexuality was firmly in train in Krafft-Ebing's work, especially regarding theories of sexual degeneration. In this respect, Freud can thus be better understood to have continued a critique of biological theories of psychiatry rather than introduced such critique into psychoanalysis. Conversely, despite his focus on the everyday role of the unconscious and desire, for which Freud is famous, biological considerations clearly remained central to his theories of sexual and psychological development, as illustrated by his identification of stages of psychical development associated with zones of the body, such as the oral, the anal, and the phallic.

Yet where Krafft-Ebing continued to engage with biological theories of sexual degeneration and with the anthropometric studies of Lombroso, Freud argued much more categorically that biological factors such as an abnormal neurology could be definitively disregarded as the determining factor in the development of certain conditions. His analysis of sexual inversion was a prime example of this. While "the first attention bestowed upon [sexual] inversion gave rise to the conception that it was a congenital sign of nervous degeneration," Freud argued, this hypothesis was demonstrably unsound.[60] It was disproved by the fact that "inversion is found among persons who otherwise show no marked deviation from the normal" and "is found also among persons whose capabilities are not disturbed, who on the contrary are distinguished by especially high intellectual development and ethical culture."[61] Although Freud's starting point here was opposed to Krafft-Ebing, especially on the point of degeneration, he comes to the same conclusion, and this made analysis of inversion exemplary of the increasing problematization and redefinition of the normal in the context of writing about sex at the turn of the century in both sexology and psychoanalysis. Havelock Ellis epitomized the increasing tendency to accord inversion a place within the range of normal sexual variation when he argued: "the law and public opinion combine to place a heavy penal burden and a severe social stigma on the manifestations of an instinct which to those persons who possess it frequently appears natural and normal."[62] As inversion claimed a place within normal sexuality, the normal and abnormal came increasingly to be seen as interlinked. Abnormality was not opposed to the normal state but was a constitutive element of it, Freud argued. This was amply demonstrated by how commonly perversions could be found in the normal psyche: "the very wide dissemination of perversions urged us to assume that the predisposition to perversions is no rare peculiarity but must form a part of the normally accepted constitution."[63]

Where Krafft-Ebing saw the normal drive in the perverse act, then, Freud saw perversions under the veneer of normality, and although both men concluded that the normal and abnormal needed to be understood in relational terms, as entwined conditions or states, the implications of their work were nonetheless quite different. Krafft-Ebing's argument that perverse or pathological states could be extreme manifestations of normal drives left the category of the normal itself intact and uninterrogated. Freud's assertion that even apparently normal people could be revealed to have pathological or perverse tendencies, on the other hand, problematized the category of the normal itself. As his English translator A. A. Brill recognized, Freud's work revealed "how faint the line of demarcation was between the normal and neurotic person, and that the psychopathologic mechanisms so glaringly observed in the psychoneuroses and psychoses could usually be demonstrated in a lesser degree in normal persons."[64] Despite framing psychoanalysis as a medical treatment for "nervous diseases,"[65] Freud's theories of sexual development served not to normalize (or make healthy) the neurotic or the hysteric, but to redefine the normal psyche as inherently and inescapably perverse. Even more than in sexology, psychoanalysis saw the normal as a precarious and easily disturbed state. The normal, emerging as a key concept in the field of psychosexuality, was conceptualized as a fragile ideal, not a fixed standard or dominant force. It was indeed an elusive state that no individual could fully embody or achieve.

In understanding the normal this way, Freud both problematized the taken-for-grantedness of the normal and established it as a key concept within psychoanalysis, just as sexology had done before. The centrality of the normal was further reinforced when, in the first decade of the twentieth century, Freud began more decidedly to turn his primary attention from nervous disorders to normal sexuality itself. In order to understand abnormal conditions fully, he argued, it was first necessary to understand the normal processes of sexual development. He claimed that his work on children's sexuality constituted a "discovery" of normal childhood sexuality in the face of a paucity of existing medical research in this area, as he announced at the 1908 meeting of the Vienna Psychoanalytic Society.[66] Within a few short years, a whole field of medical research focusing on normal adult sexuality had developed,[67] along with a specialist field of normal childhood sexuality.[68] In this way, even as it was destabilized as a distinct category, the concept of the normal came to be elaborated in ever more detail in the early twentieth century and was more deeply embedded in theories of subjectivity and everyday life. Normal sexuality was increasingly understood as something precarious, something difficult to cultivate and maintain, requiring constant monitoring and assessment. It became a focus for psychoanalytic research not because there were so many normal people hitherto overlooked by medical research, but because there were so few.

One of the most influential features of the normal as it developed in psychoanalysis was that, following Krafft-Ebing's focus on the whole person, it was extended to encompass consideration of subjectivity as a whole. All aspects of people's lives were governed by sexual impulses and unconscious drives, Freud argued, and this meant that the changing configurations of the normal and the abnormal shaped all psychosocial experiences. In the *Psychopathology of Everyday Life*, Freud showed how everyday errors—a slip of the tongue or momentary forgetfulness—were an effect of the same systematic but unconscious processes that governed sexual pathologies: practices of forgetting or slips of the tongue were not random expressions of "psychic arbitrariness but follow[ed] lawful and rational paths."[69] Minor slips of the tongue or memory were recognizable manifestations of an internal disturbance, and part of a regular pattern of behavior. This expanded domain of psychoanalytic investigation gave the idea of the normal much greater room in which to develop. The increasing centrality and authority of the idea of the normal at this time did not rely on its quantitative significance: Freud's approach, like Krafft-Ebing's was through the case study rather than statistics, narrative rather than numbers. His concern was not with the average incidence of sexual practices but with what their manifestations revealed about the psychology of the individual subject. The medical and mathematical meanings of the normal had not yet converged, and despite their normative pronouncements, neither Krafft-Ebing nor Freud described the normal as a fixed standard. For Jonathan Ned Katz, Freud's writing played a key role in the normalization of heterosexuality and the entrenchment of heteronormativity as a key dynamic in contemporary culture: "though the word *heterosexual* is not much employed by Freud, the term *normal* is repeated *over and over* in reference to the sex-love of women and men for each other."[70] What we see in Freud's representation of the normal is something whose meaning is more complex and volatile than this, however. The normal as invoked in Freud's work was the site of constant examination and adjustment, of negotiation between the individual and the group to which all subjects must attend. Thus, despite his problematization of the normal—or rather, his construction of the normal as a problematic and precarious state—Freud's work did much to establish the concept of the normal and to extend its range of applications into all aspects of everyday life.

SEXUAL HEALTH AND ADVICE LITERATURE: GUIDES FOR THE "NEARLY NORMAL"

Although Freud's ideas were widely known by the first decades of the twentieth century, the first books on human sexuality written especially for a general audience were advice books on marriage and sexual health, educational guides on sexual anatomy and reproduction, and works of popular psychology. Most texts in this field were written from a broadly medical perspective, by physicians, public health campaigners, members of the purity movement, psychiatrists, biologists, and even, in the case of Marie Stopes, a paleo-botanist. Together, their work constituted a new public discourse on sex within the framework of health education in the first decades of the twentieth century. That discourse was quite different from the one found in sexology and psychoanalysis, and these books were part of a very different project from the professional developments out of which sexology and psychoanalysis emerged. Where those fields focused on the perverse and pathological individual, the exceptional case, these books considered the normal person, to whom many were directly addressed. As William Robie argued: while "a great many histories have been obtained from, and numerous facts have been recorded concerning, individuals clearly abnormal, and much philosophizing has been inspired by such data, very little study has been made of normal human sexuality."[71] C. W. Malchow, an American professor of proctology, made a similar claim at the start of *The Sexual Life, Embracing the Natural Sexual Impulse, Normal Sexual Habits and Propagation* (1907): "Much has been written about sexual debility, perversion and pathological conditions affecting manhood, and very little attention has been given to the elucidation of what should be considered the normal condition of the masculine portion of humanity."[72] Malchow's pairing of natural sexual impulses and normal sexual behavior is telling. The idea of normal sexuality here can be seen to emerge from an earlier understanding of the natural, in much the way it emerged from the pathological and perverse in sexology and psychoanalysis.

This emergence can be seen further in one of the earliest appearances of the word "normal": *Sexual Science; or, Manhood and Womanhood, Including Perfect Husbands, Wives, Mothers, and Infants*, written by the famed American phrenologist Orson Fowler and published in 1870. Although Fowler included a chapter on "A Vigorous and Normal Love Element" in his text, he mentioned the word "normal" itself just once, instead referring repeatedly to the natural. Sex, Fowler argued, was a "part of Nature," and as such "must needs have its governing *laws*. These laws establish a love *science* over this part, the same as mathematical laws establish a mathematical science; because each is equally a part and parcel of Nature."[73] It was precisely by framing sexuality as a natural impulse, as Fowler did, that writers of guides on sexual behavior sought to establish a public discourse on sex at the end of the nineteenth century. Sex was a natural impulse, but it was also a volatile one that required strict supervision. For this reason, education programs were imperative, public health campaigners argued. Public sex education campaigns were thus one of the key sites at which a new public discourse on sex emerged at the turn of the century, and reference to the natural and the normal gave rhetorical force to their arguments.

Lavinia Dock's *Hygiene and Morality*: *A Manual for Nurses and Others*

(1910) exemplified the rhetoric of these texts. Dock, a purity movement campaigner and president of the American Nurses' Association, contended that it was imperative that sexually transmitted diseases be treated like any other communicable disease: "as in combating typhoid fever and the plague the first thing needful is that all shall know that there are such diseases, whence their origin, and how they may be cut off at their source, so it is essential that every citizen shall know that there are venereal diseases, where they arise, and how they may be exterminated. Therefore, a wide-spread campaign of popular education must be the first movement made."[74] A distinguishing feature of these guides on sexual health and behavior, whether written for a professional audience (like Robie's and Dock's) or a popular one (like Fowler's), was their contextualization within public health discourses. The field of "sexual hygiene," emergent at the end of the nineteenth century, was an expression of this, as David Pivar shows in *Purity Crusade: Sexual Morality and Social Control, 1868-1900*.[75]

Sexual hygiene was an attempt to destigmatize the medical treatment and understanding of sexually transmitted diseases by treating them, as Dock recommended, just like any other infectious disease. Dock and others urged this approach in a context in which venereal disease was seen as punishment for vice, and fear of disease as an effective deterrent. This belief had significant medical and social consequences, and it was in protest against it that Dock wrote *Hygiene and Morality*. As Haller and Haller note in their study of sexual medicine in nineteenth-century America: "many hospitals in New York and elsewhere had rules prohibiting the treatment of gonorrhoea or syphilis" on moral grounds.[76] Against such attitudes, campaigners for reform in sex education and research argued that education was the most effective means by which to halt the spread of such diseases, which also afflicted otherwise normal people: "Who are responsible for the introduction of venereal diseases into marriage and the consequent wreckage of the lives of innocent wives and children?" Robie asked rather ominously in *Rational Sex Ethics*. "Not, as a rule, the practiced libertine or confirmed debauchee; but, for the most part, men who have presented a fair exterior of regular and correct living . . . who, indulging in what they regard as the harmless dissipation of 'sowing their wild oats' have entrapped the gonococci or the germs of syphilis."[77] The greatest threat to the existing social and familial order was not the perverse or pathological person but on the contrary men who, like Krafft-Ebing's masochist correspondent, appeared to be perfectly normal.

The need for education to protect the innocent—especially women and children—was one that drove much of the publication in this field. It was argued that ignorance was a much more perilous state than that afforded by a careful education, and so it was imperative that children be informed about sexual health and proper sexual behavior at a young age. Such education was perfectly consistent with the standards of polite society, argued Wood-Allen in *What a Young Girl Ought to Know*: "We do not simply believe that it may be a good thing to forestall evil knowledge by wise, parental instruction, but we know that a pure knowledge is a most sure and trustworthy safeguard of present innocence and reliable prophecy of future virtue."[78] Her book was designed to provide a trustworthy guide by which parents could instruct their children, by showing "how the physiological facts of reproduction may be clothed in delicate language and surrounded by an atmosphere of sacredness and self-reverence that will render the knowledge thus obtained a guarantee of right conduct."[79] Much of the literature on sexual health and behavior written for the general public in the early twentieth century was characterized by this combination of delicacy and frankness. It was also characterized by an increasing tendency to frame education and proper conduct as a social responsibility or civic duty. This can be seen in Frank Lydston's popular guide *Sex Hygiene for the Male and What to Say to the Boy* (1912). Lydston, an American urologist and author of texts on sexual and social diseases, argued that public health relied on individual behavior: "such knowledge should not be restricted to the physician—the layman should enjoy his share of enlightenment and do his part in prevention," he wrote.[80] That was Wood-Allen's view also: "Public health is made up of the health of individuals, and the health of grown people is very much what the children, in their care of themselves, have made it," she wrote. "It is the little children of today who are deciding what will be the health and vigor of the nation in years to come."[81] Guides such as Wood-Allen's and Lydston's were designed to provide the requisite knowledge by which individuals would be able to make socially responsible choices about private behavior.

The popularity of such guides attested to a widespread public desire for the information they provided and an equally widespread concern that natural sexual impulses and normal sexual behavior were fleeting and elusive states. The normal as it was understood in sexual health and hygiene literature was in this respect like the normal as described in sexology and psychoanalysis: a precarious and volatile condition that needed much monitoring and management to maintain. Many of these popular guides were written in the context of the perceived epidemic of nervous exhaustion described in Beard's *American Nervousness*. Like Beard, they argued that the unnatural conditions of modern life had corrupted normal sexual behavior. This is particularly evident in popular health guides written by well-known figures like the nutritionist and vegetarian John Kellogg and the physical culturist Bernarr Macfadden. Both men were ardent health reformers and published health guides designed to redress the debilitating conditions of urban life, often by sleeping with windows open, bathing in cold water, and eating plain food. For both Kellogg and Macfadden, sexual behavior was a key component in overall health. Kellogg's *Plain Facts for Old and Young: Embracing the Natural History of Hygiene*, published in 1882, was one of the most popular health manuals of its time, selling 100,000 copies in its first decade of publication. Kellogg's concern was with natural sexuality rather than normal sexuality, but much of what he had to say about it accorded with what we have seen in sexology and psychoanalysis: sexuality was a site of volatility and vulnerability, which made the natural a fragile, precarious state: "If properly educated, and surrounded by the proper influences, a boy of [fifteen] will know nothing of the overwhelming excitements of the sexual functions," Kellogg asserted, "no natural demand for their use occurring until after the body has achieved full maturity. Unfortunately, however, the natural order of things is too frequently interfered with through the influence of evil companions, and the majority of boys become more or less contaminated morally long before this period."[82] As a result, Kellogg lamented, "real boys are scarce nowadays,"[83] and girls fared no better: "natural little girls are almost as scarce as real boys."[84]

Like the normal, the natural was surprisingly elusive, requiring vigilant supervision and self-discipline to maintain. Bernarr Macfadden was one of the best-known advocates of his day of disciplinary practices aimed at cultivating the natural. Macfadden was widely photographed in a loincloth to show off his famed musculature, and recommended fresh air and exercise as a remedy for the unnatural conditions of modern life. Like the sexual hygiene campaigners, Macfadden recommended widespread sex education. In his popular *The Virile Powers of Superb Manhood: How Developed, How Lost, How Regained* (1900), Macfadden reiterated Kellogg's conviction that natural manly vigor was being perilously undermined by urban culture:

> Ignorance of the facts in reference to the sexual instinct that should be as plain as the noonday sun to every human being, this, together with lack of knowledge of the great laws of health, so necessary in order to build vigor and symmetry of the body, have resulted in filling civilized countries with a host of pygmy men. Immediately after birth they come in contact with abnormal

influences. They are encumbered with clothing that discourages rather than encourages muscular movements; they are compelled to breathe foul air when the weather is cold; they are always overfed; the bottle often does duty for the female breast, and they come in contact with all sorts of conditions that tend to depreciate vitality. Of course over half are killed by all this, and those that survive are greatly weakened, and never attain the superb manhood that should be their inalienable right.[85]

Like men, women were corrupted by the "deplorable ... unnatural conditions" of urban life, so that "few of them grow into normal womanhood."[86] As a result, "the proportion of women diseased sexually exceeds twenty-nine in every thirty. [N]ot one woman in every hundred has a fair amount of sexual vigor, and at least nine in every ten, if not nineteen in every twenty, are more or less prostrated, or else actually diseased sexually."[87] Being natural was something that did not come naturally: like the normal it had to be worked at. Statistically, it was extremely rare. At the same time, it was also a site of change, able to be actively worked on and improved, not biologically inherited and determined, as Lombroso and Galton had claimed.

This interweaving of the natural and the normal remained foundational to the first large-scale public sex education campaign in the United States, which was run by the American Social Hygiene Association in the 1910s and 1920s.[88] The American Social Hygiene Association sought to make education about sex freely available, experimenting with popular media that could be exhibited in public places like YMCAs and YWCAs. The ASHA produced two poster series, *Keeping Fit* (1919), aimed at boys, and *Youth and Life* (1922), aimed at girls. The former was based on an earlier campaign run during the First World War for soldiers using a series of lantern slides and a film called *Fit to Fight*, which warned men they had a patriotic duty to avoid venereal disease while deployed: "a man killed by Germs is just as dead as a man killed by Germans," warned one slide.[89] The *Keeping Fit* and *Youth and Life* series, made up of 24 posters each, moved sex education into the public sphere, as they were exhibited free in public places, unlike the books of Freud and Krafft-Ebing, which had to be individually purchased. On the one hand, the ASHA posters represented sex as a natural part of life. For instance, poster 26 of *Youth and Life* read: "Sex endows the girl with beauty of body, vivacity, and charm of manner. It is the sex or creative impulse which inspires her warmth of affection, her intensity of purpose, her desire to devote herself to the welfare of humanity." Poster 20 of *Keeping Fit* read: "The sex instinct in a boy or man makes him want to act, dare, possess, strive. When controlled and directed, it gives ENERGY, ENDURANCE, FITNESS" (see figure 3). The ASHA campaign represented the "sex instinct" as something that was both natural and needing to be disciplined and channeled into productive pursuits. This advice recalled the kind found in the guides of Kellogg and Macfadden: boys and girls should eat well, exercise often, enjoy the outdoors, read good books, and avoid "self-abuse." At the same time, however, sexual hygiene insisted that health and fitness were not just personal qualities but also a social responsibility. Each individual was *personally* responsible for the health of the race, the ASHA posters stated. For boys, sex was natural but needed cultivation; for girls, running a home was the key source of happiness, but must be approached as a science (see figure 4). The ASHA exemplified a popular approach to sex education, representing sex as a natural force whose power was to be both celebrated and strictly controlled.

Campaigns like that of the ASHA, which were circulated among a mass audience, played a pivotal role in elaborating and popularizing a new concept of the normal in the first decades of the twentieth century, providing an important channel through which the idea of the normal moved from professional discourses into the public sphere. Guides and studies of human sexual behavior written at this time framed sex as a natural part of normal life but also as something that required constant monitoring and discipline. The "normal person" who emerged from such a context was one whose normality was a site of continual adjustment and negotiation. Malchow's *The Sexual Life* provides a striking example of how the normal person was understood. His description is one of the most detailed and extensive of this period:

> A thing is normal, or in its normal state, when strictly conformed to those principles of its constitution which make it what it is, and abnormal when it departs from those principles. . . .
>
> To be a normal man does mean to be of a certain height, weight, strength or mental caliber, but implies a condition which permits of the performance of those functions which are designated to and are in compliance with, the natural order of his being.
>
> To comply with these requirements he must be in both such a physical and mental state as will enable him to be adapted to his surroundings and live in accordance with social, as well as natural, laws.
>
> The standard must vary with the conditions, for the abilities of a man at the age of sixty vary greatly from those of a youth of sixteen, yet both should be considered normal. . . .
>
> It is difficult to formulate the rules whereby to definitively measure the normal standard, and if one is to be governed by what appears in the daily press and elsewhere . . . one would be forced to the conclusion that there are few, if any, normal men.[90]

The key characteristic of the normal person in Malchow's account was the ability to function in present circumstances and to be able to adapt when those circumstances changed. Normality was explicitly not a fixed standard or measure but rather a state of flexibility and balance. That state could be enhanced—or damaged—by a wide range of variable factors, making it an extremely difficult one to examine and evaluate. Or indeed to achieve: Malchow reluctantly concluded, as Kellogg and Macfadden had before him, that there were probably few normal men. The normal was an ideal most people fell short of. It did not represent a statistical majority or a quantitative measure. In fact Malchow explicitly dissociated the normal from the numerical.

In this way, texts such as Malchow's did not so much describe the normal person as invent that person as a new cultural figure. And while that figure was the product of disciplinary institutions and regulatory systems, they were not those of the legal system or asylum as they had been for sexology and psychoanalysis but those of public health reform, self-improvement and fitness cultures. Guides from Fowler's *Sexual Science* to Macfadden's *The Virile Powers of Superb Manhood* promoted forms of self-discipline that were also practices of self-cultivation, designed to provide personal fulfillment as well as improve fitness. One of the most popular guides of the late 1910s and 1920s was Marie Stopes's *Married Love: A New Contribution to the Solution of Sex Difficulties*, which aimed to improve and celebrate marital happiness. Stopes's book, like many such guides, was addressed to normal couples, which she defined as those without obvious sexual problems: "In the following pages I speak to those—and in spite of all our neurotic literature and plays they are in the great majority—who are nearly normal, and who are married or about to be married, and hope, but do not know how, to make their marriages beautiful and happy."[91] Unlike many of the other figures examined above, Stopes was not a psychiatrist or a physician, but rather a paleo-botanist.[92] She was a well-known popularizer of science, and one of the few women in Britain to occupy an academic position in her field. Stopes produced *Married Love* as her own marriage was being annulled on the grounds of impotence. Having initially failed to find a publisher for the book, she eventually managed to publish it in 1918 with the financial support of her then future husband, Humphrey Roe. The book was immediately and astonishingly successful, selling thousands of copies each week for the first few years of its publication. It made Stopes a household name in Britain, and she was inundated with letters from readers seeking advice on their own marital difficulties.

Although Stopes addressed her guide to the "nearly normal," the fact that so many readers required such a guide suggested that the space between the nearly normal and the normal was a significant one that was difficult to bridge. Normality was hard to achieve. Just as no one was perfectly normal or natural, no marriages were so perfectly happy that they could not benefit from advice and improvement. Stopes was thus able to address herself to readers "who are reckoned, and still reckon themselves, happy, but who yet unawares reveal the secret disappointment which clouds their inward peace."[93] Nearly normal sexual subjects were those who suffered from a sexual disappointment so vague and secret that they might even be unaware of it, assuming themselves happy even as their peace of mind was cast into an almost imperceptible shadow. This is why guides like Stopes's were required: to make truly happy those who only imagined themselves happy, or perhaps to suggest to those who believed themselves truly happy that they only imagined themselves so. In this way, although Stopes distanced her own work from sexology and psychoanalysis, whose focus on perversion and pathology she found distasteful, she nonetheless represented normal sexuality much as Krafft-Ebing and Freud had before her, as a precarious and elusive state that must be actively cultivated. Stopes's method for increasing marital happiness was to anchor sexual practice in natural biological processes. In so doing, Stopes introduced a quantitative and biometric element to the scientific study of sex that, as we have seen, did not play a prominent role in previous research in this field. The significance of this could be seen in her research on sex tides. As Laura Doan notes in her wonderfully illuminating study of Stopes's sex tide charts, Stopes created a series of time charts on graph paper by "entering her assessment of the daily values of her sexual arousal based on close and methodical self-observation. Using a statistical method called time series analysis, the botanist-cum-sex-researcher placed an x along a horizontal axis to measure the highs and lows from a point designated as a "datum line," a technique that, she would later argue, made "graphically clear" the regularity of the "fundamental rhythm of feeling."[94] Stopes believed that sexual feeling in women was periodic, and the regularity of its ebbs and flows could be charted and used as a marital guide. As Doan argues: "In formulating the expression 'sex-tide' Stopes created a neologism that wonderfully captured the uniqueness of a project at the nexus of biometry and sex research, construing the 'manifestations' of 'sex-feelings' as observable, measurable and, above all, natural—as natural as the rhythmic tidal flows of the lunar cycle. To correct what she perceived as the misconceptions of medical practitioners regarding the true nature of women's sex feelings, Stopes applied a rigorous statistical approach to the study of the body and populations."[95]

Stopes introduced statistical and quantitative research methods to the scientific study of sex that did not yet play a prominent role in sex research in the 1920s. Indeed, the role of statistics in sex research remained highly contentious until well into the 1930s, when the gynecologist and sex educator Robert Dickinson published his coauthored study with Lura Beam, *The Single Woman: A Medical Study in Sex Education* (1934). Dickinson contended that: "Studies involving human values cannot be settled as a yes or a no, objective or not-objective. Individuals cannot be added together and discussed as totals because they are not comparable."[96] However, "if statistics are taken with the necessary grain of salt, the simpler mathematical processes are helpful."[97] As late as the 1930s, studies of normal sexuality did not rely significantly on statistical methodologies, and the idea of the normal they elaborated was not primarily quantitative in meaning. Even texts that did refer to statistics as an authoritative source of truth, such as Robie's *Rational Sex Ethics*, did not refer to specific statistical data. Robie's chapter "Consideration of Statistics" did not cite a single statistic but merely asserted that the majority of men masturbated, and that there was no empirical evidence that the practice was damaging to health.[98] Rather, the normal as it was understood here referred to a state of overall balance and sustainability. It was not a stable condition or a standard but a site of constant adjustment undertaken during a period of rapid cultural change. The normal person, as imagined at the start of the twentieth century, was an adaptable and functional one, whose state was generally healthy and happy. If the "concept of the normal is itself normative," as Canguilhem argued, the regulatory systems it perpetuated in this context, that of public health at the start of the twentieth century, were very different from those of the large nineteenth-century disciplinary institutions examined by Foucault, or the legal-administrative systems within which Krafft-Ebing and, to a lesser extent, Freud, worked.[99] The idea of the normal that emerged from the context of public health literature and sex advice guides addressed the "nearly normal" subject for whom the cultivation of normality was a form of self-improvement or personal fulfillment.

CONCLUSION

Becoming normal was a difficult business at the start of the twentieth century: it required a great deal of work and attention, and it could never be ultimately or definitively achieved. No one could be perfectly or consistently normal. Everyone was somewhat perverse, or hysterical, or nervous, or secretly unhappy. At the same time, habits could be worked on, fitness improved, selves cultivated. The normal here did not operate as a standard, although a number of leading figures in this field were concerned it would appear so. Havelock Ellis worried that the insights of social hygiene would make managing sexuality seem bureaucratic and dull: "I fear," he wrote, "that by many persons social hygiene is vaguely regarded either as a mere extension of sanitary science, or else as an effort to set up an intolerable bureaucracy to oversee every action of our lives, and perhaps even to breed us as cattle are bred."[100] Lillian Gilbreth, the productivity and efficiency expert, articulated a similar concern about domestic science, which might impede its potential benefits for home-makers: "Housekeeping is an industrial practice," Gilbreth argued. "Industry has reduced much of its procedure to standard practice, and this is equally applicable to and available to housekeeping. . . . The home-maker may at first feel that to acknowledge that much of her work parallels work in industry is to make it mechanical, 'machine-like,' monotonous or uninteresting."[101] However, the application of such practices and the time efficiencies they produced allowed more time for other pursuits, Gilbreth argued. She recommended measuring time efficiency in terms of "happiness minutes."

Efficiency measures, such as those developed through the time-and-motion studies Gilbreth undertook with her husband Frank, allowed productivity to be maximized in sustainable ways in an increasingly industrialized workplace. The Gilbreths filmed the sort of repetitive motions seen in the early twentieth-century factory and office and identified redundant or inefficient movement to maximize productivity while limiting fatigue. In this respect, efficiency studies were a product of the same culture that produced *American Nervousness* and nervous exhaustion, in which the fatigue of modern life was a health risk that required constant management. Gilbreth's reference to industrial standards and her use of quantified analysis does seem to further attest to the increasing importance of standardization and quantified forms of knowledge within the human sciences at the time, as did Stopes's sex tide charts. The Gilbreths promoted their "One Best Way as a standard, a norm from which individual differences may be calculated."[102] Yet even here the normal was not a fixed measure but a dynamic and interactive system subject to a wide range of factors: "Balance seems to be the thing to keep in mind," Gilbreth reflected of efficient home-making, "a chance to let strained, tense nerves relax, to let relaxed nerves tense up, ready to function."[103] The conceptual elaboration of the normal that took place at the turn of the century and especially during the first decades of the twentieth century needs to be positioned in the context of these discussions about nerves and exhaustion as well as medical studies of perversity and pathology. The normal understood in this context refers to the condition of the subject under industrialization, with its increasing urbanization and mass production, rather than the nineteenth-century disciplinary institutions on which Foucault focused.

It is in this heterogeneous cultural space that a more popular idea of the normal began to emerge at the start of the twentieth century. As the term came into broader use, it also acquired more cultural authority. This authority was not, however, informed in any significant way by statistics and did not draw primarily on the authority of numbers until after the 1930s. Studies in sexology, psychoanalysis, and sexual hygiene hence occupy a pivotal place in the critical genealogy of normality because they were written in the period immediately before the convergence of the average and the healthy, of scientific quantification and medical examination. Where physical and criminal anthropologists had focused on the collection and analysis of anthropometric data, using quantitative methods of analysis, the texts examined in this chapter were largely narrative in approach. They relied not on the statistical chart but on the case study. Their concern was not primarily with average incidence—what proportion of the population engaged in particular sexual practices—but individual experience—how one person's history or biology shaped one's sexual practices and general psychology. We have seen the beginnings of this change when first Lombroso, and then to a much greater extent Galton, began to take the measurement and identification of statistically normal physical features as an indexical measure of character and temperament. Our genealogy of physical and criminal anthropology, and then eugenics, has shown how anthropometric analysis increasingly came to be used to identify not only racial types, understood as distinguishable by a set of fixed physical characteristics, but also intelligence, talent, and other mental faculties. Research in the field of sexual science thus represented a distinct shift away from the measurement of skulls that had preoccupied Broca and Lombroso, and away from the measurement of fixed physical characters, to the assessment of dynamic states of health and balance.

Studies in sexual science played a pivotal role in establishing sex education as integral to the management of everyday life, and were an important site for the construction of the idea of the normal person, allowing that figure to become established in the popular imaginary of the early twentieth century. Such studies were necessary because even the sexuality of ordinary people—those suffering no obvious sexual problems or pathologies—was itself increasingly understood to be at least potentially pathological and a source of social or personal problems. Middle-class men frequenting prostitutes, young women susceptible to seduction, wives unknowingly giving birth to syphilitic children, and adolescent boys weakening their constitutions by excessive indulgence in masturbation all took the place of the pervert and the criminal at the center of sexual hygiene literature. The normal, as it was understood in this context, was less a standard than a system; it was repeatedly described as a condition characterized by adaptability and moderation, by flexibility and ease of adjustment. The subject caught up in the dynamics of normality would be the adaptable subject called for by the mass production and consumer capitalism of the twentieth century, rather than the docile subject required and shaped by nineteenth-century social discipline.

FIGURE 3. "A Natural Experience." *Keeping Fit* American Social Hygiene Association poster series, 1919. Courtesy of the Social Welfare History Archives, University of Minnesota Libraries.

FIGURE 4. "Home-Making a Science." *Youth and Life* American Social Hygiene Association poster series, 1922. Courtesy of the Social Welfare History Archives, University of Minnesota Libraries.

CHAPTER EIGHT

The Object of Normality: Composite Statues of the Statistically Average American Man and Woman, 1890–1945

The star exhibit in the Hall of Anthropology at the 1893 World's Fair in Chicago was a pair of composite statues labeled "The Typical American: Male and Female" (see figure 5). Commissioned by Dudley Allen Sargent, the director of physical education at Harvard, and made by the artists Henry Kitson and Theodora Ruggles, these statues had been modeled by calculating the averaged physical dimensions/proportions of the vast collection of anthropometrics collected from college students, which Sargent had acquired in the course of his work.[1] Sargent's aim was to develop a "scientific" approach to physical education, and he saw anthropometrics as the best way to achieve this. During the 1870s and 1880s, he measured the bodily size and strength of thousands of (young, white, male) college students. He then compiled these individual results into a single chart, using the statistical mean—the apex of the bell curve—as the most typical for each measurement. He took the measurements of forty individual parts of the body, comparing left and right sides to assess their symmetry. He also tested for strength and stamina, using new devices he spoke of as his own inventions, calling them such names as the spirometer, the manometer, and the dynamometer.[2] In this respect, Sargent's anthropometric practice continued the history we examined above, where we saw that anthropometricians like Broca and Lombroso regularly developed new instruments and practices of measurement, quite unlike the later work in sexual psychology and hygiene. Yet Sargent's work also represented a significant shift in the history of anthropometrics. Where Francis Galton's composite portraits were seen as a record of the biologically inherited attributes of a type, which was fundamentally fixed and correlated to physical features, Sargent intended his charts and statues to provide information on how the body could be worked on and reshaped or changed through programs of (self-)discipline and efficient training. In 1887, Sargent published his charts, along with an explanation of their application, in the popular magazine *Scribner's*. The purpose of his charts, he explained, was "to furnish the youth of both sexes with a laudable incentive to systematic and judicious physical training by showing them, at a glance, their relation in size, strength, symmetry and development to the normal standard."[3] Anthropometric data about the statistically average body could be used to provide a template by which to assess and correct individual bodies, Sargent argued. In this respect, the purpose of Sargent's charts was more instrumental than diagnostic. Where earlier forms of anthropometric research had used statistical charts and composites to identify types understood to be relatively stable and to categorize individuals within them, Sargent saw his charts as providing a template for improving bodies by making them conform more closely to this statistical norm. The issues surrounding heredity, which had been such a vexed question in later nineteenth-century anthropology, had here receded in importance. Instead, Sargent's charts were designed to allow individuals to transform their bodies by measuring their distance from those normal standards of which they fell short, and engaging in practices by which they might move more closely toward them.

In this way, Sargent intended his statistical charts to have an explicitly normalizing function, in the Foucauldian sense, in that they allowed individuals to chart their position in relation to statistical norms and to try to move closer to those norms. We have seen that, for Foucault, normalizing technologies served a twofold purpose. In the first instance, their function was "primarily regulatory,"[4] since they were implemented through "infinitesimal surveillances, permanent controls, extremely meticulous orderings of space, indeterminate medical or psychological examinations" as well as "comprehensive measures [and] statistical assessments."[5] In the second instance, however, they were also productive. Thus, in the formulation we have previously examined, while normalization "imposes homogeneity," it is also what constitutes the modern subject as an "individual."[6] This is exactly how normal standards were conceptualized in Sargent's anthropometrics. For those whose measurements fell below the fiftieth percentile, the charts could be used to encourage self-improvement and correction. For those with measurements above the fiftieth percentile, Sargent's charts could provide a means by which to optimize their health and fitness and to measure their superiority in quantified ways. Sargent actively promoted this use of his charts at the 1893 World's Fair, where he ran a participatory Anthropometric Laboratory at which members of the public were measured and their anthropometric data recorded. Each participant was provided with an anthropometric record to take home. Sargent was not of course the first anthropometrician to establish a public laboratory of this sort: Galton had done the same thing at the 1889 London International Exposition, and in the last decades of the nineteenth century, being measured in public anthropometric laboratories was something of a fad. However, Sargent's charts and the statues he modeled from them were intended to serve a new purpose. Previously, anthropometric data had been collected for the anthropologist or other professional to interpret — primarily so that individual subjects could be identified according to their particular type. Sargent, however, wanted to train the general public to apply these anthropometric charts to programs of self-improvement that they monitored and managed themselves. As Catherine Howe notes in her study of Sargent's statues: "the hope was that the contrast between the physiques displayed by the statues and their own would inspire visitors to the fair to improve their fitness after returning home."[7]

From the first decades of the twentieth century, then, anthropometric data would no longer be seen as the province of scientists and other professionals, but were rather reconceptualised as part of an emergent culture of self-improvement and self-management. As we have seen, until the very end of the nineteenth century, use of the word "normal" was still largely confined to professional discourses, in which it had a quite specialized meaning. We have also seen that the word began to appear with great frequency in medical writing on sexuality at the turn of the century. In the 1890s, when Sargent was exhibiting and promoting his work, the word "normal" was still rarely used in popular contexts. This is reflected in Sargent's naming of his composite statues "The Typical American Male and Female": for Sargent, typicality, rather than normality, was the lens through which these statistically average bodies were understood. At the turn of the twentieth century, the normal remained an emergent, rather than established, idea in popular culture. The development whereby anthropometrics moved out of the prisons and hospitals and into the college further allowed use of the word "normal" to expand from closed professional spaces into the public sphere.[8]

This chapter will trace the development of a new concept of the normal within early twentieth-century anthropometrics through the history of composite statues, focusing on two further sets of statues exhibited in America during the first half of the twentieth century. The first of these is Jane Davenport's eugenic "Average Young American Male" statue, made in 1921, modeled from the vast collection of First World War soldiers' anthropometric data undertaken by her father, the eugenicist Charles Davenport (see figure 6). The second is a pair of statues made in 1945 by the gynecologist Robert Latou Dickinson in collaboration with the artist Abram Belskie, named Norma and Normman (see figure 7). While composite statuary, like composite portraiture, might seem to occupy a very marginal position in the histories of both scientific and popular cultures in the first half of the twentieth century, these three sets of statues are nonetheless highly representative of the history of the normal during this period. In the first instance, it is important to recognize that composite statues were valued as part of what Tony Bennett has termed the "exhibitionary complex" of their day—that network of cultural institutions, such as museums, galleries, and public exhibitions designed to improve those who saw them, thereby constituting the disciplinary subject and society.[9] Moreover, the very names of these statues—the "Typical American," the "Average Young Male," and "Norma and Normman"—exemplify the chronological and conceptual transformation undergone by the statistically average body during this period. This moment in our critical genealogy of normality is particularly important because the typical, the average, and the normal have now come to be thought of, and used as, synonymous terms. For instance, Sarah Igo's excellent *The Averaged American: Surveys, Citizens, and the Making of a Mass Public*, which examines the way participation in the large-scale social surveys that emerged in the early twentieth century helped to constitute new forms of citizenship, uses the terms "typical," "average," and "normal" interchangeably. Until the midcentury, however, those terms had been quite distinct. The three sets of composite statues examined here allow us to see these differences more clearly and thereby to recognize the specificity of the idea of the normal emergent over this period and in this context. As popular objects of display in anthropology, eugenic, and public health exhibitions, they also allow us to trace the extension of the term "normal" from scientific discourses into the popular sphere, and thus exemplify the much wider range of cultural contexts in which the concept of normality came to be found during the first half of the twentieth century. Thus, where we previously have focused on very concentrated sites of professional debate or activity, the circulation of composite statues requires us to consider a much more diverse range of cultural activity and spaces, as we follow the discursive and institutional emergence of a popular idea of the normal into the popular sphere.

After their exhibition at the 1893 World's Fair, Sargent's "Typical American Male and Female" were briefly exhibited at the Peabody Museum at Harvard University and have been in storage ever since. Despite their ephemeral display, however, the exhibition of Sargent's composite statues established a pattern that would also characterize the exhibition of the two following sets of statues: they established the statistically average body as central to discourses of public health and connected these to emergent cultures of self-improvement in a way that would make the idea of normality so resonant in mid-twentieth-century America, even as they drew attention to the insurmountable gap between the statistical ideal of an average and the variability of individual bodies. Indeed, following Foucault, it is precisely because the normal was conceived as a state of perfect averageness impossible to embody that it became so entrenched in the practices of everyday life at this time. We see this in Sargent's own instrumental approach to the identification of normal standards. Everyone must participate in practices of self-measurement and self-monitoring, Sargent argued, because no person could hope to be perfectly normal. Rather, "the same man may be above the normal in one measurement and below the normal in another."[10] Sargent's statistical composites thus produced an ideal body rather than a real one: "less than one per cent" of men had the properly symmetrical physiques indicative of the "harmonious development of all parts of the human economy essential to robust, vigorous health"[11] Because nobody could be perfectly normal, everybody needed to engage in practices of normalization based on self-measurement, self-monitoring, and self-management.

The "normal standards" that most concerned Sargent, precisely because he saw them as hardest to achieve, were those related to the physical symmetry of the body. In 1890, he launched a competition to find the most "perfectly symmetrical" man and woman in America, funding a $500 prize for the winners. While the results were enthusiastically announced in the *Boston Sunday Herald* ("The Human Form Divine: Prizes for Symmetry Awarded by Dr. Sargent"), it was noted that no living subjects could be found with perfectly symmetrical bodies.[12] This was the reason Sargent commissioned the "Typical American Male and Female" statues for the 1893 World's Fair: to provide a visible and concrete representation of bodies too statistically perfect to exist in the natural state. Yet what the statues revealed about typical Americans was the subject of debate rather than consensus. For Theodora Ruggles, sculptor of the "Typical American Female" statue, if this figure did not quite match the idealized form of classic Greek statuary, it was nonetheless representative of a healthy and attractive generation of young American women.[13] However, many contemporary reviewers did not agree with this assessment, criticizing the figure for her "thick thighs" and "weak spine," which were enough to make "believers in feminine loveliness shudder."[14] Such differences in opinion about what composite statues revealed about the overall state of the typical, or average, or normal American would be repeated in the public reception of Davenport's statue and then the statues by Belskie and Dickinson. In this respect, the history of composite statuary in the first half of the twentieth century reproduces the dynamic evident in earlier periods in the history of normality, one that may indeed be one of the defining characteristics of this history: at key moments of emergence, the term "normal" was repeatedly challenged and contested rather than embraced or widely agreed upon.

"THE AVERAGE YOUNG AMERICAN MALE"

Two new composite statues were featured in the exhibition staged as part of the Second International Congress of Eugenics at the American Museum of Natural History in New York in September and October 1921.[15] These statues were positioned at either end of a long hall. At the entrance, visible on arrival, was the figure of "the Composite Athlete, 30 Strongest Men of Harvard."[16] Opposite this, at the far end of the hall, a second figure could be seen: "the Average Young American Male, 100,000 White Veterans 1919."[17] The display of these two statues was intended to be a comparative one and was designed to prove a eugenic point. Where the statue of the Harvard "Composite Athlete" was presented as an aspirational figure, the idealized embodiment of the physical and mental elite of young white American masculinity, the "Average Young American Male" was presented as proof of the racial decline of the white American male population. Despite—or perhaps because of—this, it was "the Average Young American Male" statue that attracted all the attention during the exhibition. Very little was recorded about the design and manufacture of "the Composite Athlete," and no known images of this piece survive.[18] The "Average Young American Male," on the other hand, was the focus of a great deal of commentary in its own time and has recently been the subject of a number of scholarly studies on eugenics exhibitions.[19] This statue was modeled using the averages of the vast collection of anthropometric records of drafted and demobilized American soldiers undertaken during the First World War by Charles Davenport and

Albert G. Love, commissioned by the Office of the Surgeon General. Davenport and Love's first published report was called *Defects Found in Drafted Men: Statistical Information Compiled from the Draft Records* (1919). A second, expanded version of this was produced in 1921: *Army Anthropology: Based on Observations Made on Draft Recruits, 1917-1918, and on Veterans at Demobilization, 1919.*

Jane Davenport's "Average Young American Male" statue was produced the same year and was seen to demonstrate a worrying decline in white American masculinity: "Contrasted with the vigorous and idealised body of the composite Harvard athlete," Mary Coffey notes, "the average male's slight shoulders, distended belly, and lack of firm musculature implied that the national (white) body was degenerating as a result of an improvident mixing with inferior European stocks."[20] The "average man" of the 1920s was thus a very different figure from that of Quetelet's *homme moyen* of the previous century. Whereas for Quetelet the "average man" was "ideal" in more than one sense—not just an abstract outcome of mathematical calculation, but a model that deserved to be an object of admiration for a given population, just like Sargent's Typical Americans—eugenicists saw in the average man signs of degeneration and racial decline requiring intervention.

This is reflected in the wider context in which the "Average Young American Male" statue was exhibited. Although the data used to model the statue were drawn from enlisted and decommissioned soldiers, the influence of the Great War on the state of the average young American man was never mentioned during the Eugenics Congress when the significance of this figure was being discussed. Rather, the supposed decline was attributed to biological inheritance and immigration. This view was articulated in the public lectures delivered alongside the exhibition as part of the Congress. The opening address by Henry Fairfield Osborn (cofounder of the Galton society) exemplified this, exhorting the men in the room to recognize they were "engaged in a serious struggle to maintain our historic republican institutions through barring the entrance of those who are unfit to share the duties and responsibilities of our well-founded government."[21] As we have seen, such debates had been conducted in the closed spaces of learned societies and royal academies for some time, but in the 1920s they were increasingly disseminated to a general public through popular sites like museums and in public exhibitions. The poster designed to introduce eugenics to exhibition visitors, "What Is Eugenics?," was representative of this. Eugenics was defined there as "that science which studies the inborn qualities—physical, mental, and spiritual—in man, with a view to their improvement. Nothing is more evident in the history of families, communities and nations than that, in the change of individuals from generation to generation, some families, some races, and the people of some nations, improve greatly in physical soundness, in intelligence and in character, industry, leadership, and other qualities which make for human breed improvement; while other racial, national, and family stocks die out—they decline in physical stamina, in intellectual capacity and in moral force."[22] The two composite statues on display in the entry hall of the American Museum of Natural History were intended to illustrate this account of eugenics. The "Composite Athlete" exemplified "physical soundness, intelligence and character," while the "Average Young American Male" evidenced a "decline in physical stamina, intellectual capacity and moral force." The "average," in the 1920s, in this respect meant something very different from what the "typical" meant for Sargent in the 1890s: whereas the typical American was a figure to aspire to, a standard to normalize oneself in relation to, the average American was a figure of concern, requiring intervention.

That said, it is also important not to attribute a false and retrospective homogeneity to the way terms like the "typical" and the "average" were used in the first decades of the twentieth century, as their meaning was often a source of discussion and debate. As with Sargent's "Typical American" statues, debate about the significance of the "Average Young American Male" was played out in newspapers and magazines rather than academic journals, in popular sites rather than professional spaces. In the case of Davenport's statue, there was less debate about the aesthetics of the figure—there was a negative consensus on this point—rather, doubt was actually thrown on the project of composite statuary itself. One reviewer at the 1932 exhibition of the statue, during the Third International Congress on Eugenics (held once again at the American Museum of Natural History), described the statue as a "perfectly lifeless" and "bloodless figure." It was not a work of art and bore no resemblance to life, being simply "a chart in the round."[23] That is, its statistical representativeness did not make it a representative portrait of the average American. Rather, it was a purely imaginary figure: a statistical ideal that captured nothing essential of the embodied reality.

Such criticism of the project of composite statuary is important because it draws our attention to the fact that eugenics was not universally convincing or persuasive when it was introduced to the wider public in the sphere of public exhibitions during the 1920s and 1930s. While eugenics did play an important role in establishing anthropometrics within scientific and public health discourses in the United States at this time, and while this obviously shaped the conditions in which the popular concept of the normal would emerge, eugenics was not always accepted, much less embraced by the public, neither did it find a ready uptake in state or institutional administrative practices. Efforts prior to the Second International Congress of Eugenics to implement a eugenic program of involuntary sterilization in the United States had met with resistance, and in a series of legal challenges had been ruled unconstitutional. The year before the Congress, Harry Laughlin, who was Charles Davenport's assistant director at the Eugenics Records Office at Cold Spring Harbor, wrote the "Model Eugenical Sterilisation Law," which was designed to provide a constitutionally sound template for an easily adaptable and applicable eugenic sterilization program. The law targeted anyone who "regardless of etiology or prognosis, fails chronically in comparison with normal persons, to maintain himself or herself as a useful member of the organized social life of the state."[24] This law, whose application hinged on an interpretation of the "normal," at no point defined the word, while nonetheless moving to enshrine its value in law. After the Congress it did indeed come to be applied: the first sterilization law was passed in Virginia in 1924, and eventually laws of this kind were in effect in thirty states. Some 65,000 involuntary sterilization procedures were undertaken between 1927 and the mid-1960s.[25] While the implementation of eugenic policy is an important part of the history of anthropometrics in the twentieth century, and thus of normality, the extreme violence enacted against certain kinds of bodies has tended to overshadow the much more culturally pervasive and insidious ways in which anthropometrics informed the emergent disciplinary technologies of the public health and self-improvement cultures that had vastly more cultural reach over this period. This is why the history of twentieth-century anthropometrics is so important to a critical genealogy of normality: the large-scale projects of the mid-twentieth-century grew out of this history of the administration of public institutions and management of bodies inside them while growing toward something else—an increasingly consumer-oriented commercial world that emerged in the 1930s and 1940s. Anthropometrics, and the idea of the normal it carried, entered the American popular cultural imaginary not primarily through negative, overtly disciplinary programs or institutions, but through participatory and individually managed programs of self-improvement.

REIMAGINING THE AVERAGE: THE MIDDLE AMERICAN AS AN OBJECT OF STUDY

Throughout the 1920s and 1930s, the figure of the "average man" would become an object of widespread interest. However, while the figure of "the

Average Man" would become a regular topic of public discussion and professional debate from the late 1920s into the 1930s, this interest was routinely spoken of as a novel development. An article in the *New York Times* in 1927, reporting on Columbia University psychologist Harry Hollingworth's study of the "average man," exemplified this: "Science now turns its attention to the average man. Not so long ago it was the mentally defective, the diseased or the criminal whose characteristics were most industriously studied. Then the attention of science turned to the person with very high intelligence and unusual abilities. Now science gives us a picture of the man who does most of the work of the world, fights most of the battles, likes the movies, believes in stories with a happy ending and becomes the father of the generations of the future."[26] That historical overview was not an entirely accurate one, as this book has shown. First, the "average man" had been an object of scientific study since the mid-nineteenth century, as the work of Quetelet demonstrates. Second, studies of the exceptional and talented, such as Galton's *Hereditary Genius* (1869), were published alongside — or even before — well-known studies of the "defective" and "degenerate," such as Richard Dugdale's *"The Jukes": A Study of Crime, Pauperism, Disease and Heredity* (1877). Finally, Lombroso made a kind of conceptual composite out of criminals, mad people, and geniuses.

Yet it is certainly the case that in the 1920s and 1930s, the "average man" suddenly became a figure of great scientific and popular interest, in a way that represented a significant departure from the predominant nineteenth-century focus on social problems, such as that which motivated Charles Booth's landmark study of urban poverty in *Life and Labour of the People in London* (first published in 1889, and ballooning to seventeen volumes by 1903). Sarah Igo deftly describes this shift in *The Averaged American*: "Scientific characterisations of 'average' or 'typical' Americans were a striking phenomenon of the new century. This constituted a shift away from the almost exclusive study of 'degenerates, delinquents, and defectives' that had marked nineteenth-century social investigation . . . statisticians across the nineteenth century plied their tools to establish demographic medians and outliers. But rigorous inquiry for its own sake into the typicality of everyday practices and opinions was a twentieth-century enterprise."[27] In the nineteenth century, this kind of statistical work had often been undertaken to identify classes of people prone to illness, madness, or criminality. Here, the focus appeared to be on the ordinary person. And yet the habits and physique of the "average man" became the focus of scientific interest only when they were themselves understood to constitute a problem that needed to be solved. Both Davenport's statue of the "Average Young American Male" and Hollingworth's "textual composite portrait" of the "average man" were seen to reveal worrying signs of incipient degeneracy. As the newspaper's headline announced of Hollingworth's study: "The Average Man Found by Science. He Is Shown to Be Superstitious, Ill Educated, Conventional and Possessing the Mind of a Boy of 14 Years."[28] Thus, while much was made of the novelty and significance of Hollingworth's study, under closer scrutiny the figure of the "average man" was transformed into something much more familiar: a problem requiring social and scientific intervention. The distinction between the "average" and the degenerate or defective was frequently invoked in popular writing in the early twentieth century, only to be immediately undermined.

This is a consistent feature of how the average was understood in popular culture in the first decades of the twentieth century. On the one hand, there was a clear and growing interest in studies of such things as the typical and the average person. But on the other, there was no clear consensus about what those terms meant, so that they were often the subject of debate rather than agreement. Thus, as the typical and the average became the focus of scientific and social research, they also came to be understood as terms that described highly problematic categories, and as inherently precarious cultural formations. We see this played out in the emergence of the figure of the Middle American, which, like the idea of "everyday life," was a new one in the 1920s and 1930s. The Middle American found an important site of popularization in Robert S. and Helen Merrell Lynd's 1929 study *Middletown: A Study in Modern American Culture*, one of the most popular and successful social studies of the first half of the twentieth century. The Lynds' project was funded by the Institute for Social and Religious Research, and their aim, as they explained in their introduction, was to identify and study the average American amid the historical and cultural transformations currently taking place as a result of rapid urbanization and industrialization. In this respect, the Lynds' research was motivated by the same set of concerns that drove Sargent's and the eugenicists' work. However, their methods and approach were very different. In order to study the everyday practices of "average" or "ordinary" Americans, the Lynds used not the quantitative methods of physical anthropology, based on anthropometrics, but rather a new method of "cultural anthropology" by which the distinguishing features of ordinary American culture could be thrown into sharp relief, so that observers might "gain precisely that degree of objectivity and perspective with which we view 'savage' peoples."[29] While the

Lynds sought to study the inhabitants of Middletown scientifically, they also sought to measure something beyond the anthropometrics of physical anthropology, focusing instead on shared cultural practices.

To examine those practices, the Lynds undertook immersive fieldwork: for an extended period they lived in a medium-sized town they did not name but that was immediately identified on the publication of *Middletown* as Muncie, Indiana. They examined the inhabitants' work, leisure, family, community, and religious practices through interviews and participation in daily activities, including attending town meetings, religious services, community events, and so on. The Lynds had selected Muncie for their research because it seemed to them most representative of "ordinary" American life: of average size, located in the Midwest, inhabited by a fairly homogenous population of white Protestants, and in the process of rapid industrialization. The Lynds did not see this change as positive, and they were critical of what they saw as Middletowners' overly enthusiastic adoption of a consumer culture, which threatened to disrupt the economic and social patterns of their lives. They noted, for instance, that the purchase of (then very expensive) motorcars and radios was transforming traditional practices and social structures in ways that served to fragment what had previously been a cohesive community, establishing new divisions between rich and poor, old and young. Middle America, at its point of emergence, was already a nostalgic concept, evoking a "typical American" way of life that was under threat from consumer culture and urbanization.

The Lynds acknowledged that finding a suitably Middle American town, with a predominantly "native white" population, was already difficult in the 1920s, and that this made Middletown statistically nonrepresentative.[30] In addition, the Lynds then excluded Middletown's African-American population from their survey entirely, arguing that studying "racial change" would complicate and compromise their primary focus on "cultural change."[31] In this way, while this focus on cultural practices distinguished the Lynds' approach to anthropology from the focus on anthropometrics and biological typing that dominated physical anthropology and eugenics, assumptions about race nonetheless played a structuring role in shaping their study. On the one hand, then, the Lynds' methodology was consistent with statistical analysis as it was practiced at the time. As we have seen, researchers in scientific and statistical analysis had maintained that it was possible to marginalize certain people without formulating explicit judgments about their supposed inferiority, since it was taken as a requirement of method that statistics properly conducted required the elimination of individuals who would distort the calculations. Because nonwhite Americans were supposed on the basis of research in physical anthropology to be of a different type from that of whites, the inclusion of these different populations could be expected to skew the results of statistical inquiry. If nonwhite populations were studied along with whites, researchers tended to suppose, there would be more than one median and more than one cluster, with a consequent distortion of the distributional curve. As a result, although the concept of Middle America that emerged from the Lynds' study was an exclusively white one, this could be represented as the product of scientific impartiality and in the interests of statistical accuracy, rather than political exclusion.[32] At the same time, however, the Lynds also acknowledged that Middletown was a largely imaginary place and the idea of the Middle American a cultural composite. As Sarah Igo argues: "*Middletown* made 'the typical' visible, and empirically real, at a moment when any sense of American commonality was difficult to discern and national culture seemed deeply fragmented and unstable."[33] Thus, while the Lynds' work did much to introduce and establish the figure of the Middle American in the cultural imaginary of this time, it did so in a way that foregrounded, precisely, the imaginary qualities of this figure. The Middle American who emerged from the pages of *Middletown* was as fictional as the ghostly portraits that emerged from Galton's composite photographs.

However, the cultural impact of the Middle American was exponentially greater than Galton's composite photographs of anthropological types, or his books on eugenic theory, and circulated among a much wider and more heterogeneous readership. Where Galton's work had been read primarily by a small community of professional men, or by the educated readership of specialized journals, *Middletown* was a genuine blockbuster. It was one of the first studies of this kind to appeal to a mass audience, selling a quarter of a million copies in the first year alone. Clearly readers in 1929 were fascinated by this portrait of the experiences of ordinary Americans in the midst of changing cultural conditions. As one reviewer put it: "Nothing is so interesting as ourselves, and [reading *Middletown*] was like looking at yourself in a mirror."[34] One indication of the cultural influence and reach of the Lynds' study can be seen in the rapid expansion of the concept of Middle America it popularized. As Igo notes, both *Middletown* and the extensive public discussion about its significance represented a "repeated emphasis on the middle, whether the middle class, the 'middle-of-the-road,' or the largely implicit Middle West."[35] A further sign of the rapid expansion of this term was the fact that, within a decade, it could be used to refer to an entirely different sort of subject from that described in the Lynds' study. This is exemplified in the film-length advertisement made by Westinghouse to promote its exhibits at the 1939 World's Fair in New York and titled *The Middleton Family at the New York World's Fair*. This film explicitly addressed contemporaneous concerns about the cultural and economic impact of industrialization on Middle American life, such as those articulated in *Middletown*. The narrative begins with young son Bud, sulky and disaffected, complaining that the automation of work and the rise of mass-produced products meant that boys of his generation would be unable to find meaningful work or social purpose. Daughter Barb, an art student in New York, has fallen under the influence of her socialist European tutor and is also critical of the consumer culture promoted by companies like Westinghouse. Reunited in New York to visit the World's Fair, the Middleton family is given a private tour by a rival for Barb's affection, a Westinghouse engineer called Jim, who is also from their hometown. Jim is the voice of reason in the film, repeatedly consulting his notebook to recite statistics that prove how industrialization will quantitatively and qualitatively increase both affluence and leisure. In one scene, he takes the Middleton women to see a dishwashing competition that pits "Mrs. Drudge," washing by hand and looking increasingly harried and dishevelled, against "Mrs. Modern," who stands serenely beside her dishwasher as it does all the work. (The men go to see the Westinghouse Robot, a life-size figure able to do addition, answer general knowledge questions, and smoke a cigarette.)

This figure of the Middle American, celebrated and reassuringly confirmed in the Westinghouse film, had not eclipsed the figure of the typical American celebrated fifty years earlier by Sargent, however, but rather had developed alongside and in relation to it. That becomes apparent in the staging of another contest designed to find the living embodiment of the typical American, also held at the 1939 World's Fair. The competition in question was run by sponsors of the Fair, including Westinghouse, and was designed to identify the most "Typical Families" from states across the country. Winners were offered an all-expenses-paid trip to the fair for one week. Entrants were "invited to write essays explaining why they were typical."[36] Although families of all sizes could enter the competition, however, only two parents and two children were able to take the winning trip. The families were photographed meeting admiring crowds behind a white picket fence, and interviewed at the Westinghouse radio booth. Each family member was asked one personal question — e.g., "What is your favorite exhibit at the Fair?" — and required to make one sponsor endorsement — "And how did you get to the Fair?" "By Ford motorcar, Sir!"[37] Here the typical American family member has become a spokesperson for commercial organizations, and is defined almost exclusively in terms of his/her product choices. The interviews included no questions or reflections about the typicality of the family itself.

The Middle American, as a new cultural figure emergent in the 1930s, was one enmeshed in a developing consumer culture, but once again the significance of this was a subject of debate rather than consensus, a debate increasingly taking place in the mass media of advertising, radio, and cinema: in the context of *The Middleton Family at the New York World's Fair* and the "Typical Families" contest, consumer culture was seen as a positive productive influence; in the context of *Middletown*, it was understood as a force that undermines traditional American ways of life. This latter view was still in evidence well into the 1940s, as seen in the popular film *Magic Town* (1947). *Magic Town*, like *Middletown*, reflected ambivalence about the changing economic and social organization of midsized American towns during the first half of the twentieth century. The film follows a struggling opinion surveyor, Rip Smith, played by Jimmy Stewart, as he discovers a town whose inhabitants' views are perfectly representative of the national population's—a quality Smith promptly and fully exploits. In so doing, however, he brings the small town and its inhabitants to national attention. Under the glare of public scrutiny, the residents' opinions and preferences skew, and their lives are exposed not as representative but ridiculous. The town itself becomes a national laughing stock—although the object of ridicule in the film is less the actual inhabitants of Magic Town than the commercial belief in this magical space of pure statistical representativeness. While the source of mockery here is different from that of the *New York Times*'s mockery of Hollingworth's "average man" twenty years earlier, then, the extent to which the average was a potentially risible category during the first half of the twentieth century is again apparent. The average had been mocked in the French Academy of Medicine in 1837, and it was still vulnerable to attack. So while *Magic Town* and *Middleton Family at the New York World's Fair* both attest to the widespread adoption of the term "Middle America" after the Lynds' study, their very different perspectives on the significance of this figure also draw attention to the continual debate that arose around this concept as it came to circulate more broadly in the public sphere.

Given how central *Middletown* was to the way the typical American was understood in the 1930s and 1940s, it should be emphasized that the object of the Lynds' study was, precisely, the "Middle American," rather than the "typical" American, as it had been for Sargent, or the "average American," for Davenport. While the "average" or "typical" American" referred to a statistical measurement, the Middle American suggested something more vague but also more integral to the cultural imaginary of the time: a set of shared cultural practices and values. The Lynds did not physically measure "Middle Americans" and base their findings on biologically or anthropometrically determined results: rather, they recorded the feelings, experiences, and opinions of the residents of Middletown to find those most representative of its culture. This shift would have far-reaching effects in American anthropology and its emergent survey culture, as Igo has shown: from this time onward, what would be measured and assessed in social studies of large populations was not just the physical body but subjective states and opinions, mined for their commercial utility. This context would be constitutive of the normal as it moved into popular use in the midcentury. However, it is significant that, as late as the 1930s, this subject was not yet identified as a normal one. Although Igo argues that the Lynds' work revealed "an unwavering concern with the normal as an object of study," albeit one grounded in "a very peculiar vision of the normal,"[38] the text of *Middletown* itself does not include the term "normal," and Igo's citations from Robert Lynd on the subject of normality come from letters written much later than the text of *Middletown*. The Lynds' use of the term "middle" is an important precursor to that of the "normal," and the concept of the "Middle American" continued to occupy a central place in the American cultural imaginary of the midcentury in ways that shaped the distinctive meaning taken on by normality at that time. Yet the two terms are not identical, and the difference between them allows us to identify more clearly the distinguishing features of "normality" during a period when this term was just beginning to enter into popular use. The concept was being used much more broadly than previously, and with a signification that was decidedly less precise.

THE GRANT STUDY OF NORMAL YOUNG MEN

We see that change played out in the first study to take the "normal" person explicitly—rather than the typical, average, or Middle American—as its object of study. The research in question began at Harvard in 1938, nine years after the Lynds' *Middletown* was published. It was known as the Grant Study of Normal Young Men, after its financial patron, dime store magnate W. T. Grant. The study was directed by Arlie V. Bock, professor of hygiene, and was intended to identify "that combination of sentiments and physiological factors which *in toto* is commonly interpreted as successful living."[39] A team of researchers collected a broad range of anthropometric, physiological, social, and psychological data from the participants, with the aim of identifying the behaviors and biological indicators of successful, well-adapted men. They conducted a range of physical and psychological tests, studying the participants' "personality" and "adjustment patterns," as well as "socio-economic," "morphological," "physiological," "medical," and "mental" measurements.[40] The results appeared in two separate books published simultaneously in 1945: one by Earnest A. Hooton, written for a popular audience, and reassuringly titled entitled *Young Man, You are Normal: Findings from a Study of Students*, and a second, more academic, report, written by Clark W. Heath, called *What People Are: A Study of Normal Young Men*. This double publication is itself significant. It shows the means by which the discursive and cultural expansion of anthropometrics from professional discourses into popular culture was being achieved. It is also an important reminder of how late this expansion occurred, and how recently the word "normal" came into more popular, everyday use. Both Hooton and Heath felt called on to provide detailed, highly qualified definitions of "the normal" at the start of their reports, which suggests that the word was not yet in common use, nor could its significance be taken for granted even when addressing an educated audience. Hooton refers his readers to the more precise definitions provided by Heath, who begins his report by explaining: "For our present purposes, 'normal' is defined as the *balanced*, harmonious blending of functions that produce good integration. Many kinds of such integrations are reflected in widely divergent types of personality and behaviour. The 'normal' individual, therefore, here is regarded as the *balanced* person whose combination of traits of all sorts allows him to function effectively in a variety of ways."[41] Normal was being identified not as a fixed state—or a statistical measure—but as a condition of physiological and psychological balance, associated with optimal adaptability and self-management. This understanding of the normal as a relational property recalls the one we first encountered in Étienne Geoffroy Saint-Hilaire's work, which saw the normal state as a condition of harmony that enabled the smooth functioning of the body. It also recalls that provided by sexual hygienists and popular psychologists like Malchow, for whom it was a state of balance and moderation. This is no coincidence. Unlike the anthropometricians examined earlier, Heath's approach to normality was a medical, rather than statistical, one. Heath's primary training was in medicine, and he would later become a professor of hygiene at Tufts, specializing in hematology.

Heath himself was conscious of a difference in perspectives, arguing that it was medicine, not statistics, that provided the most "strategic position [from which] to attack the problem of the diagnosis of the normal person."[42] A key reason for this was that the medical understanding of the normal allowed for far more flexibility—in conceptualization and application—than statistics could provide: "This idea of 'normal' is accordingly an elastic one. It does not restrict our studies to the statistically average person, nor does it project us into a vain search for the normal person in the sense of a 'perfect' one."[43] Unlike the perfect person, or the type, the normal person was not identifiable as a fixed or static state. This made the normal a particularly difficult and challenging area of study: "the varieties of traits of medically 'normal' people are fully as complex and numerous as those of the manifestations of disease."[44] For this reason, the study of normal young men was no less challenging, nor any less important, than that of the diseased. So it was both curious and concerning, Heath noted, that the subject of normal men remained so underexamined: "While large endowments have been given and schemes put into effect for the study of the ill, the mentally and physically handicapped, very few have thought it pertinent to make a systematic inquiry into the kinds of people who are well and do well. Study of the normal person, who is not seriously afflicted with some disease or defect and whose behaviour does not get him into difficulties in society, has been comparatively neglected."[45] We have seen this assertion made repeatedly, in relation to the typical, the average, and the Middle American: here it was the lives of normal men that had been neglected by science in favor of the diseased and defective. It might have seemed that this was because normal men did not require intervention and assistance; however, it had to be recognized that "the normal person who does not have problems for which he would welcome assistance is a rarity."[46] That was why normality and abnormality, or normality and disease, ought not to be thought of as binary opposites, Heath argued. They were, rather, possibilities ranged along a spectrum, so that "one may find the 'normal' within the person having mental or physical abnormality"[47] and vice versa. This is a medical concept of normality that we have examined at length in the context of writing on sex. However, Heath's understanding of the normal does more than simply reiterate the view we identified in a previous generation of work. Undertaking his research in the context of a multidisciplinary study that drew extensively on quantitative research methods, Heath conceptualized the normal as a quality of adaptability. Indeed, the follow-up study to Heath's report, published in 1977 was titled *Adaptation to Life*.[48] For Heath, normality was quite precisely the capacity to adapt successfully to one's environment. What was required for the "successful living" the Grant study set out to examine was flexibility and adaptability to the changing conditions and circumstances of one's life.

While Heath proposed a different understanding of the normal from the one found in anthropometrics—for which it was a numerical measure or standard range—the Grant study itself did much to entrench a statistical, quantitative understanding of normal in popular culture. One way it did so was by commissioning Hooton to write the popular account of the study. Hooton was a well-known physical anthropologist at Harvard before becoming in the 1930s curator of somatology at the Anthropometric and Statistical Laboratory at the Peabody Museum. He was also a popularizer of eugenics, concluding his 1936 article "What Is an American?" with a proposal for involuntary sterilization: "Every racial strain in our country should be purified by the sterilization of its insane, diseased and criminalistic elements. The candidates for such biological extinction would not be selected on the basis of Aryan or Semitic descent, blond hair or black skin, but solely on the score of their individual physical, mental and moral bankruptcy."[49] While such eugenicist positions clearly informed the concept of the normal as it emerged from Hooton's account of the Grant Study of Normal Young men, it should be recognized that such arguments were not necessarily widely persuasive in the 1930s. Hooton's book *Apes, Men and Morons*, published in 1937, was criticized in book reviews for being biologically determinist: "It is Hooton's view that mental functions are the functions of a biological organism, that mental functions are biologically determined, and that the ability to function up to and beyond a certain level is biologically determined," pointed out one critic.[50] Hooton's 1939 anthropometric study of fourteen thousand criminal bodies, *The American Criminal*, was similarly criticized for recalling the work of Lombroso in supposing that the existence of criminal types could be identified by biological markers.

The belief that types could be recognized by particular biological markers also informed the Grant study, especially in Hooton's report. It informed the manner in which the researchers sought to identify the relationship between particular kinds of physique and health, personality, social capabilities, and intelligence. Many of the anthropometric and psychological tests for the Grant study were overseen by the "constitutional psychologist" William Herbert Sheldon, who also undertook a large-scale study of posture and health across Ivy League and "Seven Sisters" universities from the 1930s to 1960s. From this research, just as Sargent had developed a template of the statistically average American, Sheldon developed a theory of somatotypes, in which bodies could be identified as one of three types: ectomorphic, predisposed not to store fat or muscle; mesomorphic, muscular; or endomorphic, predisposed to store fat.[51] Sheldon's primary interests were psychological, and he attempted to demonstrate a statistical correlation between somatotype and psychology. He referred to this as "constitutional psychology." Drawing on a modified version of Sheldon's somatotypes, the Grant study measured the "masculine component" of each participant, identifying his "type" of masculinity.

The reliance on biological markers and the extent to which particular body "types" were correlated to leadership potential and success in adult life suggest a much more fixed and quantitative understanding of normality than that proposed by Heath, and one much closer to a physical anthropological or eugenic approach. The difference between Hooton's and Heath's concepts of the normal—found in the two publications commissioned by the Grant study itself—is representative of the broadening range of significance the word "normal" itself took on in the 1940s, and indeed betrays the fundamental incompatibility of some of these meanings. Heath's scientific report and Hooton's popular book thus exemplified the way the normal moved from the context of scientific research, like that of the Grant study itself, into popular culture. Within the context of the study, the normal served as the point of conceptual intersection at which the different methodologies of measurement making up the Grant study converged, without being reducible to a single coherent definition. It is precisely because it drew on both medical and statistical understandings of the normal, and not always coherently, that the Grant study contributed to the conceptual expansion and increasing cultural uptake of the normal. Conceptual elasticity—in which normality could be used to signify even incompatibly different states—enabled it to expand rapidly and widely.

BETTER THAN AVERAGE: THE MAKING OF NORMA AND NORMMAN

The two composite statues made by Dickinson and Belskie, Norma and Normman, exemplified this expanded and heterogeneous context in which the word "normal" moved into more widespread popular use in the middle of the twentieth century (see figure 7). They were produced in 1945, the same year that the findings of the Grant Study of Normal Men were published. The statues were the product of an earlier successful collaboration between Dickinson and Belskie, who had produced a collection of obstetric models, called the "Birthing Series," for the 1939 New York World's Fair, models designed to make medical knowledge both appealing and comprehensible for a broad audience.[52] Dickinson was an energetic campaigner for public sex education who had published well-received books, such as *The Single Woman: A Medical Study in Sex Education* (1934), coauthored with Lura Beam, based on over five thousand case studies.[53] The exhibition at the New York World's Fair was enormously popular, and afterward, Dickinson's correspondence reveals, he was inundated with requests for photographic plates and plaster models. Norma and Normman were intended to complement this collection of obstetric models, representing the reproductive adult male and female. Normman, like the "Average Young American Male," was modeled on Charles Davenport's vast collection of First World War anthropometrics, augmented by a number of smaller data sets: the records from an anthropometric laboratory Dickinson set up at the Chicago World's Fair in 1933 and insurance company data, as well as results from surveys of male college students undertaken by physical anthropologists, like Hooton, working in university contexts.[54] Norma, however, was modeled from a new data set: the averaged measurements of "native white" American women recorded and standardized by the Bureau of Home Economics in 1940, in an attempt to devise the first standardized system of sizing for ready-made clothes.

Dickinson's first reference to the statues was in a letter to his publisher written in 1945, regarding the preparation of *Human Sex Anatomy: A Topographical Hand Atlas* (1949), which helps to date the manufacture of the statues to late 1944 or early 1945.[55] Referring to the (much delayed) completion of the book, Dickinson wrote: "Now it has a noble opening—the perfect woman, the average American, in a Belskie sculpture in four aspects.... Over it will be a transparency of four outlines, front, back and side, with every measurement.... The man is as fine. I am urged to put him in, also."[56] In a subsequent letter to Hooton, written three months later, Dickinson suggested that he was attempting to produce the figures as anatomical models: "These figures will be opened up to show their interior organs and also made in plastics with individual systems, like nerves and blood vessels and lymphatics. Attempts—life-size—were found too large, and one-half dimension is adopted."[57] While Dickinson did include the anthropometric charts in *Human Sex Anatomy* when it was finally published in 1949, the statues were never made to include the internal anatomy.

Of the three sets of composite statues exhibited in the first half of the twentieth century, Norma and Normman were by far the most significant. They would remain on display for an extended period, and copies can still be seen today at the Warren Museum of Anatomy, at Harvard University. They are also representative of the moment in which the word normal first began to move into widespread popular use and acquire its contemporary significance. Norma and Normman were normal in the medical sense of the term, in that they were designed to make visible the "normal anatomy" and "normal state of health" of the average young white American, a project Dickinson argued (incorrectly) "ha[d] never been undertaken before."[58] They were also "normal" in the statistical sense, with their dimensions calculated from the average anthropometrics of (young white) American bodies.[59] Beyond this, however, Norma and Normman were also "normal" in another, newer way. We can see this by comparing the figure of Normman with that of the "Average Young American Male." Where the "average male" was slouched, Normman stood tall; where the "average male's" muscles were soft, Normman's were clearly defined. These differences are particularly striking given that the two statues were modeled from what was essentially the same data set.[60] Anthropometric measurements may have provided a great deal of precise information about the average dimensions of the body, but they did not provide a record of musculature or posture; this was left to the discretion of the artist. The difference between the two statues was thus a manifestation of the role interpretation played in their modeling, and of the difference between the average and the normal body in the scientific imaginary of the time. Where Davenport associated averageness with degeneracy, for Dickinson normal was a mark of privilege. Like Sargent, he believed that the exemplarity of both Normman and Norma made them the most beautiful and "perfect" of their type. He described Norma to his publisher as "the perfect woman, the average American."[61] To Hooton, he described her as the "perfect lady."[62] The notion of perfection had been given scientific value in the work of nineteenth-century anatomists such as Cuvier, for whom the type was not to be thought of as flawless but as perfect in a formal, categorical sense: it stood perfectly for the whole species. It appears there was something of this in play in Dickinson's work a century later. Normman and Norma, like the typical American, were represented with balanced and harmonious physiques, free from extremes: they were fit but not too muscular, neither too stocky nor too thin. Their perfection derived, precisely, from their "normal" or typical states. This was a much more positive view of the statistically average body than the one represented in Davenport's "Average Male" statue.

Norma and Normman first went on display at the American Museum of Natural History in New York, the location of the Second and Third International Congresses on Eugenics, in June 1945. An article written by curator Harry L. Shapiro, entitled "Portrait of the American People," appeared in the Museum's journal, *Natural History Magazine*, in the same month. Shapiro, who had been Hooton's student at Harvard, was an associate curator during the eugenics congresses and would serve from 1956 to 1963 as the president of the American Eugenics Association. He argued that Norma and Normman were important for two key reasons. First, they contributed to wider anthropological studies of the "native white" American type, which he argued was a relatively recent area of research and subject of contemporary fascination. Norma and Normman provided a rare, scientifically accurate portrait of this type.[63] Second, the statues prompted the viewer to think about what statistics had to tell us about that type and the relation between the "average" and the individual, in a way that drew attention to the differences between these terms:

> Norma and Normman exhibit a harmony of proportion that seems far indeed from the usual or average. One might well look at a multitude of young men and women before finding an approximation to those normal standards. We have to do here then with apparent paradoxes. Let us state it in this way: the average American figure approaches a kind of perfection of bodily form and proportion; the average is excessively rare.
>
> Ordinarily, when we think of perfection . . . we place it at one extreme of a curve of frequency, whose middle range or average is equivalent to mediocrity. Virtuosity, for example, is never at the middle range of a curve of frequency. . . . Why, then, should the average of our bodily proportions strike us as a form of perfection, and if it is average, why is it rare? I shall confine myself to statistical explanations. The extremes of any single physical character are generally statistically rare, whereas the average is frequent. . . . But the combination of many [individual] averages in one individual is rare and unusual.[64]

Whereas for Quetelet there was no inherent tension between the typical and the ideal, for Shapiro the coexistence of the typical and the ideal had become an "apparent paradox" to be explained. How was it, he asked, that these models of the average American young man and woman seemed so much better and more perfect than the actual average American? The answer, he explained, came from understanding the relationship between statistics and individuals. The "perfectly average" bodies represented by Norma and Normman were "excessively rare" because they embodied an anthropometric average in not just one part of their bodies, but in every single aspect. Any individual body was statistically unlikely to coincide with so many averages. Like Sargent's Typical Americans, they were ideal and imaginary figures.

Shapiro's article set the tone for the reception of the Norma and Normman statues in the popular press, which was primarily positive, whereas those for the "Average Young American Male" in 1921 had been predominantly negative. After their short exhibition at the American Museum of Natural History, the statues were purchased by the Cleveland Museum of Health, where they went on display in July 1945.[65] The museum also bought the "Birthing Series" at the same time, along with reproduction rights for photographic plates and plaster models for all Dickinson and Belskie's models. The collection was exhibited together as "The Wonders of Life."[66] The Director of the Museum, Bruno Gebhard, had known of Dickinson and Belskie's work for some time, having overseen the public health exhibitions at the 1939 World's Fair in New York. Gebhard's involvement is another reminder of how important eugenics was to the exhibitory culture of the time: Gebhard had been the curator of the Deutsches Hygiene Museum in Dresden from 1927 to 1935. He was well known for his eugenics exhibition at the Second International Hygiene Exhibition in Dresden, which was attended by a number of American health workers, including representatives of the American Public Health Association.[67] This exhibition had featured the popular Transparent Man and Transparent Woman exhibitions: large plastic anatomical models that allowed the internal organs to be illuminated with the press of a switch. The Transparent Man was also exhibited at the 1933 World's Fair in Chicago, where Dickinson had run an Anthropometric Laboratory, and may have influenced his desire to produce Normman and Norma as anatomical figures that could be "opened up to show their interior organs." In 1934, a group of American eugenics supporters had arranged for a Gebhard-curated exhibition to be brought to the United States. Titled "Eugenics in New Germany," it toured the United States in 1934 before eventually finding a permanent home at the Buffalo Museum of Science.[68] Gebhard's role in public health education through exhibitions from the 1920s to 1960s thus reminds us of the interconnectedness of this field with that of eugenics, but also draws attention to the complexities and differences within the field of eugenics itself. Gebhard was an enthusiastic supporter of eugenics but was also opposed to Nazi eugenics: he lost his position at the Deutsches Hygiene Museum for refusing to join the Nazi party and relocated to the United States shortly afterward. None of the exhibitions he organized or curated in America were explicitly eugenicist;

rather, they reinterpreted eugenic ideals of racial improvement within the American context of public education and self-improvement seen in Sargent's work. Gebhard spoke of Norma and Normman as Sargent spoke of the Typical Americans: as templates for physical correction.

Although Dickinson himself had many close professional relationships with well-known eugenicists, such as Hooton, he did not view the Norma and Normman statues through the lens of eugenics, nor in the anthropological context that interested Shapiro. Dickinson was a gynecologist and sexologist, and he saw the statues quite differently, as part of his campaign to encourage more frank and open discussion about human sexuality. At an official luncheon to celebrate the purchase of the statues in July 1945, Dickinson congratulated the Cleveland Museum for taking the bold stance to display the statues: "No revolution is greater than the sex revolution that is in progress," he said in his address. "To hide, to hush, to hurry past is giving place to a directness of approach. But a vocabulary is needed and an anatomy to give language to the new attitude. . . . We have never studied sex behavior until now. Your museum has become a leader in this field."[69] That Norma and Normman were taken to be such important figures in three key areas of early twentieth-century thinking — anthropology, eugenics, and sexology — shows how much these statues were loaded with cultural significance. That they could be taken to exemplify such different things in each of these areas attests to the increasingly wide range of cultural spaces and discursive networks in which the term "normal" itself was beginning to circulate at this time.

This overloading of meanings has been recognized in two recent cultural histories of normality that both open with detailed discussions of the Norma and Normman statues: Anna Creadick's *Perfectly Average: The Pursuit of Normality in Postwar America* and Julian B. Carter's *The Heart of Whiteness: Normal Sexuality and Race in America, 1880–1940*. For Carter, the statues are representative of the new formations of heterosexuality and whiteness emergent in the United States at the start of the twentieth century. He argues that whereas whiteness and heterosexuality had often been seen as precarious cultural formations in the late nineteenth century, by the twentieth they had become so institutionally and discursively entrenched that their specificity had disappeared. In this way, whiteness and heterosexuality were dissociated from their history of disciplinary authority, acquiring "the appearance of blank emptiness and innocence."[70] Normality was the name for that disappearance or erasure of specificity: a scientific and quantitative term that could be used as proof of epistemic neutrality. Like

Carter, Anna Creadick argues that the cultural force and authority of normality as it emerged in the middle of the twentieth century derived in large part from its claims to mathematical rationalism and scientific objectivity. For Creadick too, Norma and Normman are representative of the manner in which ostensibly objective facts—statistical averages—were also shaped by interpretive frameworks and evaluative judgments. Creadick, however, argues that even at its point of emergence, the concept of normality was the subject of contention rather than agreement, and that it fragmented as much as consolidated emergent ideas about gender, sexuality, race, and embodiment. The postwar American culture of conformity so closely associated with the popular concept of normality was, even at its inception, a focal point for resistance and dissatisfaction, Creadick argues, perceived by some as stultifying and oppressive: "Even as it was being employed at this time, however, normality was also being questioned and critiqued. The concept shifted from . . . claiming the authoritative discourse of scientific rationality to voicing a contradictory discourse of popular psychology; from offering a source of security to being a metonym for conformity and danger to 'progress.'"[71]

Both Carter and Creadick take Norma and Normman as representative figures of a postwar American culture in which new cultural understandings of whiteness, heterosexuality, and conventionality were emergent. Both are concerned to trace the extension of this new concept into a broad range of popular cultural domains. So it is unproblematic within the framework of their own histories that the case studies on which they focus—ranging from marital advice literature and sex education programs to popular fiction and fashion—hardly use the term "normality" itself at all, but rather focus on the typical, the ordinary, the average, the usual, the conventional, or the standard. What we now refer to as normality, Creadick and Carter show, is a recognizable structuring force within such texts and a result of the cumulative effects of these terms. At the same time, normality was greater than the sum of its parts, greater than the numerical measures of health, race, and youth it described. It was something that no actual, individual subject could possess: "we cannot adequately theorise normality through the lens of ability, class, gender, sexuality, or race alone—or even in its combination—because normality's epistemological sweep and capacity to organise experience extends far beyond these categories," says Creadick.[72] As she and Carter both show, it was where midcentury privileging of whiteness, heterosexuality, and conventionality met quantification that the healthy, able-bodied, young white body came to represent both a statistical average and an aspirational ideal whose specificity had been entirely erased. Yet in 1945, as we have seen, "normal" was still not the most commonly used word to describe this body. The Norma and Normman statues were thus produced at the very moment this word began to move into the popular sphere, and have much to tell us about the conditions in which it did so.

THE SEARCH FOR NORMA

In September 1945, Gebhard announced that the Cleveland Museum of Health, in collaboration with the local newspaper, the *Cleveland Plain Dealer*, would hold a competition to find the living embodiment of Norma. There was no parallel competition to find a Normman, thereby reinforcing Dickinson's assessment of his male figure as a "mere afterthought," and suggesting that the shift between the "average American" and the "normal person" also marked a corresponding shift in the gender of this representative figure. The competition was scheduled to run for ten days, and a "Norma editor," Josephine Robertson, was tasked with writing Norma-themed daily stories on topical issues, keeping the competition in the news. She announced the launch of the competition in a front-page article, headlined: "Are You Norma, Typical Woman?" The article announced: "A search for Norma, the typical American woman, will begin today in Ohio in order to discover whether there is actually a woman whose measurements coincide with those of an average computed from the measurements of 18,000 women all over the United States and represented by the statue Norma at the Cleveland Health Museum."[73] Robertson drew heavily on the Harry L. Shapiro article to explain the significance of the Norma statue to the *Cleveland Plain Dealer* readership. Norma was described as the embodiment of an "American type," that of the "native white" young American female:

> Norma is the product of the American melting pot. In the beginning of this country's history, there was no truly typical American woman unless it was Pocahontas. Settlements were made largely according to European nationalities and there was not much intermingling. In the latter part of the 18th century, certain European scientists began to think of the Americans as a distinctive type, but as an inferior one due to what they termed an inferior environment....
>
> It wasn't until after the American Revolution that inhabitants of this country began to think of themselves as one people and of being of a distinctive type. With the progressive intermarriage of the citizens of varied origins the American type began to emerge, and the American woman became easily recognizable wherever she travelled.

However, statistics indicate that the typical American woman is still in the process of evolution, that Norma is growing taller and heavier and that her bust measurement is increasing. She exhibits "a harmony of proportion that seems far indeed from the usual or the average," according to Harry L. Shapiro, curator of physical anthropology of the American Museum of Natural History in a recent article. He calls this average "excessively rare."

Are you Norma, that rare individual?[74]

Here what was still most commonly defined as the "native white" American type was in the process of being transformed into the much more general category of the "normal American." The newspaper included an application form and detailed instructions on how to measure and record one's anthropometric data, which it reprinted daily for the duration of the competition. The caption for the photograph of the statue of Norma, pictured on the front page was: "The *Plain Dealer* Search for Norma seeks to learn if such a figure exists in life as well as statistics."

For the next ten days, the *Cleveland Plain Dealer* ran front-page stories about the Norma competition. Robertson interviewed local prominent figures in medicine, health, and education. "Norma" appeared in articles about fashion and sports, and in cartoons. The extensiveness and comprehensiveness of this media coverage is particularly remarkable if it is placed in its wider historical and cultural context. Norma first went on display in New York a month after the Second World War had ended in Britain, and the "Search for Norma" competition was launched just a few weeks after that war had come to an end for the United States, in the immediate aftermath of the two atomic bombs dropped on Nagasaki and Hiroshima. For ten days during this particular historical moment, the front page of the *Cleveland Plain Dealer* was occupied not by reportage of these epochal events, but by daily updates on a local competition to find the physical embodiment of the "perfectly average" woman. Yet this was perhaps not so surprising, since the competition was used as a means by which to discuss the possible roles ahead for women, especially those who had been engaged as "War Workers." It was precisely in such a volatile historical and cultural moment, perhaps, that questions about "normal femininity" were so pressing. The Norma statue thus attests to that transformation and the uncertainty about the normal at this time. It appears to have been constitutive of the new ideas about the normal entering into the public sphere, rather than representative of an established and familiar concept of normality.

Debate about the significance of Norma was again conducted primarily in the popular press and the public sphere. This was reflected in the *Cleveland Plain Dealer*'s own coverage of the competition. For Robertson, Norma was a model of postwar American self-sufficiency and hard work, the new woman who had found a modern independence through her years as a War Worker:

> Norma is not attenuated like the ethereal women of Botticelli . . . but being nearer the earth she is more practical and useful in a work-a-day world. . . . Her hands are strong and skilful from pounding the typewriter, milking the cow, playing the piano, stringing beans, washing dishes, operating the punch press, manipulating the forceps, employing the beaker and test tube, the mortar and pestle.
>
> The muscles of her arms and back are well developed from driving the car, the tractor, the jeep, swinging golf clubs, hoeing the corn, lifting the baby, rubbing the stains out of greasy overalls.
>
> No longer circumscribed by conservatism, she is free to choose her own work in the world. She is the physical expression of that new freedom. Perhaps she is also an expression of the political tendencies of the times which glorify the worker.[75]

However, other references to Norma in the *Cleveland Plain Dealer* understood her significance in a very different way—as a representative figure for a generation of female War Workers happily returning to the domestic sphere to resume their traditional role as wife and mother. The September 13, 1945, edition of the *Cleveland Plain Dealer* included a cartoon of a woman identified as a "War Worker" cooking in the kitchen, with the caption "Gee Norma, We're Glad You're Home Again!"[76] The different perspectives offered here by Robertson—for whom Norma was a model of the new woman, capable and independent—and the anonymous cartoonist—for whom she represented a welcome and reassuring return to domestic femininity—indicate an uncertainty about what "normal" femininity might mean at the end of the exceptional conditions produced by the war. It is just this uncertainty that made both Norma, and the new concept of normality she embodied, so important, and it is perhaps for this reason that the most exemplary figure of normality in the immediate postwar period is female.

Yet again, however, finding a living embodiment of the statistically average body would prove elusive, as seen in the announcement of the winner of the "Search for Norma" competition. When the contest closed on Wednesday, September 19, the *Cleveland Plain Dealer* reported that over 3,700 women had sent in their measurements. On Friday, September 21, shortlisted finalists were invited to an Open House Day held at the local YWCA. Over a thousand people attended the open day, and several hundred finalists were measured by a panel of experts drawn from the YWCA, the

Academy of Medicine of Cleveland, Flora Stone Mather College, the School of Medicine, Western Reserve University, and the Cleveland Board of Education. The statue of Norma was also on display, and the eventual winner, Martha Skidmore, was photographed standing beside her (see figure 8). The article announcing the winner, "Theatre Cashier, 23, Wins Title of 'Norma,' Besting 3,863 Entries," ran in the Sunday edition of the paper. Skidmore was described as "a former war worker who now sells tickets at the Park Theatre."[77] In occupation as well as physique, Skidmore was perfectly representative. In the interview celebrating her win, Skidmore "indicated that she was an average woman in her tastes and that nothing out of the ordinary had ever happened to her until the Norma search came along."[78] Ironically, the only exceptional thing that had ever happened to Skidmore, by her own account, was being identified as the most typical of her peers. Skidmore was exceptionally unexceptional: she was, as Shapiro had foreseen she would be, unusual in her very typicality. However, as Robertson noted, while Skidmore's measurements came closest to those of Norma, "they did not coincide with those of the statue.... After assessment of the measurements of 3,863 women who entered the search, Norma remained a hypothetical individual."[79]

The space of nonidenticality between Skidmore and Norma was highly consequential. In the first instance, the margin of difference between real bodies and statistical ideals was identified as a source of aspiration and potential (self-)improvement. For Gebhard, it reinforced the importance of cultivating one's body and health as a social as well as personal responsibility—a position closer to Sargent's focus on self-improvement than that of eugenics: "'In this nation where we are so rich in all facilities,' Dr. Gebhard said, 'we haven't nearly realized our health needs nor that health is something that must be worked for. We cannot buy it from the druggist nor from the doctor. People must achieve physical fitness by their own efforts. The unfit are both bad producers and bad consumers. One of the outstanding needs in this country is more emphasis on physical fitness. And you can't make that statement too strong.'"[80] Here we see how a recognizably eugenic focus on un/fitness, by which Gebhard was certainly influenced, was reinscribed within midcentury discourses of public health education and self-improvement practices. The space between individual bodies and the new ideal of normality embodied by Norma was a gap to be narrowed, but one that could never be closed.

A second consequence of the nonidenticality of Norma and Skidmore was that it encouraged new practices of measurement in which every aspect of the self could be quantified. For the Baptist Minister Dr. B. C. Clausen, at a sermon delivered at the Euclid Avenue Baptist Church on the Sunday after the "Search for Norma" competition had concluded, anthropometric principles could be applied to spiritual life. Clausen took "Norma's spiritual measurements" and urged his congregation to "measure yourself today.... Where have you failed? What dimensions do you lack? Use the experience of prayer for daily measurement in comparison with your ideals. Then exercise, practice, restrict and discipline yourself, noting the improvement each day. Let your church serve as a constant incentive, as well as gymnasium, for your correction."[81]

The context in which Norma appeared was one of constant measurement and assessment, of vigilant self-scrutiny and cheerful self-improvement, but this was an endeavor in which falling short of one's goals was assumed to be inevitable ("Where have you failed? What dimensions do you lack?"). It is precisely this combination of the inspirational and the unachievable, of the evaluative and the quantitative, that would come to characterize the concept of the normal and the more general category of normality as they entered popular discourse and culture during the midcentury.

CONCLUSION

The gap between the averaged measurements of Norma and the individual measurements of Skidmore represents the space of oscillation between the individual and the aggregate that Foucault identifies as the key dynamic of normalization.[82] But the point of emergence of contemporary ideas of normality was not, as Foucault claims, within the large disciplinary institutions and biopolitical systems that came together at the start of the nineteenth century. It was only in the first half of the twentieth, as the history of exhibited composite statues reveals, that contemporary ideas of normality emerged in the context of self-improvement and consumer cultures. Anthropometrics continued to play a decisive role, but that role was transformed, as were the cultural networks of which they were a part. Whereas the large-scale anthropometric collections of the nineteenth century had been undertaken in the context of scientific research or administrative practices of various kinds, anthropometric studies in the mid-twentieth were increasingly undertaken for commercial purposes, to build and sustain the mass marketing of consumer products. That can clearly be seen in the collection of the data used to make the Norma statue. It was carried out by a government agency, as it had been for Normman, in this case the Bureau of Home Economics, part of the US Department of Agriculture. However, this project was always intended to have a commercial outcome.[83] It was undertaken at the request of the Mail Order Association of America in order to develop an industry-wide standard for sizing in the ready-made clothing industry. In the 1930s, the Bureau of Home Economics was provided with a Federal Work Projects Administration grant to "develop the measuring techniques and to train the measurers."[84] The aim of the project was to remove obstacles to expansion of the industry caused by the irregularity of its existing sizing: "Garments labelled the same size but made by different manufacturers vary greatly, and few of them fit without alteration."[85] Because some garments could not be altered, this led to "a large volume of returns," causing particular hardship for "rural homemakers who frequently must buy by mail."[86] In 1939 and 1940, (native white) women were invited to participate in a national survey of women's bodily measurements.

Almost 15,000 women volunteered and had their measurements professionally recorded. Fifty-nine measurements were recorded for each woman (see figure 9). This research—including a detailed account of its methodologies and findings—was published as *Women's Measurements for Garment and Pattern Construction* (Miscellaneous Publication No. 454) under the auspices of the US Department of Agriculture in 1941. The authors noted that younger women, especially those between 18 and 24 years of age, were disproportionately represented among the participants: "it became apparent early in the study that too many young women were being measured, because older women were reluctant to participate."[87] Several thousand schedules were discarded to redress this. Further schedules were later adjusted by being "edited to eliminate gross errors, that is, combinations of measurements which are anthropologically impossible."[88] A further selection criterion, stated but not explained by the authors, was that only the measurements of white women were sought.[89] This fact was hidden from the volunteers themselves, who were recruited through local women's organizations and clubs: "When it was found necessary, for the sake of good feeling within a group, to measure a few women of other than the Caucasian race, this fact was entered under remarks and the schedule later discarded."[90] In the end, just over 10,000 schedules were assessed.[91]

The results of the survey were disappointing. The authors concluded that any system of standardized sizing was unlikely to accommodate the majority of bodies, given the diversity of their anthropometrics: "the measurements of the average woman are of limited usefulness. The Nation's women vary far too much in size to be properly fitted by garments made for the average woman."[92] Instead, the "almost bewildering variety of shapes and sizes" of women's bodies[93] would require at least three permutations of garment length—"regular," "long," and "short"—and three standard girths—"regular," "stout," and "slim."[94] This system was never adopted. We find here again the tension between statistical averages and actual individual bodies that we have seen throughout this chapter. It makes clear that the production of mass-produced objects cannot take for granted the existence of standardized bodies: on the contrary, the system of standardization itself must find a careful balance between uniformity and adaptability, especially in the context of the consumer capitalism emergent in the postwar period. Mass-produced objects, like clothing, must be designed to fit as many bodies as possible in order to maximise commercial efficiency and potential. The collection of anthropometric data for commercial purposes was driven by different imperatives from the ones that motivated scientific and governmental collections: in a commercial context the aim was not necessarily to normalize bodies, but rather to identify and respond to the normal distribution of characteristics among them in order to maximize the value of the data. Hooton, for instance, conducted an anthropometric survey of commuters at Boston's North Station in order to develop more comfortable seats for the Heywood Wakefield Company (publishing the report as *A Survey of Seating* in 1945).[95] He also worked with various government agencies during the war to develop better-fitting military uniforms, aircraft seating, and gas masks. He extended capacity for mass data analysis by working with new technologies, such as IBM's punch card and the Hollerith mechanical sorter. Like the collection of the data needed to create a system of standardized clothing, Hooton's work was designed to make mass-produced products and objects that better suited the average or typical human body. In other words, anthropometrics could be used and were in fact used to enable objects, rather than bodies themselves, to be standardized. Consumer clothing was standardized in ways that were responsive to, but not restricted to, the dimensions of the average body. In the strict sense, then, the ready-to-wear clothing market did not act normatively. It did not impose a single size on all consumers. In order to develop and flourish, it needed to allow for a (normal) range of variations.

At the same time, certain bodies were privileged in this process—mostly young, white, fit, and able bodies—and others expected to adapt. The history of standardized clothing is a key cultural site at which this privilege has been entrenched, as recognized by cultural historians, since it encoded assumptions about who and what counted as a typical or normal body.

Joyce L. Huff makes this point in her study of Daniel Lambert, the most famous professional fat man of the late eighteenth and early nineteenth centuries. Huff observes that fascination with bodies of unusual dimensions at this time accompanied the rise of industrialization and mass-produced goods: "Mass production assumes a consumer who possesses an adaptable body, a body that can and will adapt to fit into preconstructed spaces. As the [nineteenth] century progressed, the public sphere was slowly standardised and, increasingly, those with bodies that did not fit the norms found themselves out of place in an environment built to meet the needs of the 'average' body."[96] It was certainly the case that nontypical bodies were neglected and effectively excluded by an increasingly standardized environment while somehow being expected to accommodate themselves to it. But it was also true that mass-produced goods were increasingly designed in the twentieth century to suit a variety of body types in order to maximize their commercial value.

Such commercial projects were an essential part of the culture that produced composite statues. Together, they are exemplary of the history of normality in the first half of the twentieth century, not despite but because of their rather eccentric cultural position: their claim to representativity draws attention to what it excludes, to the characteristics beyond the numerical by which normality is constituted, and the schism between the statistically average ideal and embodied individuality at the heart of normality as a set of cultural practices. The story of the normal in the first half of the twentieth century is thus the story of its gradual expansion into everyday speech. Yet even as the term became increasingly ubiquitous and powerful, it also grew much more vague and difficult to define. That is, as the idea of the normal became more generally applied, its significance became more diffuse, disappearing into the cultural background even as its cultural and material effects became more pronounced. And while the space of nonidenticality between statistically ideal and real bodies could be, and sometimes was, a site of political violence or exclusion, it was also a space of self-cultivation and self-improvement, and it is in this capacity that the more generalized concept of normality would become so culturally ubiquitous from the midcentury onward.

This concept of the normal emerged not in the context of nineteenth-century disciplinary institutions but of twentieth-century consumer capitalism, shifting its point of popular emergence out of the prison and into the office or suburban home, and from compulsory regulatory regimes to voluntary practices of self-management. The normal subject in the midcentury, as the Grant Study of Normal Young Men emphasized, was the adaptable, flexible subject of twentieth- century capitalism. In this way, Norma and Normman, like the other composite statues examined here, exemplified the tensions and uncertainty that coalesced around the figure of the statistically average American. Argument about its cultural status would continue in the decades that followed.

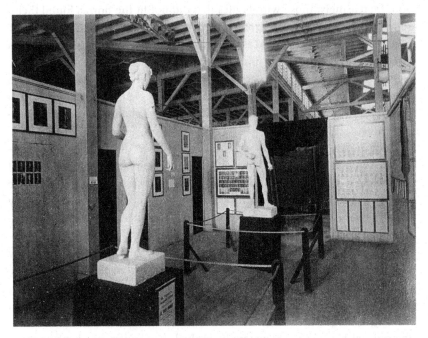

FIGURE 5. Dudley Allen Sargent's "Typical American Male and Female" statues, on display in the Hall of Anthropology at the Chicago World's Fair, 1893. Courtesy of the Peabody Museum of Archaeology and Ethnology, Harvard University, historical photograph, PM# 93-1-10/100264.1.1 (digital file# 99010037)

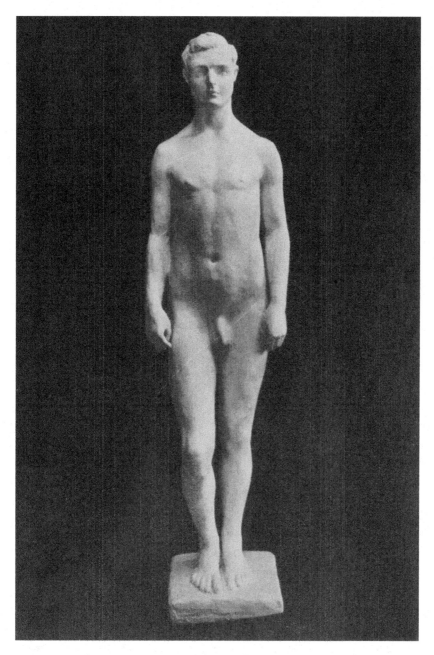

FIGURE 6. "The Average Young American Male," Jane Davenport, 1921. In Harry Laughlin's *The Second International Exhibition of Eugenics*, 1923. Courtesy of the Wellcome Library.

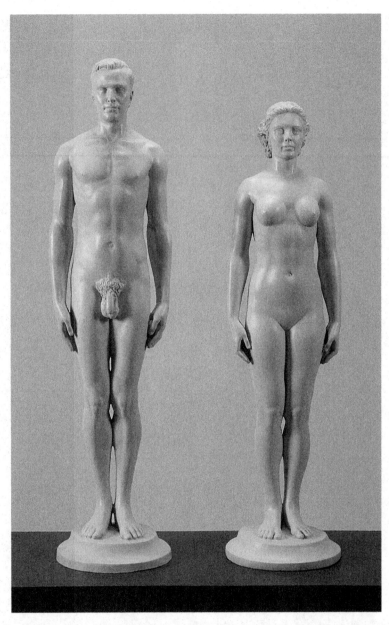

FIGURE 7. Dickinson and Belskie's Normman and Norma statues, 1945. Courtesy of the Warren Anatomical Museum in the Francis A. Countway Library of Medicine. Photograph by Samantha van Gerbig, Collection of Historical Scientific Instruments, Harvard University.

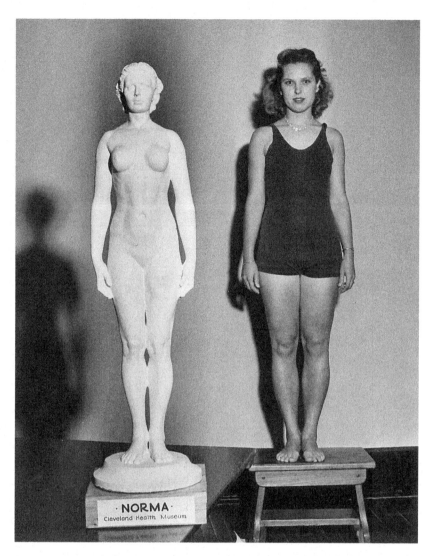

FIGURE 8. Search for Norma competition winner Martha Skidmore, next to the Norma statue, 1945. Image courtesy of The Cleveland Museum of Natural History.

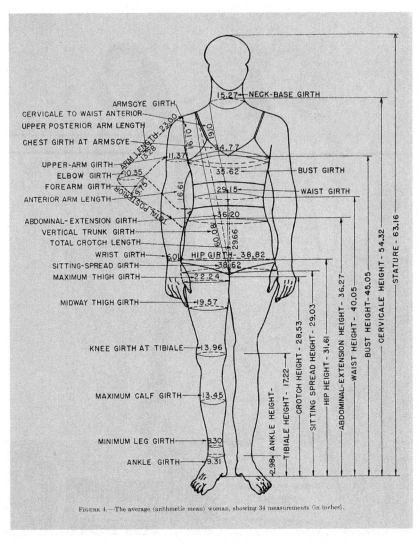

FIGURE 4.—The average (arithmetic mean) woman, showing 34 measurements (in inches).

FIGURE 9. From *Women's Measurements for Garment and Pattern Construction*, US Department of Agriculture, 1941

CHAPTER NINE

Sex and Statistics: The End of Normality

The publication of Alfred Kinsey's *Sexual Behavior in the Human Male* in 1948 was widely hailed as a landmark event: "after decades of hush-hush" on the subject of sex, announced the *New York Times* in its review, "comes a book that is sure to create an explosion and to be bitterly controversial."[1] Kinsey's report certainly produced a discursive explosion, generating an enormous amount of cultural commentary and a good deal of public controversy. The book was reviewed in every major newspaper as well as in academic journals and popular magazines. Even people who had not read Kinsey's report were rapidly exposed to the cultural sensation around it. Within a few months of its release, the report was already so well known it could be casually referred to in the lyrics of a popular song: Cole Porter's "Too Darn Hot," something highly unusual for a scientific report.[2] While medical studies of sexuality had been increasingly marketed to general audiences from the start of the twentieth century, and public exhibitions on sexual health had featured in high-profile events like the 1939 World's Fair, the publication of *Sexual Behavior in the Human Male* nonetheless represented something genuinely new: a scientific study of sex that was also a blockbuster publication. A quarter of a million copies were sold within the first weeks of its release, and the book spent six months on the *New York Times* best-seller list: "Not since *Gone With the Wind* had booksellers seen anything like it," wrote *Time* magazine on the eve of the publication of the highly anticipated follow-up, *Sexual Behavior in the Human Female*.[3] By the time this second report was released, in 1953, Kinsey was a celebrity, his name widely synonymous in popular culture with the study of sex. The release of the second report was preceded by a media campaign, largely stage-managed by Kinsey himself, during which he was featured on the cover of *Time*, surrounded by Disneyesque illustrations of birds, bees, and flowers. In the feature article he was hailed as the great discoverer of modern sexuality: "Kinsey has done for sex what Columbus did for geography."[4] The findings of the second report, like the first, were the subject of extended commentary in professional journals and the popular press alike. An illustrated book of photographic responses to the second report, titled *Oh! Dr Kinsey!* and featuring photographs of readers' startled reactions to the text, was released the same year.[5] Kinsey was even featured as a character in a 1953 episode of the popular *Jack Benny* television show. All this sensational coverage in the popular media did not undermine Kinsey's status as a scientist, however, and his work did not lose its cultural or professional prestige as a scientific report, although a number of academic and professional journals dedicated special sections to reviewing his findings and critiquing his methodology. At Kinsey's request, the American Statistical Association undertook a detailed review of his methods in the second report, having criticized the methods used in the first.[6] Kinsey's two reports in this way enjoyed a vastly larger circulation and had a much wider cultural impact than any of the texts examined previously while, as scientific studies, remaining strongly tied to this earlier history.

Kinsey's reports brought the concept of the normal into a much more central position than ever before. Nearly all reviews of the two books identified the subject of the reports and their key point of innovation as their focus on "normal" sexuality. Without using that term himself, Kinsey noted a relevant point of distinction between his report and the work of a previous generation of sexologists and psychoanalysts: "the monumental studies of Havelock Ellis and Freud . . . did not involve a general survey of persons who did not have sexual problems which would lead them to professional sources for help," Kinsey explained at the start of *Sexual Behavior in the Human Female*.[7] By contrast, and like earlier guides to sexual health and happiness, Kinsey's work aimed to focus on the "more typical of the general population."[8] Attention to the sexuality of the normal person contributed greatly to the popularity of his work in the postwar period. Kinsey was seen to be casting light into the hidden recesses of the respectable family home, illuminating the sexual desires and practices of that great majority of the population who, by virtue of their very ordinariness, had remained outside the sphere of medical or legal attention. His first report was widely praised for its focus on the sexuality of "average men, who are encountered daily on the street or the bus."[9] Even Kinsey's detractors felt compelled to admire him for undertaking "the greatest mass survey of normal sex activity in our society" and providing "the first work in all the mountain of writing on the subject which presents an authenticated picture of normal sexual behavior."[10] As Sarah Igo recognized, the opinion pollster George Gallop was an early supporter of Kinsey's work, commending his study of "the sexual behavior of normal individuals" and his impressive "skill at extracting opinions on private matters from ordinary Americans."[11] Yet Kinsey's decision to focus on the sexual behavior of "ordinary" or "average" Americans was not without precedent, as Kinsey himself acknowledged. He recognized that his research owed a debt to the pioneering work of earlier researchers such as Katherine Davis and Robert Dickinson who, as we have seen, had undertaken studies on the sexual behavior of "average people" in the 1920s and 1930s.[12] Moreover, and despite Kinsey's characterization of Freud's work as focused on those with "sexual problems," Freud himself had claimed to be undertaking the first study of normal sexuality as early as 1908, while the sexual hygienist William Robie made the same claim in *Rational Sex Ethics*, published in 1916. By the time Kinsey published *Sexual Behavior in the Human Male* and *Sexual Behavior in the Human Female* in the middle of the twentieth century, claims to be undertaking the first study of normal sexuality were already fifty years old.

While Kinsey's two reports may not have been the first studies of normal sexuality, then, the fact that they were so widely heralded as such is itself significant. It has much to tell us about the historical conditions in which his two reports were produced, and the reasons they caused such a cultural sensation. As Kinsey explained in *Sexual Behavior in the Human Male*, he was neither a psychologist nor a sociologist but a biologist, and his methods were accordingly different from those used in previous forms of sex research.[13] Unlike the methods used by the sexologists, psychoanalysts, and sexual hygiene writers, Kinsey's methods were numerical and data-driven. His reports contained information from over 5,000 interviews with white men and women[14] and should be understood as "progress reports" on the much larger study that was underway,[15] for which he aimed to amass 100,000 interviews each with men and women. Kinsey claimed that his reports represented the greatest number of interviews about sex with "average people" ever collected, and while this claim was not strictly accurate, it is indicative of how important the volume of data being collected was to Kinsey's view of his research, and how central quantitative methods were to his approach to the study of sexuality.[16] In this he was departing markedly from earlier approaches to the study of sex. Each of Kinsey's interviews involved between 300 and 500 questions and lasted several hours. The interview data were stored on punch cards, which could then be mechanically sorted to determine the statistical frequencies of various categories of sexual behavior, including anal sex, bisexuality, masturbation, oral sex, premarital sex, homosexuality, sex with prostitutes, extramarital sex, and sadomasochism. In this way, Kinsey transformed the sexual histories and experiences of his research participants into numerical data, rather than recording them as narrative case studies. He then collated the initial findings into a series of statistical tables and charts. The inclusion of these long and detailed numerical lists throughout Kinsey's reports—which promised to quantify accurately the frequency and incidence of various sexual behaviors in hard numbers, over a very large sample—did much to add cultural heft and gravitas to his research and to his subsequent claims about sexual behavior. Thus, although there were occasionally droll newspaper reports about readers returning copies of Kinsey's book because its contents were insufficiently racy, it was precisely this combination of sex and science, rather than the focus on sex itself, that made Kinsey's reports such a sensation in the middle of the twentieth century.[17] Their dry tone and statistical charts densely packed with columns of figures enhanced their cultural authority and increased popular interest in them, rather than limiting it, by giving his study of sex an air of scientific seriousness and empirical truth.

So while Kinsey's reports were in some respect a continuation of the tradition of medical writing on sex, they also represented an important methodological and conceptual break with it, one that had important consequences for the concept of the normal. Into what had previously been a predominantly qualitative medical study of sex, Kinsey introduced quantitative and statistical methods.[18] This, as Howard Chiang has argued, was part of a larger quantitative turn across the life sciences, in which Kinsey's research played a significant role. This turn, Chiang argues, gave rise to a "shift in the norms of clinical truth—from one that found the case-studies method sufficient for distinguishing the pathological from the normal to one that became increasingly grounded in the statistical notion of normalcy and socio-populational approaches."[19] This shift is very evident in Kinsey's work. Where sexology and psychoanalysis had focused on the case study, Kinsey's research focused on the statistical chart; where they had examined individual experience, Kinsey calculated statistical incidence. As we have seen, late nineteenth-century and early twentieth-century studies of sexuality made very little use of statistics. Their approach was narrative, not quantitative, and the concept of the normal that underlay their work was medical rather than mathematical. Kinsey's work changed this by applying to the study of sexuality the statistical methods developed as part of the anthropometric projects that can be traced back to the mid-nineteenth century in the work of Quetelet and Bertillon, to the final decades of the century in the work of Lombroso and Galton, and into the twentieth century in the work of Davenport and Davidson. By applying quantitative methods of this kind to the study of sex, Kinsey transformed the critical and interpretive frameworks through which sexuality was examined and understood, as well as the concept of the normal that underpinned them.

Kinsey's work can thus be said to draw together the two key strands in the critical genealogy of the normal, which until then had remained largely separate, as our history indicates. We began with the work of Étienne Geoffroy Saint-Hilaire, who put forward a broadly medical conception in which the normal referred to the achievement of stability in a dynamic environment. That made it possible to talk anatomically and physiologically of the normal state, a notion that quickly claimed a central place in medical discourse. Alongside talk of the normal state, and quite often in competition with it, we have seen the uneven development of scientific quantification, of anthropometrics and vital statistics. We have been careful to show that the medical and mathematical meanings of the normal—of the healthy and the average, to put it as succinctly as possible—remained largely distinct throughout the nineteenth century, a period in which they are often assumed to have been closely entwined. It is in Kinsey's work that we see most clearly the convergence of medical and mathematical meanings of the normal that Ian Hacking identifies as the source of its cultural ubiquity.[20] The cultural sensation caused by the publication of Kinsey's reports tends to prove Hacking's point: the combination of sex and statistics drove discussion and debate about the normal into a central position within popular culture, professional discourse, and the larger public sphere in an entirely new way. This is reflected in the fact that the period in which Kinsey's two reports were published—the first in 1948 and the second in 1953—has been widely identified as the period in which the word "normal" became firmly entrenched in popular culture. The 1950s, in particular, have come to be seen as the high-water mark of the popular concept of the normal. This is the normal of postwar conventionality, which Creadick examines in key 1950s texts like *The Man in the Grey Flannel Suit* and *Organization Man*; it is the normal that Michael Warner associates with American respectability and assimilationist politics in *The Trouble with Normal*, the one that David

Halperin takes in *Saint Foucault* as synonymous with the "legitimate" and the "dominant." Kinsey's reports played a significant role in establishing this concept of the normal in the cultural imagination and academic research of the 1950s. Whereas earlier, the sexologists and psychoanalysts found the normal to be an elusive and precarious quality that dissipated under closer examination, and the anthropometricians created idealized figures of the typical or normal body too perfect to be found in embodied form, Kinsey promised to provide a factual and accurate account of normal sexual practice, pinned down with scientific precision and hard data.

STATISTICS AND SEXUAL SCIENCE

The idea of the normal established in popular culture by Kinsey's reports derived in large part from the application of statistical methods to the study of sex, an approach that was, as the *New York Times* had predicted, both enormously influential and bitterly controversial. Kinsey devoted the first third of both reports to providing a detailed account of his statistical methods, including interview techniques, coding of data, sampling methods, and measures taken to ensure the validity of the results. The second report also noted variations from the methodologies used in the first one.[21] The vast majority of the interviews were undertaken by Kinsey himself, or by one of his two key collaborators, Wardell Pomeroy and Clyde Martin. They were conducted in a fairly concentrated geographical area focused on Indiana, although the area extended to a number of adjacent northeastern states. This restriction was seen by critics to limit the representativeness of the sample, although Kinsey's reports did provide a detailed account of the steps taken to manage sample sizes and included data so as to make the figures more representative. The first wave of data, collected single-handedly by Kinsey in the 1930s, was coded onto gridded charts, but by the 1940s all data were recorded and stored on punch cards. The development of this new technology for recording, storing, and mechanically processing data played a pivotal role in Kinsey's research, as Donna Drucker demonstrates in *The Classification of Sex: Alfred Kinsey and the Organization of Knowledge*.[22] Where previous generations of anthropometricians, such as Galton and Charles Davenport, had also amassed vast quantities of data, the lack of technologies with which to store and analyze this information had limited their ability to undertake statistical analyses across different categories of data and with different variables: the majority of Galton's data, for instance, remained unanalyzed on his death, and it was not until the 1980s that computing technologies allowed his anthropometric records to be compiled and examined in their entirety. Kinsey, on the other hand, had available to him—through the development of the new punch card—a technology that would allow automatic statistical and quantitative processing to keep pace with his collection of data, even if his data set continually expanded to include hundreds of thousands of items.

The enormous scale and scope of Kinsey's research and the sheer magnitude of the data set he collected met with universal praise and frequent expressions of awe from his contemporaries, even those who otherwise profoundly disagreed with his project or methods. The psychologists Hyman and Barmack wrote in their "Special Review" of *Sexual Behavior in the Human Female* that the text "provides the greatest wealth of data ever accumulated on the sexual behavior of a particular population, American women, and therefore is a factual contribution of great interest to psychologists."[23] That prominent psychologists like Hyman and Barmack, whose case study-focused and qualitative research methods were challenged by Kinsey's data-driven approach, accepted his statistical findings as a "factual contribution" to psychology says much about the persuasive power attached to the sheer magnitude of Kinsey's database. Even trenchant critics of Kinsey's research like Lewis Terman, who wrote a twenty-five-page account of the statistical and other methodological flaws he discerned in *Sexual Behavior in the Human Male*, began by acknowledging that "like others, the reviewer has been deeply impressed by the magnitude and potential significance of Kinsey's research. From the first volume of his projected series it is obvious that no one has ever obtained so much information from so many persons regarding the most secret phases of their sexual histories."[24] While the rigor and validity of Kinsey's actual statistical methods were the subject of extensive debate—including a 300-page critique of the first report produced by the American Statistical Association—the size of his data set nonetheless gave it strong persuasive force.

For Kinsey, what was most important about a quantitative method was that it enabled an objective and scientific approach to the study of sex, an approach he identified as that of a biologist rather than a sociologist or psychologist. Where the latter disciplines were, by their history and methodology, evaluative and thus subjective, Kinsey saw statistics as a way to record the neutral empirical facts about sex in a manner free from bias or personal judgment. His reports simply documented sexual behavior in a disinterested and impartial way, he claimed: "The present study, then, represents an attempt to accumulate an objectively determined body of fact about sex which strictly avoids social or moral interpretations of the fact," Kinsey wrote at the start of *Sexual Behavior in the Human Male*. "Each person who reads this report will want to make interpretations in accordance with his understanding of moral values and social significances; but that is not part of the scientific method and, indeed, scientists have no special capacities for making such evaluations."[25] Previous studies of sex had been limited by the critical and interpretive frameworks of the researchers, and by their social and moral preconceptions, Kinsey argued. This had compromised their impartiality and ability to acquire a properly scientific knowledge of sexual behavior. Kinsey's aim, in contrast, was simply to provide a neutral "report on what people do, which raises no question of what they should do, or what kinds of people do it."[26] His studies were not designed for legal or therapeutic application, as Krafft-Ebing's and Freud's had been, nor were they manuals of practical advice, like Stopes's. Kinsey's work had no such agenda, he repeatedly stated. As Heike Bauer has shown, Kinsey believed that a political agenda or instrumental purpose undermined the validity of sex research.[27] This was evident in Kinsey's critique of Hirschfeld's work, which Kinsey saw as scientifically compromised by Hirschfeld's political commitment to changing public perceptions about homosexuality.[28] Kinsey intended his own reports to be free from any such political agenda or instrumental purpose. Indeed, as Bauer shows, the lack of such an agenda was, for Kinsey, precisely what made his reports "scientific."[29]

However, this association of numbers with facts and the privileging of a quantitative approach to the study of sexuality did not go uncontested in the reception to Kinsey's work. On the contrary, in both professional and popular contexts it was the subject of debate on a wide range of fronts. Hyman and Barmack were among the many critics who raised concerns about the representativeness of Kinsey's sample: while only 13 percent of American women had gone to college, this group made up 75 percent of his sample; similarly, 40 percent of American women had not been educated beyond the eighth grade, but this group made up only 3 percent of those Kinsey studied.[30] The 300-page American Statistical Association report criticized just about every aspect of Kinsey's method, from his sampling techniques to his development of variables to a wide range of what were judged to be systemic errors.[31] Kinsey's work was also challenged for assuming that information about sexual behavior could be reduced to numerical data, and that this quantitative approach provided a meaningful way to examine sexual practice and experience. Lewis Terman, an academic psychologist at Stanford, questioned the reliance in Kinsey's research on memory and reported experience, and the assumption that the information provided by interviewees was accurate and honest: people forgot things or evaded the truth, he argued, and the kind of people who would divulge information about their intimate lives to complete strangers were by definition not representative of the population as a whole.[32] A similar concern was raised by Hyman and Barmack in their critique of the representativeness of Kinsey's sample: "those women who agree to an interview might be less inhibited and consequently may engage in more different types of sexual activity, and may do so more frequently."[33] A further source of academic criticism regarding the suitability of quantitative methods of analysis to the study of sex came from within the social and life sciences, disciplines that used primarily qualitative methods.[34] For Columbia University anthropologist Ruth Benedict, Kinsey's application of statistical methods provided the appearance of a properly scientific method but was fundamentally ill suited to the actual subject. As Donna Drucker writes, "Benedict saw Kinsey's focus on quantitative method as a way to assure that the scientific community would accept his work as 'real science' instead of the qualitative and humanist method she believed revealed deeper truths about sexuality."[35] Turning sexual histories into numerical data, said Benedict, stripped them of the emotional drives and social context that motivated actual sexual behavior, thus undermining the value of the research. In Benedict's view, Kinsey's focus on quantitative method "lessened, rather than improved, the work's possible impact."[36] Statistical analysis could not account for the complexities of the human psyche, which is shaped by emotional and social circumstances that Kinsey, distancing his own approach from that of psychology, disregarded.

In popular reviews and commentary, too, Kinsey's quantitative approach met with considerable skepticism. Like his academic critics, popular writers protested that his interviewees were, by definition, not a representative sample of ordinary Americans. It was believed that "no decent woman would have taken part in such an immodest study," as Weber argues.[37] Reviewers found it "questionable that the six thousand or so women who allowed themselves to be quizzed by men on such intimate matters as sex experience would constitute a normal cross-section of American womanhood."[38] Lionel Trilling, whom Pomeroy would later dismiss as "a literary critic . . . quite unable to grasp the scientific nature of the book,"[39] found Kinsey's reliance on statistics a reflection of the unsettled conditions of postwar American culture. Trilling, like so many of Kinsey's critics, recognized the publication of his report as "an event of great importance in our culture."[40] However, its significance for Trilling was largely that of a symptom of the changing conditions of knowledge and life. Contemporary Americans had become so isolated from one another that statistics had arisen to fill a cultural void, enabling new forms of community building. For Trilling, it was a rather melancholy state of affairs when "the community of sexuality requires now to be established in explicit quantitative terms,"[41] so that "we must assure ourselves by statistical science that [our] solitude is imaginary."[42] Trilling contextualized Kinsey's quantitative approach to the study of sex as representative of a wider positivism that encouraged an overly biological understanding of sex. As a zoologist, Kinsey was, in Trilling's view, placing too much importance on the physical aspects of orgasm and on orgasm itself as the primary quantifiable unit of sexual experience.[43] In so doing, he was revealing his own biases and the determining effects on his study of his own interpretive frameworks, despite his claims to scientific objectivity and the neutral recording of facts.

As these varied critical responses to his work demonstrate, while Kinsey took the impartiality and factuality of numerical data for granted, quantification and statistical analysis were still not securely established as uncontested sources of truth and knowledge as late as the middle of the twentieth century. Rather, Kinsey's application of statistics was often critiqued as methodologically flawed or ill suited to the study of sex. At the same time, however, his first report actually helped to change the status of statistics in both professional contexts and popular culture, contributing to its increasingly privileged status as an accepted source of objective and impartial scientific knowledge. In this respect, as Weber argues, the reception of Kinsey's work constituted "a public referendum on the reliability of science and statistics," and it was one that had a decisive result.[44] Sarah Igo provides a detailed account of this process. Between the publication of his first report on male sexual behavior in 1948 and the follow-up volume on female sexual behavior in 1953, Igo demonstrates, the status of numerical data as neutral empirical facts became much more securely established: "by the time Kinsey's second volume arrived in 1953, the quantification of sexual behavior was not up for debate in the same way it had been just five years earlier," she argues. "Americans in the brief time between 1948 and 1953 had become accustomed, no matter how grudgingly, to the social scientific takeover of intimate acts—that is, the right to think about sex in the language of numbers."[45] In the middle of the twentieth century, Kinsey's reports changed attitudes toward both sex and statistics, entrenching both within the public sphere in a way, and to an extent, that had not previously been the case.

Yet this did not produce unified effects. It was not a single, coherent idea of the normal that emerged into popular culture through Kinsey's reports. On the contrary, the normal was a focal point of discussion, debate, and often disagreement. In this way, although Kinsey's work can be seen to represent a key site at which the medical and mathematical concepts of the normal, for so long developed in parallel, finally came together, closer inspection challenges the assumption that this conceptual coexistence securely established the normal as a dominant cultural concept at this time. The normal, as it was discussed in Kinsey's work and in the cultural responses to it, did not function in a unified way. Its position was not one of easy or assured dominance. Instead, it was constituted as—and at the intersection of—competing theories, some of which were incompatible. If the concept of the normal came to occupy a place of cultural ubiquity at this time, it was one established through discursive proliferation and disagreement, rather than consensus. This was nowhere more evident than in Kinsey's actual findings about sexual behavior. His first report showed that nearly 11 percent of men had experience of anal sex within marriage, that 46 percent had experience of both heterosexual and homosexual activities, that almost 21 percent had had intercourse before 16 years of age, that half of all married men had engaged in extramarital activities, that 92 percent reported they had masturbated, and that 10 percent had performed oral sex with female partners. His second report did not provide data on anal sex and found a much lower incidence of homosexuality among women. However, a third of the women interviewed had had intercourse by the age of 19, and a quarter of married women had engaged in extramarital activities. 62 percent reported that they had masturbated, while almost half of the married women reported they had performed oral sex with their husband.

Despite the praise he received for undertaking the first scientific study of normal sexual behavior, then, the key finding of Kinsey's reports was that the sex undertaken by the average American did not align very closely with cultural assumptions about what constituted "normal" sexuality, which, as we have seen, was largely understood as reproductive heterosexuality. Popular assumptions and cultural expectations about what constituted average sexual behavior were quite "remote from the actual behavior ... of the average citizen," Kinsey remarked.[46] The significance of this finding was the subject of much debate, quite a lot of which centered explicitly on its implications for what was generally taken to constitute normal sexual behavior. For some commentators, it meant that the idea of the normal as currently understood was too narrow, and needed to change to correlate more closely with actual sexual practice as evidenced in Kinsey's statistical data: "to be of any use at all, a standard of sexual normality must be based on what we, the majority of Americans, *do*, rather than what we say," wrote one popular commentator.[47] While a number of writers argued that the normal could, and should, be statistically determined, an equally large number of cultural commentators argued forcefully that a high statistical frequency of "perverse" behavior did not make it socially acceptable: if "premarital and extramarital relations, which are morally and legally wrong by Anglo-American standards, are nonetheless widely practiced by the American male," one radio broadcast announced, and "other practices are indulged by the American male in large though varying degrees held to be abnormal, queer, and immoral," these were causes for grave concern about the state of American masculinity.[48] They were prompts for intervention and prevention, not for resignation or indifference. Moreover, quantitative thinking itself was seen to play a nefarious role in the state in which America found itself. M. F. Montagu protested: "in America, quantity is a moral value which makes acceptable and normalizes what in lesser quantities would be unacceptable and abnormal."[49] As a result, it was being argued, and indeed expected, that "what has hitherto been thought unacceptable and abnormal must now be accepted and regarded as normal."[50] Montagu was opposed to this development and to the amorality he saw as inscribed in quantitative thinking. So we can see in Kinsey's expansion and popularization of the category of the normal not its secure establishment at the heart of midcentury American life, but rather a profound problematization in and around the very concept. As Michael Warner has succinctly argued: "Kinsey exploited the confusion at the heart of the normal" by demonstrating that "nonnormative sex acts are, in fact, the statistical norm."[51]

THE END OF NORMALITY

Kinsey's most famous example of the discrepancy between statistically normal behavior and cultural assumptions about what constituted normal sex—and the likely reason his first report found its way into a Cole Porter lyric—concerned the same-sex experiences of men. While it was commonly assumed that average male sexual behavior clustered around a cultural norm of heterosexual monogamy, Kinsey's data revealed a much greater range of sexual behavior. Almost half the married men interviewed reported extramarital activities, Kinsey found, while 37 percent of all men interviewed reported that they had engaged in homosexual activity to the point of orgasm.[52] The significance of this finding was an important focus of Kinsey's report, bolstering the theory of sexual behavior he developed from it. For Kinsey, this statistical revelation was exemplary of the way in which moral and social interpretive frameworks had hindered the development of objective scientific knowledge about sexual behavior: what his data revealed was not simply that a greater percentage of men were homosexual or bisexual than had been previously recognized, but that the range of sexual activity undertaken by the average man was much broader and more varied than was commonly assumed. This, in turn, challenged a popular assumption about the structure of sexuality, specifically that it was organized as a binary: "Males do not represent two discrete populations, heterosexual and homosexual. The world is not to be divided into sheep and goats. . . . The living world is a continuum in each and every one of its aspects."[53] Kinsey's view of this matter was not unprecedented. We have seen a similar critique of a binary concept of normal and abnormal sexuality in the work of Krafft-Ebing and Freud, elaborated alongside a similar finding that the normal and the abnormal do not constitute a pair of binary opposites but are rather interconnected parts of a dynamic system or condition. Kinsey's work developed that view further, making it the cornerstone of his theory of sexuality. In his famous conclusion, there were seven identifiable "stages" along a continuum of sex. These were: "0: Exclusively heterosexual with no homosexual inclinations; 1: Predominantly heterosexual, only incidentally homosexual; 2: Predominantly heterosexual, but more than incidentally homosexual; 3: Equally heterosexual and homosexual; 4: Predominantly homosexual, but more than incidentally heterosexual; 5: Predominantly homosexual, only incidentally heterosexual; 6: Exclusively homosexual."[54] While the development of the "Kinsey Scale," as it became widely known, is now considered one of the most influential contributions to twentieth-century theories of sex, here Kinsey's position largely reiterated, rather than substantively revised, the understanding of sexual psychology and behavior that had been dominant in the first half of the twentieth century — one that was based on medical rather than mathematical forms of knowledge.

However, Kinsey's problematization of the normal did in fact go further than that of Krafft-Ebing and Freud. Where the sexologist and the psychoanalyst had argued that the normal and the abnormal were interrelational states rather than binary opposites, Kinsey sought to do away entirely with the concepts of the normal and the abnormal. Despite all the praise he received in the media for his focus on "normal" sexuality, Kinsey himself was very careful to specify that the subject of his research was, specifically, "average" or "typical" sexuality rather than "normal" sexuality. The former terms, in Kinsey's view, were properly quantitative, and referred to neutral and numerical values. He contrasted them to the meaning of the word "normal" itself, arguing that this was mired in the kind of social and moral assumptions he sought to avoid. For Kinsey, "normal" was not primarily a statistical term, nor was it a properly medical one. On the contrary, it was an inherently moralizing expression. In explaining his own approach to the study of sex, Kinsey offered the following clarification: "No preconception of what is rare or what is common, what is moral or socially significant, or what is normal and what is abnormal has entered into the choice of the histories or into the selection of the items recorded on them."[55] To introduce such a process of subjective selection into the collection of the data would have been to determine in advance the eventual shape of the findings. As Kinsey explained, "such limitations of the material would have interfered with the determination of the fact. Nothing has done more to block the free investigation of sexual behavior than the almost universal acceptance, even among scientists, of certain aspects of that behavior as abnormal."[56] In this way, although defenders and detractors alike routinely described his two reports as studies of normal sex, Kinsey repeatedly and explicitly opposed the use of the normal as a conceptual framework through which to undertake a scientific study of sexual behavior. He emphasized how antithetical the concept of the normal as he understood it was to his understanding of scientific research as neutral and objective:

> The term "abnormal" is applied in medical psychology to conditions which interfere with the physical well-being of a living body. In a social sense, the term might apply to sexual activities which cause social maladjustment. Such an application, however, involves subjective determinations of what is good personal living, or good social adjustment; and these things are not as readily determined as physiologic well-being in an organic body. It is not possible to insist that any departure from the sexual mores, or any participation in socially taboo activities, always, or even usually, involves a neurosis or psychosis, for the case histories abundantly demonstrate that most individuals who engage in tabooed activities make satisfactory adjustments.[57]

For this reason, he concluded: "The terms normal and abnormal have no place in scientific thinking."[58] Kinsey's extensive critique of the term "normal" and its companion term "abnormal" demonstrates in effect that he was not preoccupied with any possible tension between medical and mathematical concepts of the term. For Kinsey, the ab/normal had no place in either mathematical or medical thinking.

By the middle of the twentieth century, it would appear, both the medical and the mathematical meaning of the term "normal" had been entirely eclipsed, at least for Kinsey, by what he saw as its primarily moral significance: "The similarity of the distinctions between the terms normal and abnormal, and the terms right and wrong, amply demonstrates the philosophic, religious, and cultural origins of these concepts," he asserted.[59] That this claim was historically inaccurate is less significant than what it tells us about the status of the normal in Kinsey's work, and in the scientific and popular cultures of the mid-twentieth century. One of the most striking things about this is that in the middle of the twentieth century the word was still very new in popular discourse. The two Grant Study reports, Heath's *What People Are: A Study of Normal Young Men* and Hooton's *Young Man, You Are Normal*, began with detailed explanations of the meaning of the term "normal," taking it to be unfamiliar to general readers. These reports were published in 1945, a mere three years before Kinsey's first book. In the space of three years, it seems, the word "normal" had transitioned from that of a new term requiring careful introduction into popular discourse to one associated with cultural prestige and ubiquity. While this suggests a rapid transformation in the significance and circulation of the term in the midcentury—in which Kinsey's own reports played a key role—it is also the case that Kinsey's reports attributed a retrospective dominance and meaning to the normal that it did not enjoy during the first half of the twentieth century. The association of the normal with the moral was not an issue of historical fact but of conceptual necessity in Kinsey's work. It played a foundational role in his theory of sexuality by providing a focal point for his critique of binarized concepts of sexual behavior: "It is a characteristic of the human mind that it tries to dichotomize in its classification of phenomena," Kinsey wrote. "Sexual behavior is either normal or abnormal, socially acceptable or unacceptable, heterosexual or homosexual; and many persons do not want to believe that there are gradations in these matters from one to the other extreme."[60]

Given this insistent critique of the normal, it is surprising how many of Kinsey's reviewers and readers continued to commend his reports for their innovative and important focus on "normal sexual behavior." The point here is not that this aspect of Kinsey's work was so widely misunderstood—although it certainly was, and it is indeed remarkable how thoroughly Kinsey's careful comments about the problematic status of the normal were overlooked. What is more significant in the context of our genealogy is the extent to which this oversight is itself indicative of the culturally imperative force of the term "normal" at that time. The word, as Kinsey's reports demonstrate, came to enter popular usage in the middle of the twentieth century through the pages of a book explicitly designed to critique it. For, despite his intentions, Kinsey's books were one of the main conduits by which the term "normal" moved into widespread circulation. As Trilling notes in his critique of the first report, "although the Report directs the harshest language toward the idea of the Normal, saying it has stood in the way of any true scientific knowledge of sex," Kinsey's own approach stealthily shores up its cultural status by using the concept of the "Natural" in a very similar way, and allowing this to "quietly develop into the idea of the Normal."[61] Trilling's point here was well made. He was correct, for instance, to note that Kinsey retained a concept of normal biology throughout his first report. Trilling thus provided an accurate account of the way in which the idea of the normal remained conceptually integral to Kinsey's study of sexuality even as he explicitly disavowed it. A similar dynamic can be seen in the extensive reviews that warmly praised Kinsey for the publication of a landmark study of "normal sexuality" he was very explicitly not undertaking. Such contradictions are exemplary of the cultural processes by which the word "normal" came to occupy a central place in popular discourse in the middle of the twentieth century. The increasing dominance of the idea of normality at that time was driven not by a consensus about its meaning, nor by its cultural coherence, but by the sustained debate and enduring irresolution surrounding and underlying it.

For Kinsey, his data revealed that the range and variety of sexual practices and preferences found among the supposedly "more typical of the general population" were much wider and more diverse than was commonly taken to be the case: "The data show that divergences in a sexual pattern of social groups may be as great as those which the anthropologists have found between the sexual patterns of different racial groups in remote parts of the world. There is no American pattern of behavior, but scores of patterns," he wrote.[62] Thus the effect of his work was not to consolidate a unified or coherent category of normal sexuality, but on the contrary to disperse it, charting a much greater range of sexual behaviors than had previously been assumed to be the case. For some commentators, this was a positive finding. The reviewer of *Sexual Behavior in the Human Male* for the *Journal of the American Medical Association* noted that Kinsey's "figures indicate the present concepts of normality regarding sex must be greatly broadened," and that doing so would "be helpful towards partially relieving many persons of guilt complexes relating to feelings of sexual inferiority and abnormality."[63] It was reassuring for men to discover how common experiences of homosexuality or masturbation were, or to see how varied was the frequency of marital intercourse. The reports were, for this reason, taken by many as progressive texts, likely to encourage a more inclusive approach to varied sexual behaviors. This view was evident in the *New York Times*, which commended Kinsey for the way in which "the facts are presented with scientific objectivity, and without moralizing—but they provide the knowledge with which we can rebuild our concepts with tolerance and understanding."[64]

In the years following the publication of *Sexual Behavior in the Human Male*, the word "normal" came to be used with far greater frequency, and in a much wider range of popular contexts, than ever before. Thus, while "Kinsey had hoped to avoid enshrining any single definition of 'normal,'" Sarah Igo notes, his work had cultural effects that Kinsey himself could not control: "the same process of tabulation had the power, in others' hands, to make the category [of the normal] all the more legible."[65] This is evident in the cultural reception of Kinsey's work, in which few of the popular commentators who congratulated Kinsey for his groundbreaking study of "normal sexuality" felt the need to explain or qualify their own use of this term. In the space of three years, between 1945 and 1948, the word "normal" had moved from its position as an unfamiliar term of scientific jargon to one whose meaning could be taken for granted by popular readers, with no clear trajectory in between. The popular emergence and the professional disassemblage of the term "normal" were thus historically synchronous events. This has much to tell us about the complex and sometimes surprising way in which the normal came to be established as culturally pervasive at just that time, doing so through debate rather than agreement, and through dispersal rather than coherence. There is no moment in the genealogy we have traced throughout our book in which the normal becomes a decidedly persuasive force or a focus of critical or cultural consensus. The 1950s did not represent the emergence of a new and dominant concept of normality, neither was it a watershed moment in the historical transformation of the term. Rather, the midcentury reception of Kinsey's work allows us to recognize the significance of the key dynamic we have examined throughout our study: its conceptual and cultural emergence as an object—and to some extent a product—of debate.

CONCLUSION

Kinsey's two reports thus represent a pivotal moment in the history of normality, through which the word came into much more frequent use, occupying for the first time its current position in popular discourse as the most commonly used word for the ordinary and common. Until 1945, the idea of the normal had not yet entered popular culture in any significant way. During the first half of the twentieth century, a more complex and multifaceted idea of the normal had been elaborated in medical and scientific writing, but the word "normal" was rarely used in everyday speech. Kinsey's two reports, published in 1948 and 1953, attributed to this term a ubiquity it did not yet possess. This was a characteristic feature of the popular idea of the normal emergent in the postwar period: once it appeared, it seemed as though it had been always already there. Ironically, Kinsey's stringent critique of its validity as a scientific term was one of the key conduits by which it came to acquire its subsequent cultural status. That the word "normal" was introduced into popular discourse in the context of such critique is highly significant. The public reception of Kinsey's reports was in this way exemplary of the pattern we have seen throughout this book, according to which the word "normal" came to be established in popular culture not through consensus but through debate. This was how the normal came to enter everyday speech, its legitimacy contested before it had been established, a precarious formation that seemed, and that seems still, to have always already been there.

The reception of Kinsey's reports exemplifies the fact that criticism does not necessarily—and perhaps does not ever—undo the dominance of the normal. We have seen that Kinsey's sustained critique of the concept of the normal served, ironically, to strengthen its cultural pervasiveness in the early 1950s by dramatically extending the range and frequency of its use. In encouraging so much debate about the normal, Kinsey did not do away with the concept, as he had intended, but lodged it all the more securely in the popular imaginary of the day. This was clearly evident, as Kinsey himself noted, in the way his interview participants responded to his research. Despite Kinsey's insistence that his report be read as "first of all a report on what people do, which raises no question of what they should do, or what kinds of people do it," many of his readers clearly used his work as a practical guide.[66] The most frequent questions Kinsey and his interviewers were asked involved "comparisons of the individual's activities with averages for the group to which he belongs"—and the question "Am I normal?"[67] Regardless of Kinsey's own intention, the cultural effect of his reports was to encourage a widespread adoption of, and identification with, the concept of the normal.

The popularity of Kinsey's reports and their authority as accurate representations of the average American's sex life were closely associated with his statistical methods—contentious as his methods might have been within professional circles. We have seen how the intellectual history of quantification is often taken as the key to normality's present cultural authority and as the dominant meaning of the term when it is used in contemporary contexts. For Ian Hacking, it was only when the mathematical understanding of the normal converged with that of the medical that the term acquired its current power and prestige. Our research has placed this moment of convergence quite late: in the immediate postwar period. This context is very different from the one in which Foucault situated the emergence of normalization. It is a period of mass marketing and public surveys, of self-help and consumer culture. What was required for such a culture was the participatory subject of democratic capitalism rather than the docile body of the disciplinary institution. This normal emerged not from the prison but the office and suburban home. In accordance with the critical genealogy we have recovered, the increasing conceptual and cultural dominance of the term during the middle of the twentieth century derived not simply from the convergence of the two main strands of the critical genealogy we have examined throughout this book—the medical and mathematical—but also from their moments of nonencounter or incoherence.

Conclusion

At each point in the genealogy we have traced, the normal emerges not as an undisputed force of dominance or compulsory homogeneity but rather as a locus of debate and contradictory assumptions. The normal was formed in controversy and was everywhere nourished by debate. It continues to be surrounded by conceptual clutter and built on contradiction. In our view, most recent scholarship fails to take proper account of that. Many have held to Canguilhem's definition of the verb "to normalize" as signifying "to impose a requirement on an existence,"[1] and aligned that understanding with Foucault's identification of normalization as "primarily regulatory."[2] This has led them to see the normal as a matter of enforced conformity to a fixed standard. We are not contending that they are simply mistaken in doing so, but histories and analyses of that kind make the normal a far simpler matter than it ought to be. Standardization is certainly a key aspect of normality, and we have shown it to be so across a period of a hundred and fifty years, especially in physical and criminal anthropology and in eugenics. But we have also seen a rather different perspective emerge alongside it. There, the normal was understood as a dynamic, interrelational state, a highly contingent individual equilibrium. This is the other side of the double movement of normalization that was identified by Foucault, one that is often overlooked in more recent work: in company with the dynamic of homogenization or standardization, Foucault identified a dynamic of individuation and differentiation. Normalization involves both. By emphasizing individuation, our history is attempting to redress an analytical imbalance and question a simplification of critique.

If we want to understand better the historical transformations and the ongoing cultural authority and ubiquity of the term "normal," we need to take the full complication of its critical genealogy into account. Our project has attempted to contribute greater historical detail to our understanding of the normal, and from that detail quite a different genealogy has emerged. In the first instance, we have discovered that the idea of the normal enters popular culture much later than is often realized: we have dated this moment quite precisely to 1945, with the publication of the two reports on the Harvard Grant Study of Normal Young Men and the production of the Norma and Normman statues by Dickinson and Belskie. The origins of the contemporary idea of the normal as it is used in everyday speech can be traced not to nineteenth-century disciplinary institutions, as Foucault has famously argued, but to the beginning of what is now called the data society, in which anthropometric measurements were increasingly used for commercial purposes such as the production of mass-produced consumer objects or the collection of information about subjective experiences or opinions. The commercial use of anthropometric data was driven by very different interests from the ones that guided scientific or governmental projects: in a commercial context the aim was not to normalize bodies, but rather to design objects that better suited the average or typical human body. In other words, anthropometrics enabled objects to be standardized in ways that were responsive to the dimensions of the average body; they did not simply impose normative standards upon those bodies.

In such a context, as we have seen, the normal body is not the standardized body but the flexible body, the one best able to adapt to constant change, to manage fatigue and maintain fitness. The importance of flexibility and its commercial implications was not in fact new in the middle of the twentieth century, but was rather an integral part of the longer history of the normal. In 1837, during the debate about numerical method in the French Académie de Médecine, the figure of the shoemaker who made all his shoes in only one size was imagined as a laughable practitioner of numerical method.[3] That was one of the most telling gibes against the average as a figure of knowledge, and it continued to resonate decades later.[4] Everyone knew in 1837 that the shoemaker was by definition a craftsman who produced only custom-made shoes. But the gibe was eventually overtaken by history as there developed a more complex relation between the statistical and the individual. Even in the 1860s, Broca found a quite different tone when discussing such matters with his anthropologist colleagues. As a connoisseur and meticulous user of craniometric instruments, he deplored the fact that some would-be scientists employed an everyday measuring instrument used by hatters—called in French the *conformateur des chapeliers*—as an expeditious means of gathering craniometric data. The principal curve in the skull measured for the purpose of fitting hats, said Broca, was not one that mattered greatly to science. True physical anthropologists had a more complex set of tasks to perform.[5] But he did recognize and respect the practical knowledge that led hatters in many towns to make bigger hats for town people than for peasants.[6] Inscribed in their commercial practice was an implicit understanding of craniometric differences corresponding to local differences in type. Making the same size for everyone was undoubtedly a commercial absurdity, but offering a measured range of variations to different types of customers showed that practical anthropometric knowledge could and did form an alliance with commercial resourcefulness. That was to be the kind of alliance on which twentieth-century prêt-à-porter was based, one in which standardization and flexibility played conjoined roles.

Particularly in the twentieth century, the contexts in which the normal emerged were not exclusively or even primarily disciplinary but most often commercial. We saw how large-scale anthropometric studies were used to develop systems of standard sizing for mass-produced clothing and seats for mass transit systems. Such mass-produced objects could not assume standard bodies; in order to be commercially viable they needed to be designed to accommodate as wide a range of bodies as possible. In a commercial culture, the system of standardization itself needed to achieve a careful balance between uniformity and individuality. As the consumer culture enabled by the manufacture of mass-produced objects expanded in the twentieth century, so did the commercial value of information about bodily measurements and the subjective preferences of target demographics increase. Such a culture required not simply the standardization of objects and bodies, but a flexible and responsive relationality between both systems and subjects.

This can be seen in Kinsey's reports. His work extended quantitative analyses of bodies to sex, transforming experiences and behaviors, opinions, and desires into quantifiable units that could be statistically examined across different categories using punch cards. That research demonstrated that even the most apparently secret aspects of social and private life could be quantified, aggregated, and statistically examined. In this way, he transformed behavior and experience into a sortable database of opinions and feelings. This method of quantifying human behavior was a key technology in the rise of the consumer culture and democratic capitalism that characterized 1950s culture, playing an important role in the development of survey cultures and opinion polls productive of new models of participatory citizenship in the midcentury, as Sarah Igo has shown. Kinsey's methods required vast quantities of data, but their operation was not simply dependent on automation or standardization. It rather required a great deal of flexibility and ready adaptability. This can be seen in the interviewing system he developed, which called for an elaborate set of skills on the part of his interviewers, who were carefully trained to establish a warm but dispassionate interpersonal relationship with participants. Interviewers were taught to adopt a manner that was agreeable and nonjudgmental, to pay attention to such things as body language and tone of voice, and to frame questions in a way calculated to encourage forthright responses. For instance, they were instructed to ask: "when was the first time you [masturbated/had sex, etc.]?" rather than "have you ever [masturbated/had sex, etc.]?" Yet no information about the actual questions asked or the raw responses to them was included in Kinsey's reports, precisely because, as he observed, the interviewing methods could not be standardized. Instead, they were designed to be flexible, adapted in situ to the exigencies of each interview as it progressed. Kinsey opposed this method of interviewing to that of the questionnaire, which, as a standardized and anonymous document, did not in his view provide reliable data.

Although a form of normalization may have occurred in the contexts of the public disciplinary institutions emergent at the start of the nineteenth century, then, the cultural networks those institutions produced were of a considerably different sort. Where Foucault's work focused on the practices of large public institutions designed to observe and manage populations, such as the prison, the hospital, and the school, this was not the context in which Stopes, Hooton, or Kinsey researched or wrote. By the middle of the twentieth century, with large-scale studies like the Grant study and Kinsey's sex research, a new generation of scientific research had emerged, and anthropometric and psychometric data were finding application in an increasingly wide range of commercial and popular contexts. Such data were used to tailor mass-produced objects to the average bodily dimensions of their users; to design spaces such as parks, offices, and public transport vehicles to maximize their accessibility and the comfort of a range of averagely sized bodies; to maximize productivity and efficiency both personally and professionally. The literature on efficiency, fatigue, and time management that flourished in the middle of the twentieth century showed that the efficiency of bodies, on the one hand, and that of factories and offices, on the other, were interdependent and needed to be carefully managed as interdependent variables requiring constant supervision and adjustment for both system and subjects to work well, as Anson Rabinach has shown in *The Human Motor: Energy, Fatigue and the Origins of Modernity*.[7] Edna Yost's *Normal Lives for the Disabled*, written in collaboration with Lillian Gilbreth and published in 1944, was intended to help people with disabilities—especially acquired disabilities, such as the loss of a limb, or sight—adapt to the requirements of the workplace and navigate the legal and medical systems, in order to lead "normal lives." Drawing on Gilbreth's expertise as a time- and motion-efficiency expert, the book provided practical advice on how to maximize functionality and minimize fatigue both at work and at home for people with a range of disabilities.[8] As the office became an increasingly important site of work, (self-)management and productivity literature like *Organization Man* (1954) and *How to Win Friends and Influence People* (1955) advised midcentury readers how to adapt to the changed conditions of labor in the midcentury, specifically to the emergence of office or corporate culture, and to maximize their productivity in that new context. The proliferation of self-help guides has continued unabated since then. Whether their focus is on how to maximize or optimize potential or how to find balance and happiness, such books claim to provide systems or insights by which cultural institutions can be navigated with greater nimbleness and agility. We are far indeed here from the early nineteenth-century prison or hospital. What contemporary Western cultures require, and produce, is not simply the docile body of the disciplinary society but the flexible body of capitalism, readily adaptable to different conditions. Individuation in this context is not to be understood as liberatory, in opposition to the violence of enforced homogenization. Individuation also allows subjects to be held personally responsible for their ability to adapt to their social conditions so that, as we have seen, twentieth-century public health discourses often held individuals responsible for the health of the nation, rather than ascribing it to governmental policies and practices. Similarly, self-help and productivity literature encourage individuals to improve themselves rather than to organize collectively. Accordingly, it should be stressed that neither individuation nor homogenization are productive or repressive forces outside the particular contexts in which they are mobilized. It is simply that individuation works in very different ways and produces different results. Those ways and those results have become more clearly visible in the course of our genealogy.

Some legacies of the way the normal entered public discourse in the middle of the twentieth century, primarily through health discourses, have been enduring. Public discussions about work and health have continued to focus on the importance of "balance," as the Grant study reports did. Discussions about fitness continue to be strongly shaped by midcentury ideas about the normal seen in the exhibition of Norma and Normman. We have seen how important public health campaigns have been to popularizing the idea of the normal over the first half of the twentieth century, and how self-improvement cultures grew alongside these. We have also traced the idea of fitness, as it is used in this context, to Galton's early work with eugenics, and the influence of Darwin on his thinking. In the contemporary period, the idea of fitness remains widely privileged, but it is also recognized as an ideal that does not represent a statistically average state of health. A popular health campaign in Australia in the 1980s featured a corpulent, beer-drinking couch potato called Norm, an Australian Everyman of bad health and habits, who was encouraged to spend thirty minutes outdoors moving around each day. Norm was the statistically average representation of normal masculinity, whose condition and behavior were not presented as desirable or normative in the least. The Norm campaign thus further embedded the normal in Australian popular and public discourse about health in the 1980s (and reinforced its whiteness) by affectionately mocking the figure of the normal man. At the same time, it demonstrated a certain cultural literacy about the contradiction at the heart of the normal: the character of Norm was average but not healthy.

For this reason, drawing attention to the incoherence of the term "normal" or adopting a position of opposition to it, as though it does in fact represent a fixed standard, do not necessarily undermine its continued function as a cultural ideal, nor are they necessarily the most effective form of political critique. Awareness of the distinction between the average and the healthy does not undermine the popular idea of the normal but is a constitutive part of its conceptualization. This can be seen in the complicated ways in which the term has continued to find a place in popular culture, surrounded by the ambivalence and debate we have observed from its outset. Current usage of the phrase "the new normal" provides one testimony to the contemporary endurance of the term in popular culture, despite half a century of skepticism and critique. Government health campaigns in Australia now promote making "healthy the new normal." The semi-ironic "normcore" movement exemplifies the extent to which the idea of the normal continues to be entwined with its critique. Adam Curtis's recent documentary *Hyper-Normalisation*, which takes its title from Alexei Yurchak's *Everything Was Forever, Until It Was No More: The Last Soviet Genera-*

tion, defines a state of hypernormalization as one in which simplified and fake media coverage replaces analysis of actual political or cultural events.[9] Critiques of the recent American election, including the widespread use of the phrase "this must not be normalized" on social media platforms, demonstrate the ongoing importance of the normal to twenty-first-century culture and the extent to which the normal continues to have a positive cultural value understood to be worth preserving.

Alongside this use of the term in the popular sphere, derivations of "normal" continue to proliferate in cultural analyses and critical theory, as seen in recent neologisms such as chrononormativity, homonormativity, neuronormativity, and so on. In each of these instances, the "normative" signifies a standard, often one that is enforced. Yet in the history from which this most contemporary form of the normal is derived, what characterizes the normal is not its function as a cultural standard or fixed type, but its role as a process of constant adjustment and rebalance in the context of changing conditions and circumstances. A critical genealogy of such a concept requires something other than a polemical opposition, even as it provides support for resistant thinking.

The normal is not monolithic. It is not an inexpugnable edifice towering over the intellectual landscape of our modernity. Its power derives from the very looseness that allows it to be everywhere available even as it continues to be questioned. Not a fortress, Foucault might have said, but an extensive network of usage across the plains of our existence. Our history has shown, moreover, that normality should not be thought of as a conceptual device malevolently assembled with the purpose of stigmatizing the deviant, disqualifying the disabled, and coercing the unfit. By following the history of its usage in nineteenth-century scientific discourse, we have been able to show that it functioned as a worthy concept with equally worthy opponents. In the course of the nineteenth century, the normal claimed its place without triumph—and indeed without actually winning any of the pitched battles fought in its name. Our genealogy has made it possible to understand both the strengths and the weaknesses of the concept and, most importantly, to understand how its power derives from the remarkable mixture of the two.

Acknowledgments

Research for this volume was supported materially by a Discovery Grant awarded by the Australian Research Council, as well as by the University of Queensland and Southern Cross University. The Huntington Foundation supported a period of archival research at the Huntington Library.

Working on this project over a period of about six years has allowed us to build a network of collegial friends. We owe a particular debt in our conception of the project to Marina Bollinger, Lisa Downing, and Karin Sellberg, all of whom helped us to think about it as a whole.

The Wellcome Trust made facilities available for us to hold a workshop at which chapter drafts were discussed by colleagues in a warmly attentive manner. Heike Bauer, Laura Doan, Kate Fisher, Jana Funke, Robert Gillett, Lesley Hall, Rhodri Hayward, Philippa Levine, Laura Salisbury, and Anna Katharina Schaffner were all part of that. This manuscript is much richer for their input.

When dealing with particular questions, we have had the benefit of conversation and feedback from Sabine Arnaud, Lennard Davis, Charlotte Epstein, Rosemarie Garland-Thomson, Stephen Gaukroger, Gerard Goggin, Ian Hesketh, Annamarie Jagose, Alison Moore, Kane Race, Nikolas Rose, Katie Sutton, Paul Turnbull, Caroline Warman, Robyn Wiegman, and Elizabeth Wilson. Our thinking on this subject has often been revised and refined in ongoing conversation with colleagues and friends, including Mark Andrejevic, Maureen Burns, Oron Catts, Sheila Cavanagh, Sean Edgecomb, Fiona Nicoll, Susan Stryker, Nikki Sullivan, and Ionat Zurr.

Our research was greatly facilitated by the patient and helpful archivists at the Francis A. Countway Library of Medicine, the Huntington Library, and the Galton archives at UCL.

Charlotte Chambers was a hard-working and reliable research assistant.

And we have had support of sustained, incalculable kinds from our partners, Wendy and Michael.

Notes

INTRODUCTION

1. Robert McRuer, *Crip Theory: Cultural Signs of Queerness and Disability* (New York: New York University Press, 2006), 7.
2. Georges Canguilhem, *The Normal and the Pathological*, trans. Carolyn R. Fawcett (New York: Zone Books, 1989), 241.
3. *The Normal and the Pathological*, 239.
4. Ian Hacking, "The Normal State," *The Taming of Chance* (Cambridge: Cambridge University Press, 1990), 160.
5. Canguilhem, *The Normal and the Pathological*, 243.
6. Hacking, *The Taming of Chance*, 169.
7. Quoted in Caroline Warman, "From Pre-Normal to Abnormal: The Emergence of a Concept in Late Eighteenth-Century France," *Psychology & Sexuality* 1, no. 3 (2010): 203.
8. Hacking, *The Taming of Chance*, 160.
9. See for instance Jürgen Habermas, *Between Facts and Norms: Contributions to a Discourse Theory of Law and Democracy* (Cambridge: Polity, 1996); Waltraud Ernst, ed., *Histories of the Normal and the Abnormal: Social and Cultural Histories of Norms and Normativity* (New York: Routledge, 2004); Stephen Turner, *Explaining the Normative* (Cambridge: Polity, 2010); Michael Warner, *The Trouble with Normal: Sex, Politics and the Ethics of Queer Life* (Cambridge, MA: Harvard University Press, 1999); Lennard Davis, *Enforcing Normalcy: Disability, Deafness and the Body* (London: Verso, 1995); and Julian Carter, *The Heart of Whiteness: Normal Sexuality and Race in America, 1880-1940* (Durham, NC: Duke University Press, 2007).
10. Lennard Davis, *Enforcing Normalcy: Disability, Deafness and the Body* (London: Verso, 1995); Rosemary Garland Thomson, *Extraordinary Bodies: Figuring Physical Disability in American Culture and Literature* (New York: Columbia University Press, 1997).
11. Elizabeth Reiss, *Bodies in Doubt: An American History of Intersex* (Baltimore: Johns Hopkins University Press, 2009); Dean Spade, *Normal Life: Administrative Violence, Critical Trans Politics, and the Limits of Law* (Brooklyn, NY: South End Press, 2011); Alice Domurat Dreger, *One of Us: Conjoined Twins and the Future of Normal* (Cambridge, MA: Harvard University Press, 2004).
12. Carter, *The Heart of Whiteness*, 31.
13. *The Heart of Whiteness*, 153-54.
14. *The Heart of Whiteness*, 4.

15. Mary Poovey, *A History of the Modern Scientific Fact: Problems of Knowledge in the Sciences of Wealth and Society* (Chicago: University of Chicago Press, 1998).

16. Jeannette Winterson, *Why Be Happy When You Can Be Normal?* (London: Jonathan Cape, 2011).

17. Warner, *The Trouble with Normal: Sex, Politics and the Ethics of Queer Life* (Cambridge, MA: Harvard University Press, 1999), 53.

18. Michael Warner, *Fear of a Queer Planet* (Minneapolis: University of Minnesota Press, 1993), xxvii (italics in the original).

19. David Halperin, *Saint Foucault: Towards a Gay Hagiography* (New York: Oxford University Press, 1995), 61–62 (italics in the original).

20. Annamarie Jagose, "The Trouble with Antinormativity," *differences: a Journal of Feminist Cultural Studies* 26 no. 1 (2015): 26. This article is part of a special issue entitled "Queer Theory without Antinormativity."

21. "The Trouble with Antinormativity," 27.

22. Michel Foucault, *The History of Sexuality, Volume One: An Introduction* (New York: Pantheon Books, 1978), 144.

23. Foucault, *The History of Sexuality, Volume One*, 144.

24. Michel Foucault, *Discipline and Punish: The Birth of the Prison* (London: Penguin, 1977), 184.

25. Foucault, *The History of Sexuality, Volume One*, 139.

26. *The History of Sexuality, Volume One*, 189–90.

27. Foucault, *Discipline and Punish*, 184.

28. *Discipline and Punish*, 193.

29. J. Jack Halberstam, *Gaga Feminism: Sex, Gender, and the End of Normal* (Boston: Beacon Press, 2012), 82.

30. Laura Doan, *Disturbing Practices: History, Sexuality, and Women's Experience of Modern War* (Chicago: University of Chicago Press, 2013), 169.

31. See, for example, Jonathan Goldberg and Madhavi Menon, "Queering History," *PMLA* 120, no. 5 (2005): 1608–17.

32. Elizabeth Freeman, *Time Binds: Queer Temporalities, Queer Histories* (Durham, NC: Duke University Press, 2010). See also "Queer Temporalities" *GLQ* 13, nos. 2–3 (2007), especially "Theorizing Queer Temporalities," 177–195, in which Carolyn Dinshaw, Lee Edelman, Elizabeth Freeman, Judith/Jack Halberstam, and Annamarie Jagose, amongst others, map out the wide range of approaches to the issue of temporality found within contemporary queer theory, and its problematization of history as a unifying sequence of events.

33. Michel Foucault, "Nietzsche, Genealogy, History," in *Language, Counter-Memory, Practice: Selected Essays and Interviews*, ed. D. F. Bouchard (Ithaca, NY: Cornell University Press, 1977), 139–164.

34. On the value of history to queer studies, see Valerie Traub, "The New Unhistoricism in Queer Studies," *PMLA* 128, no. 1 (2013): 21–39.

35. Doan, *Disturbing Practices*, 169. She refers on 167 to the "mutual benefits of an exchange between queer studies and academic history."

36. Doan, *Disturbing Practices*, 169.

37. *Disturbing Practices*, 191.

38. *Disturbing Practices*, 171.

39. *Disturbing Practices*, 172–3.

40. See also, for instance, Spade, *Normal Life*.

41. Margaret Lock, "Accounting for Disease and Distress: Morals of the Normal and Abnormal," in *The Handbook of Social Studies in Health and Medicine*, ed. Gary L. Albrecht, Ray Fitzpatrick, and Susan C. Scrimshaw (London: Sage, 2000), see especially 260–66.

42. Carter, *The Heart of Whiteness*, 4.

43. Annamarie Jagose, *Orgasmology* (Durham, NC: Duke University Press, 2012), 55.

44. Sarah E. Igo, *The Averaged American: Surveys, Citizens, and the Making of a Mass Public* (Cambridge, MA: Harvard University Press, 2007).

45. Karl Pearson, *The Life, Letters and Labours of Francis Galton* (Cambridge: Cambridge University Press, 1914), vol. 1, 111.

46. Pearson, *The Life . . . of Francis Galton*, 1, 340–41.

47. Alfred Kinsey, *Sexual Behavior in the Human Male* (Philadelphia: Saunders, 1948), 5.

CHAPTER ONE

1. Georges Canguilhem, *Le Normal et le pathologique* (Paris: Quadrige/PUF, 1966), 84–85.

2. See, for example, *Le Normal et le pathologique*, 91.

3. *Le Normal et le pathologique*, 75.

4. *Le Normal et le pathologique*, 182.

5. Warman cites this passage and comments as follows: "It would seem important here not necessarily to say that we have the thing without the word, but to look at its coming into being, the context of its emergence, and what other terms are used when it is not available." Caroline Warman, "From Pre-Normal to Abnormal: The Emergence of a Concept in Late Eighteenth-Century France," *Psychology and Sexuality* 1, no. 3 (2010): 204.

6. Michel Foucault, *Les Mots et les choses: une archéologie des sciences humaines* (Paris: Gallimard, 1966). For a discussion of this question applied to the work of historian Thomas Laqueur, see Peter Cryle, "*Les Choses et les Mots*: Missing Words and Blurry Things in the History of Sexuality," *Sexualities* 12, no. 4 (2009): 439–52.

7. Warman, "From Pre-Normal to Abnormal," 202.

8. "From Pre-Normal to Abnormal," 200.

9. Joseph Fourier, *Théorie analytique de la chaleur* (Paris: Didot, 1822), 503. Without providing the same detail as Warman, Ian Hacking states that the geometrical usage was present in a range of modern European languages. See Ian Hacking, *The Taming of Chance* (Cambridge: Cambridge University Press, 1990), 162–63. In English, he adds, it "acquired its present most common meaning only in the 1820s" (160).

10. Warman, "From Pre-Normal to Abnormal," 200.

11. Canguilhem, *Le Normal et le pathologique*, 8.

12. This view is formulated clearly in *Le Normal et le pathologique*, where Canguilhem refers to "a thesis generally adopted in the nineteenth century concerning the relations between the normal and the pathological. According to this thesis, pathological phenomena are identical to the corresponding normal phenomena, except for quantitative variations." (9).

13. In a major treatise published in 1832, Isidore Geoffroy Saint-Hilaire, the son of Étienne, lists the fields on which he expects his work to have an influence as follows: "physiology and anatomy, zoology and natural philosophy, and even diverse branches of the medical sciences." *Histoire générale et particulière des anomalies de l'organisation chez l'homme et les animaux, ouvrage comprenant des recherches sur les caractères, la classification, l'influence physiologique et pathologique, les rapports généraux, les lois et les causes des monstruosités, des variétés et vices de conformation, ou Traité de tératologie* (Paris: Baillière, 1832–1837), 1:xi. The position of medicine at the end of the list would appear to indicate the limitations he himself placed on his contribution to the field.

14. [Étienne] Geoffroy Saint-Hilaire, *Philosophie anatomique, vol. I. Pièces osseuses des organes respiratoires* (Paris: Baillière, 1818), xxvi.

15. *Philosophie anatomique*, 1:xxvi (italics in the original). See, for example, Étienne Geoffroy Saint-Hilaire, *Philosophie anatomique, vol. II. Des monstruosités humaines* (Paris: chez l'auteur, 1822), 252.

16. Isidore Geoffroy did use the word "structure" from time to time. See for example *Histoire générale et particulière des anomalies*, 1:339.
17. Étienne Geoffroy Saint-Hilaire, *Philosophie anatomique*, 1:429.
18. *Philosophie anatomique*, 1:438.
19. *Philosophie anatomique*, 1:209.
20. *Philosophie anatomique*, 1:xx.
21. *Philosophie anatomique*, 1:xx–xxi.
22. *Philosophie anatomique*, 1:xxiv.
23. *Philosophie anatomique*, 1:xviii–xix.
24. *Philosophie anatomique*, 1:xvii.
25. *Philosophie anatomique*, 1:xix.
26. *Philosophie anatomique*, 1:20.
27. *Philosophie anatomique*, 1:xxviii.
28. *Philosophie anatomique*, 1:189n.
29. *Philosophie anatomique*, 1:265.
30. *Philosophie anatomique*, 1:453–54 (italics in the original).
31. *Philosophie anatomique*, 2:27.
32. *Philosophie anatomique*, 2:184, 313.
33. See, for example, *Philosophie anatomique*, 2:14.
34. A particular organ in a certain species of fish, when compared with others of its kind, was said for example to "move into the state of a *dominant organ* when it is called on to play its role as well as the role of the absent ribs; it becomes strong and robust, far exceeding the dimensions of its normal state." *Philosophie anatomique*, 1:468 (italics in the original).
35. *Philosophie anatomique*, 1:122.
36. *Philosophie anatomique*, 2:16.
37. *Philosophie anatomique*, 2:21.
38. *Philosophie anatomique*, 2:106.
39. Isidore Geoffroy Saint-Hilaire, *Histoire générale et particulière des anomalies*.
40. Isidore Geoffroy Saint-Hilaire, *Vie, travaux et doctrine scientifique d'Étienne Geoffroy Saint-Hilaire, par son fils* (Paris: Bertrand, 1847), 260.
41. Isidore Geoffroy Saint-Hilaire, *Histoire générale et particulière des anomalies*, 2:521.
42. *Histoire générale et particulière des anomalies*, 1:17–18.
43. Étienne Geoffroy Saint-Hilaire, *Philosophie anatomique*, 2:110.
44. *Philosophie anatomique*, 2:110.
45. Even when he says of a child born with four legs that he has "toutes les qualités de l'humanité," it would be a more accurate translation of Geoffroy's scientific thinking to translate that "all the qualities of human-ness." This reference is to Étienne Geoffroy Saint-Hilaire, *Mémoire sur un enfant quadrupède, né à Paris et vivant, monstruosité déterminée par le nom générique d'iléadelphe* (Lyon: Boursy fils, 1831), 4.
46. Étienne Geoffroy Saint-Hilaire, *Philosophie anatomique*, 2:15.
47. *Philosophie anatomique*, 2:15.
48. *Philosophie anatomique*, 2:425–26.
49. *Philosophie anatomique*, 2:98.
50. *Philosophie anatomique*, 2:104.
51. *Philosophie anatomique*, 2:15.
52. John Stevens Henslow, *The Principles of Descriptive and Physiological Botany* (London: Longman, 1835), 123.
53. Henslow, *Principles of Botany*, 118.
54. See, for example, *Principles of Botany*, 9, 25, 107.
55. Canguilhem, *Le Normal et le pathologique*, 8.
56. *Le Normal et le pathologique*, 13.

57. *Le Normal et le pathologique*, 4.
58. Étienne Geoffroy Saint-Hilaire, *Philosophie anatomique*, 2:147–48.
59. Isidore Geoffroy Saint-Hilaire, *Histoire générale et particulière des anomalies*, 1:xi.
60. *Histoire générale et particulière des anomalies*, 1:45.
61. *Histoire générale et particulière des anomalies*, 1:46.
62. Canguilhem, *Le Normal et le pathologique*, 85.
63. See, for example, the text by Cuvier quoted in Étienne Geoffroy Saint-Hilaire, *Principes de philosophie zoologique, discutés en mars 1830, au sein de l'Académie royale des sciences* (Paris: Pichon et Didier, Rousseau, 1830), 67.
64. Georges Cuvier, *Tableau élémentaire de l'histoire naturelle des animaux* (Paris: Baudoin, An 6 [1797]), 84.
65. Cuvier, *Tableau élémentaire*, 93–94.
66. Georges Cuvier, *Leçons d'anatomie comparée* (Paris: Crochard, Fantin, An XIV [1805]), 1:ii. On the relative insignificance of analogies in Cuvier's view, see for example 147, 404.
67. Cuvier, *Tableau élémentaire*, 6.
68. *Tableau élémentaire*, 9.
69. *Tableau élémentaire*, 442.
70. *Tableau élémentaire*, 475. See also 484, 541.
71. In a discussion of the Geoffroy-Cuvier debate published over fifty years later, French zoologist Edmond Perrier commented that Geoffroy wished to "seek unity, not in the definitive development of animals, but in the manner in which that development comes about." Edmond Perrier, *La Philosophie zoologique avant Darwin* (Paris: Alcan, 1884), 133.
72. Étienne Geoffroy Saint-Hilaire, *Philosophie anatomique*, 1:43.
73. *Philosophie anatomique*, 1:174.
74. *Le National*, March 22, 1830 (italics in the original), quoted in Étienne Geoffroy Saint-Hilaire, *Principes de philosophie zoologique*, 215.
75. This was the second volume of the *Philosophie anatomique*, subtitled "Des monstruosités humaines" (On monstrosities in humans).
76. Isidore Geoffroy Saint-Hilaire, *Histoire des anomalies*, 1:xi.
77. *Histoire des anomalies*, 1:xii.
78. *Histoire des anomalies*, 1:123, 640.
79. *Histoire des anomalies*, 1:288.
80. *Histoire des anomalies*, 1:30n1.
81. *Histoire des anomalies*, 1:36 (italics in the original).
82. Canguilhem, *Le Normal et le pathologique*, 81–85.
83. Isidore Geoffroy Saint-Hilaire, *Histoire des anomalies*, 1:37 (italics in the original).
84. Étienne Geoffroy Saint-Hilaire, *Philosophie anatomique*, 2:18.
85. Isidore Geoffroy Saint-Hilaire, *Histoire des anomalies*, 1:30.
86. Canguilhem, *Le Normal et le pathologique*, 82.
87. *Le Normal et le pathologique*, 82.
88. Isidore Geoffroy Saint-Hilaire, *Histoire des anomalies*, 1:160.
89. *Histoire des anomalies*, 1:205.
90. *Histoire des anomalies*, 1:205.
91. *Histoire des anomalies*, 1:205.
92. *Histoire des anomalies*, 1:126.
93. *Histoire des anomalies*, 2:11.
94. *Histoire des anomalies*, 2:254.
95. *Histoire des anomalies*, 2:28.
96. Isidore's reference is to Charles Bonnet, *Considérations sur les corps organisés*, vol. 3 (Amsterdam: Rey, 1762).
97. Isidore Geoffroy Saint-Hilaire, *Histoire des anomalies*, 1:73–74.

98. *Histoire des anomalies*, 1:268.
99. *Histoire des anomalies*, 2:189.
100. The reference in Buffon's own work is Georges Louis Leclerc de Buffon, *Œuvres complètes* (Paris: Rapet, 1818), 5:369.
101. *Œuvres complètes*, 5:371.
102. Isidore Geoffroy Saint-Hilaire, *Histoire des anomalies*, 2:189.
103. *Histoire des anomalies*, 2:189.
104. Canguilhem, *Le Normal et le pathologique*, 77.
105. Isidore Geoffroy Saint-Hilaire, *Histoire des anomalies*, 1:496, 483.
106. *Histoire des anomalies*, 1:205 (italics added).
107. Xavier Bichat, *Recherches physiologiques sur la vie et la mort*, 3rd ed. (Paris: Brosson, 1805), 2–4. A note at the beginning of the book explained that Bichat's early death (in 1802) had meant that this edition was the same text as the original one (of 1799).
108. *Recherches physiologiques*, 13.
109. *Recherches physiologiques*, 15.
110. *Recherches physiologiques*, 14.
111. *Recherches physiologiques*, 14, 15.
112. *Recherches physiologiques*, 13.
113. *Recherches physiologiques*, 16.
114. *Recherches physiologiques*, 17.
115. Canguilhem, *Le Normal et le pathologique*, 29.
116. Bichat, *Recherches physiologiques*, 88. For a clear example of his use of "the natural state," see 303.
117. *Recherches physiologiques*, 270.
118. *Recherches physiologiques*, 219, 268.
119. *Recherches physiologiques*, 296.
120. *Recherches physiologiques*, 259.
121. Canguilhem, *Le Normal et le pathologique*, 29.
122. Warman, "From Pre-Normal to Abnormal," 200.
123. "From Pre-Normal to Abnormal," 200.
124. F.-J.-V. Broussais, *Recherches sur la fièvre hectique. Considérée comme dépendante d'une lésion d'action des différens systèmes, sans vice organique* (Paris: Méquignon, 1803), 7.
125. *Fièvre hectique*, 3n1 (italics in the original).
126. *Fièvre hectique*, 38, 64.
127. F.-J.-V. Broussais, *Examen de la doctrine médicale généralement adoptée, et des systèmes modernes de nosologie* (Paris: Gabon, 1816), 162.
128. *Examen de la doctrine*, 119, 221, 120, 124, 172, 173, 221.
129. *Examen de la doctrine*, 59.
130. F.-J.-V. Broussais, *Histoire des phlegmasies ou inflammations chroniques, fondée sur de nouvelles observations de clinique et d'anatomie pathologique. Ouvrage présentant un tableau raisonné des variétés et des combinaisons diverses de ces maladies, avec leur différentes modes de traitement*, 3rd ed. (Paris: Gabon, Crochard, 1822), 1:304n1.
131. Broussais, *Histoire des phlegmasies*, 1:104, 148, 110, 287.
132. F.-J.-V. Broussais, *De l'irritation et de la folie, ouvrage dans lequel les rapports du physique et du moral sont établis sur les bases de la médecine physiologique* (Paris: Delaunay, 1828), xxvi.
133. *De l'irritation et de la folie*, xxvi.
134. *De l'irritation et de la folie*, x (italics in the original).
135. *De l'irritation et de la folie*, xiv–xvi, 200.
136. The key reference here is P.-J.-G. Cabanis, *Du degré de certitude de la médecine* (Paris: F. Didot, 1798). See chapter 2 for some discussion of Cabanis's work and influence within the field of medicine.

137. Broussais, *De l'irritation et de la folie*, 229. On Broussais and materialism, see Jean-François Braunstein, *Broussais et le matérialisme. Médecine et philosophie au XIXe siècle* (Paris: Méridiens-Klincksieck, 1986).
138. *De l'irritation et de la folie*, xxviii–xxix.
139. *De l'irritation et de la folie*, 80.
140. *De l'irritation et de la folie*, 84.
141. *De l'irritation et de la folie*, 86.
142. *De l'irritation et de la folie*, 57, 84.
143. *De l'irritation et de la folie*, 4 (italics added).
144. *De l'irritation et de la folie*, 263.
145. Canguilhem, *Le Normal et le pathologique*, 23.
146. *Le Normal et le pathologique*, 19.
147. *Le Normal et le pathologique*, 26.
148. See, for example, Georges Canguilhem, *La Connaissance de la vie*, 2nd ed. (Paris: Vrin, 2009), 200, 208.

CHAPTER TWO

1. For a recent example, see Hillel D. Braude, *Intuition in Medicine: A Philosophical Defense of Clinical Reasoning* (Chicago: University of Chicago Press, 2012), 107–12.
2. Alain Desrosières, *The Politics of Large Numbers: A History of Statistical Reasoning*, trans. Camille Naish (Cambridge, MA: Harvard University Press, 1998), 2. Desrosières also refers to disagreements still ongoing today, 84.
3. Braude, *Intuition in Medicine*, 135, makes the point in general terms: "Attempts both to apply and resist the application of probabilistic methods occurred almost as soon as they were invented."
4. George Weisz, *The Medical Mandarins: The French Academy of Medicine in the Nineteenth and Early Twentieth Centuries* (Oxford: Oxford University Press, 1995), 33.
5. Weisz, *Medical Mandarins*, 76. Erwin Ackerknecht affirms that "for decades, the Paris Academy of Medicine was undoubtedly the leading medical assembly internationally, and its transactions were studied with the greatest interest everywhere." Erwin Ackerknecht, *Medicine at the Paris Hospital, 1794–1848* (Baltimore: Johns Hopkins University Press, 1967), 116–17.
6. Weisz, *Medical Mandarins*, 82.
7. Ulrich Tröhler has argued persuasively that decisive advances in the use of numbers in medicine had in fact already taken place in Britain before the great French debate. See Ulrich Tröhler, "Quantification in British Medicine and Surgery 1750–1830, with Special Reference to its Introduction into Therapeutics" (PhD diss., University of London, 1978), 2; Ulrich Tröhler, "Quantifying Experience and Beating Biases: A New Culture in Eighteenth-Century British Clinical Medicine," in *Body Counts: Medical Quantification in Historical and Sociological Perspective*, ed. Gérard Jorland, Annick Opinel, and George Weisz (Montreal: McGill-Queen's University Press, 2005), 44–45. For a critical genealogy like ours, however, public debate is at least as significant as technical advances.
8. *London Medical Gazette*, June 3, 1837, 361–64; *American Journal of the Medical Sciences* 21 (1837): 247, 525. The editorial comment was made in the *American Journal of the Medical Sciences* 21 (1837): 247.
9. Jules Guérin, "Feuilleton," *Gazette médicale de Paris* 5, no. 23 (June 1837): 356. Ackerknecht, *Medicine at the Paris Hospital, 1794–1848*, 117, describes the *Gazette* as France's most popular medical journal.
10. For the first kind, see, for example, Garabed Eknoyan, "Emergence of Quantification in Clinical Investigation and the Quest for Certainty in Therapeutics: The Road from Hammurabi to Kefauver," *Advances in Chronic Kidney Disease* 12, no. 1 (January 2005): 91; Alfred Jay Bollet,

"Pierre Louis: The Numerical Method and the Foundation of Quantitative Medicine," *American Journal of the Medical Sciences* 266, no. 2 (1973): 93; O. B. Sheynin, "On the History of Medical Statistics," *Archive for History of Exact Sciences* 26 (1982): 244. For a full-blown version of the second, see Ann F. La Berge, "Medical Statistics at the Paris School: What Was at Stake?," in *Body Counts*, ed. Jorland, Opinel, and Weisz, 90, 96. For more sophisticated versions of the second position, see Theodore Porter, *The Rise of Statistical Thinking, 1820–1900* (Princeton, NJ: Princeton University Press, 1986), 223; J. H. Warner, *The Therapeutic Perspective: Medical Practice, Knowledge and Identity in America, 1820–1885* (Cambridge, MA: Harvard University Press, 1986), 203–4; Gérard Jorland, "La Sous-détermination des théories médicales par les statistiques: le cas Semmelweis," in *Body Counts*, ed. Jorland, Opinel, and Weisz, 205.

11. J. Rosser Matthews, *Quantification and the Quest for Medical Certainty* (Princeton, NJ: Princeton University Press, 1995).

12. *Quantification*, 3–4.

13. *Quantification*, 140.

14. *Quantification*, 12–13.

15. We should mention among the influential histories of statistics Stephen M. Stigler's *The History of Statistics: The Measurement of Uncertainty before 1900* (Cambridge, MA: The Belknap Press of Harvard University Press, 1986). Stigler's work, however, has much in common with Matthews's for its progressive orientation. It tells a story of "magnificent success," 8, recounting how modern statistics, after many failures along the way, has achieved "a completion of the logic of the quantification of science," 358. Of all the published histories on the topic, Stigler's shows the greatest familiarity with mathematics and provides the fewest explanations in nonmathematical terms.

16. Lorraine Daston, *Classical Probability in the Enlightenment* (Princeton, NJ: Princeton University Press, 1988); Lorraine J. Daston, "Rational Individuals versus Laws of Society: From Probability to Statistics," *The Probabilistic Revolution. Volume 1: Ideas in History*, ed. Lorenz Krüger, Lorraine J. Daston, and Michael Heidelberger (Cambridge, MA: MIT Press, 1987), 295–304.

17. Daston, *Classical Probability*, 376–77.

18. Siméon-Denis Poisson, *Recherches sur la probabilité des jugements en matière criminelle et en matière civile, précédées des Règles générales du calcul des probabilités* (Paris: Bachelier, 1837; repr., Paris: Gabay, 2003).

19. It should be noted that, as he sets about his analysis of the 1837 debate in the Académie de Médecine, Matthews identifies the calculus of probabilities as one of the key references in the debate (*Quantification*, 23). His point is simply that none of the debaters understood the significance and applications of the calculus.

20. Daston, "Rational Individuals," 296.

21. Daston, *Classical Probability*, 253.

22. *Classical Probability*, 231.

23. *Classical Probability*, 232.

24. *Classical Probability*, 230–32.

25. Pierre-Simon Laplace, *Théorie analytique des probabilités*, vol. 1 (1847; reprinted Paris: Gabay, 1995), i, v.

26. *Théorie analytique*, ix.

27. In his introduction to the *Théorie analytique*, vii–viii, the first example is taken from astronomy.

28. *Théorie analytique*, ix.

29. *Théorie analytique*, ix.

30. *Théorie analytique*, ix.

31. *Théorie analytique*, lvi.

32. *Théorie analytique*, lx.

33. *Théorie analytique,* clxix.
34. *Théorie analytique,* clxix.
35. Desrosières, *Politics of Large Numbers,* 7.
36. *Politics of Large Numbers,* 286.
37. Daston, "Rational Individuals," 300.
38. Siméon-Denis Poisson, "Analyse mathématique. Note sur la loi des grands nombres, par M. Poisson," *Comptes rendus hebdomadaires de l'Académie des Sciences,* vol. 2 (1836), 378.
39. At the time, the range of the term "moral" covered phenomena that would today usually be called either "social" or "psychological." In another work, Poisson reached the conclusion that the law of great numbers applies just as well to moral phenomena as it does to physical ones. Poisson, *Recherches sur la probabilité,* 12.
40. Poisson, "Analyse mathématique," 378. For some discussion of this, see Hacking, *The Taming of Chance,* 95–102, and Desrosières, *Politics of Large Numbers,* 89.
41. Hacking, *The Taming of Chance,* 99.
42. Poisson, "Analyse mathématique," 378–79.
43. "Analyse mathématique," 379.
44. Poisson, *Recherches sur la probabilité,* 12.
45. Siméon-Denis Poisson, "Calcul des probabilités—Recherches sur la probabilité des jugements, principalement en matière criminelle, par M. Poisson," *Comptes rendus hebdomadaires des séances de l'Académie des Sciences,* vol. 1 (1835), 493.
46. Poisson, "Calcul des probabilités," 473–74.
47. Poisson, "Analyse mathématique," *Comptes rendus hebdomadaires des séances de l'Académie des Sciences,* vol. 2 (1836), 398 (italics in original).
48. For further discussion of this debate, see especially Daston, *Classical Probability,* 364–69. See also Stigler, *The History of Statistics,* 194.
49. Daston, *Classical Probability,* 369, goes on to speak of developments in the judicial context, but does not discuss the clash of values and perspectives that will concern us here.
50. Pierre-Jean-Georges Cabanis, *Du degré de certitude de la médecine* (Paris: Didot, 1798).
51. Daston, *Classical Probability,* 44.
52. Cabanis, *Du degré de certitude,* 15.
53. *Du degré de certitude,* 16–18.
54. *Du degré de certitude,* 26.
55. *Du degré de certitude,* 40.
56. *Du degré de certitude,* 40.
57. *Du degré de certitude,* 67.
58. *Du degré de certitude,* 67.
59. *Du degré de certitude,* 67.
60. *Du degré de certitude,* 117.
61. *Du degré de certitude,* 67–68.
62. *Du degré de certitude,* 87.
63. *Du degré de certitude,* 87.
64. See, for example, Michel Foucault, *The Birth of the Clinic: An Archeology of Medical Perception,* trans. A. M. Sheridan (London: Routledge, 1973), 61. On the difference between the glance (*coup d'œil*) and the gaze (*regard*), see 121–23.
65. Cabanis, *Du degré de certitude,* 144.
66. Foucault, *Birth of the Clinic,* 97.
67. *Birth of the Clinic,* 97.
68. *Birth of the Clinic,* 97 (italics in the original).
69. *Birth of the Clinic,* 109.
70. *Birth of the Clinic,* 97.

71. *Birth of the Clinic*, 98.
72. *Birth of the Clinic*, 109–110.
73. Philippe Pinel, *Traité médico-philosophique sur l'aliénation mentale*, 2nd ed. (Paris, 1809), 420–21.
74. *Traité médico-philosophique*, 402–3n1.
75. *Traité médico-philosophique*, 402.
76. *Traité médico-philosophique*, 402. Matthews, *Quantification*, 12, comments that Pinel's work "showed no detailed knowledge of the theory's mathematical subtleties as they were currently being developed by Laplace in his more sophisticated writings," but mathematical subtlety is not our principal concern.
77. *Traité médico-philosophique*, 404–5.
78. *Traité médico-philosophique*, 406.
79. Joseph-Henri Réveillé-Parise, "Miscellanées, no. VII," *Gazette médicale de Paris* 4, no. 12 (March 1836): 177.
80. "Miscellanées," 177.
81. "Miscellanées," 177.
82. "Miscellanées," 178.
83. When Matthews, *Quantification*, 23, begins his discussion of the 1837 debate in the Académie de Médecine, he notes accurately that the two key notions in play were the calculus of probabilities and *l'homme moyen*.
84. Adolphe Quetelet, *Sur l'homme et le développement de ses facultés ou Essai de physique sociale*, 2 vols. (Paris: Bachelier, 1835), 1:i.
85. Quetelet, *Sur l'homme*, 1:13–14. Daston speaks of "the Janus-like position of Quetelet with respect to the classical interpretation of probability," *Classical Probability*, 384.
86. *Sur l'homme*, 1:12.
87. *Sur l'homme*, 1:21.
88. *Sur l'homme*, 2:276.
89. *Sur l'homme*, 2:108.
90. Hacking, *The Taming of Chance*, 107.
91. Quetelet, *Sur l'homme*, 2:266. Daston writes: "Quetelet held up *l'homme moyen* as a literary, artistic, moral, and intellectual ideal, literally the golden mean for a given society: the great man was he who subsumed his individuality in 'all of humanity, nature, and the universal order,'" *Classical Probability*, 385.
92. Quetelet, *Sur l'homme*, 2:266–67.
93. This is our way of confirming Matthews's point that the fellows of the medical academy were not fully informed of developments in statistical mathematics. See Matthews, *Quantification*, 12–13. For a rather bland version of this line of argument, see Terence D. Murphy, "Medical Knowledge and Statistical Methods in Early-Nineteenth-Century France," *Medical History* 25 (1981): 311–13.
94. *Bulletin de l'Académie royale de médecine* (Paris: Baillière, 1836 [for 1837]), 1:593–94.
95. "Suite de la discussion sur la fièvre typhoïde," *Bulletin de l'Académie de médecine*, 1:594–95.
96. "Discussion sur la fièvre typhoïde," *Bulletin de l'Académie de médecine*, 1:595.
97. "Discussion sur la fièvre typhoïde," *Bulletin de l'Académie de médecine*, 1:594.
98. "Discussion sur la fièvre typhoïde," *Bulletin de l'Académie de médecine*, 1:595.
99. "Correspondance manuscrite," *Bulletin de l'Académie de médecine*, 1:606.
100. Risueño mentions the coincidence in the version of his speech published by the Academy to honor his contribution: Benigno Risueño d'Amador, *Mémoire sur le calcul des probabilités appliqué à la médecine. Lu à L'Académie royale de Médecine dans sa séance du 25 avril 1837* (Paris: Baillière, 1837), 6.
101. "Feuilleton," *Gazette médicale de Paris* 5, no. 17 (1837): 257.

102. Matthews, *Quantification*, 27.
103. Desrosières, *Politics of Large Numbers*, 83.
104. Desrosières, *Politics of Large Numbers*, 84 does say that the two positions maintain their opposition today, but by referring to one position as "traditional" he is in danger of missing some of the detail that interests us here.
105. George Weisz makes the point unequivocally: "The term 'calculus of probabilities' was frequently utilized in the ensuing debate and Louis's opponents sometimes spoke as if he were advocating the probability theories of Laplace, Quetelet, or Poisson. But the actual theories of Louis and his followers owed virtually nothing to the mathematical theorists of probability; they involved rather simple comparisons of proportions and averages," *The Medical Mandarins*, 164. See also O. B. Sheynin, "On the History of Medical Statistics," 252.
106. Risueño, "Mémoire sur le calcul des probabilités appliqué à la médecine," *Bulletin de l'Académie de médecine*, 1:623.
107. "Le calcul des probabilités appliqué à la médecine," *Bulletin* 1:622–23.
108. "Le calcul des probabilités appliqué à la médecine," *Bulletin* 1:624.
109. "Le calcul des probabilités appliqué à la médecine," *Bulletin* 1:624.
110. "Le calcul des probabilités appliqué à la médecine," *Bulletin* 1:624.
111. "Le calcul des probabilités appliqué à la médecine," *Bulletin* 1:648.
112. "Le calcul des probabilités appliqué à la médecine," *Bulletin* 1:635.
113. "Le calcul des probabilités appliqué à la médecine," *Bulletin* 1:636.
114. "Le calcul des probabilités appliqué à la médecine," *Bulletin* 1:634.
115. "Le calcul des probabilités appliqué à la médecine," *Bulletin* 1:667.
116. "Le calcul des probabilités appliqué à la médecine," *Bulletin* 1:659.
117. "Le calcul des probabilités appliqué à la médecine," *Bulletin* 1:659.
118. "Le calcul des probabilités appliqué à la médecine," *Bulletin* 1:634.
119. "Discussion sur la statistique médicale," *Bulletin de l'Académie de médecine*, 1:801.
120. "Discussion sur la statistique médicale," *Bulletin* 1:691.
121. "Discussion sur la statistique médicale," *Bulletin* 1: 692 (italics in the original).
122. "Discussion sur la statistique médicale," *Bulletin* 1:692 (italics in the original).
123. "Discussion sur la statistique médicale," *Bulletin* 1:692 (italics in the original).
124. "Discussion sur la statistique médicale," *Bulletin* 1:692–93 (italics in the original).
125. "Feuilleton," *Gazette médicale de Paris* 5, no. 19 (1837): 292.
126. "Discussion sur la statistique médicale," *Bulletin* 1:703.
127. "Discussion sur la statistique médicale," *Bulletin* 1:703.
128. "Discussion sur la statistique médicale," *Bulletin* 1:702.
129. "Discussion sur la statistique médicale," *Bulletin* 1:725.
130. Ackerknecht's brief report on the debate around numerical method consists essentially in a list of those who attacked Louis and a list of those who sided with him, *Medicine at the Paris Hospital*, 104. Matthews states, with greater precision, that "Louis's work had . . . framed the issue of quantification in clinical medicine in the terms in which it would be debated before the Parisian Academies of Science and Medicine in the late 1830s," *Quantification*, 20. Hacking simply calls Louis "the founder of the numerical method," *The Taming of Chance*, 84. Tröhler merely states that Louis "is viewed by all authors as the key figure of the numerical method," "Quantification in British Medicine and Surgery," 15.
131. "Feuilleton," *Gazette médicale de Paris* 5, no. 18 (1837): 274.
132. "Feuilleton": 289.
133. Pierre Charles Alexandre Louis, *Examen de l'examen de M. Broussais, relativement à la phtisie [sic] et à l'affection typhoïde* (Paris: Baillière, 1834), 155–56.
134. Louis, *Examen de l'examen*, 10.
135. Matthews, *Quantification*, 14, 16. George Weisz says simply: "The critics were right about

the limitations of counting," Weisz, "From Clinical Counting to Evidence-Based Medicine," in *Body Counts: Medical Quantification in Historical and Sociological Perspective*, ed. Gérard Jorland, Annick Opinel, and George Weisz (Montreal: McGill-Queen's University Press, 2005), 379.

136. Louis, *Examen de l'examen*, 25.
137. *Examen de l'examen*, 25.
138. "Discussion sur la statistique médicale," *Bulletin de l'Académie de médecine*, 1:722.
139. "Discussion sur la statistique médicale," *Bulletin* 1:722.
140. "Discussion sur la statistique médicale," *Bulletin* 1:722.
141. Louis, *Examen de l'examen*, 79, 80.
142. "Discussion sur la statistique médicale," *Bulletin* 1:732.
143. "Discussion sur la statistique médicale," *Bulletin* 1:732.
144. "Discussion sur la statistique médicale," *Bulletin* 1:732.
145. Jean-Baptiste Bouillaud, *Essai sur la philosophie médicale et sur les généralités de la clinique médicale, précédé d'un Résumé philosophique des principaux progrès de la médecine et suivi d'un Parallèle des résultats de la saignée coup sur coup avec ceux de l'ancienne méthode, dans le traitement des phlegma[s]ies aigües* (Paris: Rouvier and Le Bouvier, 1836), 194.
146. *Essai sur la philosophie médicale*, 187.
147. *Essai sur la philosophie médicale*, 187 (italics in the original).
148. "Feuilleton," *Gazette médicale de Paris* 5, no. 19 (May 1837): 291.
149. "Discussion sur la statistique médicale," *Bulletin de l'Académie de médecine*, 1:740.
150. Pierre-François-Olive Rayer, *Sur la statistique médicale* (Paris: Firmin-Didot, 1855).
151. "Discussion sur la statistique médicale," *Bulletin de l'Académie de médecine*, 1:783-84.
152. "Discussion sur la statistique médicale," *Bulletin* 1:784.
153. "Discussion sur la statistique médicale," *Bulletin* 1:785.
154. "Discussion sur la statistique médicale," *Bulletin* 1:785.
155. "Discussion sur la statistique médicale," *Bulletin* 1:785.
156. Rayer did make much the same point when he said that the value of statistics on mortality and the like was uncontested. What was contested, he said, was the application of this "instrument" to pathology, "Discussion sur la statistique médicale," *Bulletin de l'Académie de médecine*, 1:778.
157. "Discussion sur la statistique médicale," *Bulletin* 1:768-69.
158. "Discussion sur la statistique médicale," *Bulletin* 1:772.
159. Porter, *The Rise of Statistical Thinking*, 227.
160. F. Bérard, *Discours sur les améliorations progressives de la santé publique par l'influence de la civilisation* (Paris: Gabon, 1826).
161. Francis Bisset Hawkins, *Elements of Medical Statistics; Containing the Substance of the Gulstonian Lectures Delivered at the Royal College of Physicians: with Numerous Additions, Illustrative of the Comparative Salubrity, Longevity, Mortality and Prevalence of Diseases in the Principal Countries and Cities of the Civilized World* (London: Longman, 1829), v, vii.
162. Bernard-Pierre Lécuyer, "Probability in Vital and Social Statistics: Quetelet, Farr, and the Bertillons," in *The Probabilistic Revolution. Volume 1: Ideas in History*, ed. Lorenz Krüger, Lorraine J. Daston, and Michael Heidelberger (Cambridge, MA: MIT Press, 1987), 324.
163. Desrosières, *Politics of Large Numbers*, 82.
164. *Politics of Large Numbers*, 82.
165. Jules Guérin, "Feuilleton: Statistique médicale," *Gazette médicale de Paris* 1, no. 19 (1830): 6.
166. "Statistique du choléra. *Recherches sur la reproduction et la mortalité de l'homme aux différens âges, et sur la population de la Belgique*, par MM. A. Quetelet et Ed. Smits; rapport fait à l'Académie de médecine par M. Villermé," *Gazette médicale de Paris* 3, no. 68 (August 1832): 480-82.
167. "Statistique du choléra," *Gazette médicale de Paris*): 481 (italics in the original).

168. "Feuilleton," *Gazette médicale de Paris* 3, no. 121 (1832): 828–29.
169. Guérin, "Feuilleton: Statistique médicale," *Gazette médicale de Paris* 1, no. 1 (1830): 6.
170. "Revue clinique," *Gazette médicale de Paris* 1, no. 22 (1830): 197.
171. Hacking, *The Taming of Chance*, 81.
172. Paul Broca, "Du degré d'utilité de la statistique," *Le Moniteur des hôpitaux* 6 (1857): 47. The earlier article is found at 33–36.

CHAPTER THREE

1. Sir John F. W. Herschel, review of *Lettres à S.A.R. le duc régnant de Saxe-Cobourg et Gotha sur la théorie des probabilités appliquée aux sciences morales et statistiques*, by Adolphe Quetelet, *Edinburgh Review*, no. 185 (1850): 14. This was praise from an authoritative source. Theodore Porter comments that Herschel was "probably the most eminent man of science of his time." Theodore Porter, *The Rise of Statistical Thinking, 1820–1900* (Princeton, NJ: Princeton University Press, 1986), 119.
2. Adolphe Quetelet, "De l'emploi de la statistique dans les sciences médicales," *Lettres à S.A.R. le duc régnant de Saxe-Cobourg et Gotha sur la théorie des probabilités appliquée aux sciences morales et politiques* (Brussels: Hayez, 1845), 337–50.
3. Quetelet, *Lettres*, 337.
4. Quetelet, *Lettres*, 339.
5. Stéphane Callens, "Les Moyennes positivistes," in *Moyenne, milieu, centre: Histoires et usages*, ed. Jacqueline Feldman, Gérard Lagneau, and Benjamin Matalon (Paris: Editions de l'EHESS, 1991), 169.
6. *Gazette hebdomadaire de médecine et de chirurgie*, série 1, tome 1 (1853): 130–31. For a further example of the same kind of judgment, see M. J. Delaharpe, "Coup d'œil statistique sur la fièvre typhoïde à l'hôpital de Lausanne, de 1836 à 1850," *Gazette hebdomadaire de médecine et de chirurgie*, série 1, tome 1, no. 39 (1854): 631–32.
7. Quetelet, *Lettres*, 347.
8. Quetelet, *Lettres*, 337.
9. Quetelet, *Lettres*, 343.
10. Quetelet, *Lettres*, 343.
11. Jules Gavarret, *Principes généraux de statistique médicale, ou Développement des règles qui doivent présider à son emploi* (Paris: Bechet jeune and Labé, 1840), xiv. See also Jules Gavarret, *De l'application de la statistique à la médecine. Réponse à l'examen critique auquel M. le docteur Valleix a soumis, dans le numéro de mai 1840 des Archives générales de médecine, l'ouvrage de M. Gavarret* (n. p.: F. Locquin, 1840).
12. Gavarret, *Principes généraux*, 24.
13. Quetelet, *Lettres*, 344.
14. *Lettres*, 344.
15. See, for example, Jean Civiale, *Traité de l'affection calculeuse ou Recherches sur la formation, les caractères physiques et chimiques, les causes, les signes et les effets pathologiques de la pierre et de la gravelle, suivies d'un Essai de statistique sur cette maladie* (Paris: Crochard, 1838). For Civiale's method of data gathering by circular request, see 549. An official report on his work, in positive terms, was made to the Académie des Sciences. See *Comptes rendus hebdomadaires de l'Académie des Sciences* (1835): 167–77.
16. Quetelet, *Lettres*, 345.
17. *Lettres*, 346.
18. *Lettres*, 338.
19. *Lettres*, 339.

20. *Lettres*, 339.

21. See the *Comptes rendus hebdomadaires de l'Académie des Sciences* for sessions that took place in March and April 1837, where there was a strenuous discussion of the therapeutic value of bloodletting as opposed to purgatives, with numbers quoted by each side.

22. Quetelet, *Lettres*, 340.

23. André-Michel Guerry, *Statistique morale de l'Angleterre comparée avec la statistique morale de la France* (Paris: Baillière, 1864), vi, n1.

24. Guerry, *Statistique morale*, v–vi.

25. See, for example, Antoine-Augustin Cournot, *Exposition de la théorie des chances et des probabilités* (Paris: Hachette, 1843).

26. Herschel, review of *Lettres à S.A.R.*, 4.

27. On this point, see, for example, Gustave Rümelin, *Problèmes d'économie politique et de statistique*, trans. A. de Riedmatten (Paris: Guillaumin, 1896), 19.

28. Garabed Eknoyan, "Emergence of Quantification in Clinical Investigation and the Quest for Certainty in Therapeutics: The Road from Hammurabi to Kefauver," *Advances in Chronic Kidney Disease* 12, no. 1 (2005): 91.

29. Quetelet, *Lettres*, 341.

30. *Lettres*, 342–43.

31. *Lettres*, 342.

32. *Lettres*, 342.

33. *Lettres*, 345.

34. *Lettres*, 346–47.

35. Adolphe Quetelet, *Sur l'homme et le développement de ses facultés ou Essai de physique sociale* (Paris: Bachelier, 1835), 2:267. Alain Desrosières, *La politique des grands nombres. Histoire de la raison statistique* (Paris: La Découverte, 2000), 101, refers to the normal man and the average man as equivalent, but that seems to us not so much mistaken as unhelpfully elliptical.

36. Quetelet, *Sur l'homme*, 2:267.

37. *Sur l'homme*, 2:268.

38. *Sur l'homme*, 2:268.

39. *Sur l'homme*, 2:268.

40. *Sur l'homme*, 2:268.

41. *Sur l'homme*, 2:268.

42. *Sur l'homme*, 2:269.

43. *Sur l'homme*, 2:268.

44. *Sur l'homme*, 2:268–69.

45. *Sur l'homme*, 2:269.

46. *Sur l'homme*, 2:269–70.

47. Zheng Kang, "La société de statistique de Paris au XIXe siècle: un lieu de savoir social," *Journal de la Société Française de Statistique* 134, no. 3 (1993), 50–51.

48. Libby Schweber, *Disciplining Statistics: Demography and Vital Statistics in France and England, 1830–1885* (Durham, NC: Duke University Press, 2006), 52.

49. "Projet de classification pour l'établissement de la statistique générale d'un département," *Journal des travaux de la Société Française de Statistique Universelle*, n.s., 2, no. 19 (1837): 401.

50. César Moreau, "Procès-verbal," *Journal des travaux de la Société Française de Statistique Universelle*, n.s., 1, no. 6 (1835): 323.

51. Siméon-Denis Poisson, *Recherches sur la probabilité des jugements en matière criminelle et en matière civile; précédées des Règles générales du calcul des probabilités* (Paris: Bachelier, 1837), 7.

52. *Recherches sur la probabilité des jugements*, 13.

53. *Recherches sur la probabilité des jugements*, 318–415.

54. *Recherches sur la probabilité des jugements*, 385. Quetelet himself published a set of moral

statistics on the tendency or inclination to crime in France. Adolphe Quetelet, "Sur la statistique morale et les principes qui doivent en former la base," *Mémoires de l'Académie Royale de Belgique* 21 (Brussels: Hayez, 1848): 21.

55. Guerry, *Statistique morale*, xliii.

56. Alain Desrosières, "Masses, individus, moyennes: la statistique sociale au XIXe siècle," in *Moyenne, milieu, centre*, 258, comments that for Moreau de Jonnès statistical measurement had to be directly connected to the proper management of the apparatus of state. He was interested only in "true numbers."

57. Alexandre Moreau de Jonnès, *Eléments de statistique, comprenant les principes généraux de cette science et un aperçu historique de ses progrès* (Paris: Guillaumin, 1847), 2.

58. Moreau de Jonnès, *Eléments de statistique*, 2. Maurice Block, *Traité théorique et pratique de la statistique* (Paris: Guillaumin, 1878), 17, commented decades later that Moreau de Jonnès had always disparaged the doctrines and the production of Quetelet and others, and that he did so with "arguments of remarkable weakness."

59. *Journal des travaux de la Société Française de Statistique Universelle*, n.s., 1, no. 12 (1836): 624-725.

60. There was a reference to a study by Quetelet on life expectancy in Belgium, *Journal des travaux de la Société Française de Statistique Universelle*, n.s., 1, no. 11 (1836): 698-99.

61. Quetelet, "Sur la statistique morale," 3n1.

62. Lorraine Daston, *Classical Probability in the Enlightenment* (Princeton, NJ: Princeton University Press, 1988), 384.

63. Quetelet, "Sur la statistique morale," 4. Michel Armatte, "La Moyenne à travers les traités de statistique du XIXe siècle," in *Moyenne, milieu, centre*, 90, refers more generally to "this conflict between administrative and moral statistics that went on throughout the second third of the nineteenth century. Those who called themselves moral statisticians accused the others of blindly producing a mass of unorganized and therefore unusable figures, while they were reproached in turn with perverting statistics by engaging in improper conjecture and inference."

64. Quetelet, "Sur la statistique morale," 39.

65. Adolphe Quetelet, *Anthropométrie, ou Mesure des différentes facultés de l'homme* (Brussels: Muquardt, 1870), 19.

66. Quetelet, *Anthropométrie*, 19.

67. *Anthropométrie*, 5.

68. Quetelet, *Sur l'homme*, 1:21.

69. Desrosières, "Masses, individus, moyennes," 265.

70. Daston, *Classical Probability*, 385.

71. Quetelet, *Sur l'homme*, 2:266. Daston writes: "Quetelet held up *l'homme moyen* as a literary, artistic, moral, and intellectual ideal, literally the golden mean for a given society: the great man was he who subsumed his individuality in 'all of humanity, nature, and the universal order'" (*Classical Probability*, 385).

72. Hacking, *The Taming of Chance*, 107. Feldman, Matalon, and Lagneau comment: "Quetelet is the only thinker to have heaped such praise on the average man. That reflected his admiration for 'normal law,' which could only be of divine origin," "Introduction," in *Moyenne, milieu, centre*, 22.

73. See, for example, Quetelet, *Lettres*, 216; Quetelet, *Anthropométrie*, 22.

74. Quetelet, *Sur l'homme*, 2:108.

75. *Sur l'homme*, 2:274.

76. *Sur l'homme*, 2:266-67.

77. *Sur l'homme*, 1:21.

78. Quetelet, *Anthropométrie*, 276-77 (italics in the original).

79. *Anthropométrie*, 56.

80. *Anthropométrie*, 8.
81. *Anthropométrie*, 57.
82. Poisson, *Recherches sur la probabilité*, 204.
83. Quetelet, *Anthropométrie*, 390.
84. This example is borrowed from Hacking, *The Taming of Chance*, 106.
85. Porter, *The Rise of Statistical Thinking*, 91.
86. Quetelet, *Anthropométrie*, 269. Alain Desrosières, in his important work *La politique des grands nombres*, 18, 83, refers elliptically to the Gaussian law as "the future normal law." We are attempting to write our own history with a minimum of such anticipatory references.
87. Cournot, *Exposition de la théorie des chances*, 115–16.
88. See, for example, Quetelet, *Anthropométrie*, 7.
89. *Anthropométrie*, 8.
90. Quetelet, *Lettres*, 114–15.
91. *Lettres*, 116. Theodore Porter comments: "It is fully characteristic of Quetelet's scientific style that he interpreted his proudest discovery not as an indication that the mathematics of error had been conceived too narrowly, but as clear evidence that human variation could be understood in the same terms as errors of observation." Porter, *The Rise of Statistical Thinking*, 100.
92. Eric Brian observes that Quetelet "affirmed the equivalence of the spread of errors, that of uncertainty, and that of variability." Eric Brian, "Moyenne," in *Dictionnaire des concepts nomades en sciences humaines*, ed. Olivier Christin (Paris: Métailié, 2010), 319. Michel Armatte sums up Quetelet's position in more narrowly mathematical terms: "a series can be considered homogeneous whenever it can be considered *homologous* to a series of measurements marked by error." Michel Armatte, "Modèles statistiques de l'homogénéité et de la stabilité d'une population au XIXe siècle," in *Les Ménages. Mélanges en l'honneur de Jacques Desabie*, ed. Philippe L'Hardy and Claude Thélot. (Paris: INSEE, 1989), 90 (italics in the original).
93. Quetelet, *Anthropométrie*, 5.
94. Hacking, *The Taming of Chance*, 146–47.
95. Quetelet, *Anthropométrie*, 407. John Stuart Mill, in his discussion of the "logic of the moral sciences," discussed the apparent conflict between necessity and free will, effectively supporting the position adopted by Quetelet. John Stuart Mill, *A System of Logic, Being a Connected View of the Principles of Evidence and the Methods of Scientific Investigation*, 4th ed. (1843; repr. London: Parker, 1856), 405.
96. Adolphe Quetelet, "De l'influence du libre arbitre de l'homme sur les faits sociaux, et particulièrement sur le nombre des mariages," *Bulletin de la commission centrale de statistique* 3 [1846]: 135–55.
97. "De l'influence," 11.
98. "De l'influence," 9. For a further elaboration of Quetelet's views on "divine laws" and a "supreme being," see *Anthropométrie*, 10, 21, 380.
99. Alain Desrosières, "Homogéneité ou hétérogénéité d'une population: de Quetelet à Lexis," in *Les Ménages*, 72.
100. Poisson, *Recherches sur la probabilité des jugements*, 8.
101. Stephen Stigler, *The History of Statistics. The Measurement of Uncertainty before 1900* (Cambridge, MA: Harvard University Press, 1986), 235.
102. *The History of Statistics*, 215.
103. *Journal de la Société de Statistique de Paris*, 8e année, 1867, 280–81 (italics in the original).
104. *Journal de la Société de Statistique de Paris*, 282.
105. Jacqueline Feldman, Gérard Lagneau, and Benjamin Matalon say exactly that in "Introduction," *Moyenne, milieu, centre*, 10.
106. Quetelet, *Lettres*, 66–67.
107. *Lettres*, 66–67.

108. This was "the first mathematical proof that there really exists an average man, a man who is the type, at least as far as height is concerned. That assertion had been strongly contested by a few people, and I had not thought at the time to subject my numbers to the proofs that I have just given here" (Quetelet, *Lettres*, 216).

109. *Lettres*, 142.

110. *Lettres*, 216.

111. Quetelet, *Anthropométrie*, 17.

112. *Anthropométrie*, 15.

113. *Anthropométrie*, 15.

114. Quetelet, *Lettres*, 138.

115. Armatte, "Modèles statistiques de l'homogénéité," 92–93.

116. Quetelet, *Anthropométrie*, 15 (italics in the original).

117. *Anthropométrie*, 15.

118. Cournot, *Exposition*, 214.

119. Quetelet, *Anthropométrie*, 22.

120. Louis-Adolphe Bertillon, "Moyenne," in *Dictionnaire encyclopédique des sciences médicales*, série 2, tome 10, ed. A. Dechambre and L. Lereboullet (Paris: Masson/Asselin, 1874–79), 312.

121. *Journal de la Société de Statistique de Paris*, 15e année (1874): 115.

122. *Journal de la Société de Statistique de Paris*: 115.

123. *Journal de la Société de Statistique de Paris*: 116.

124. Jacques Bertillon, Alphonse Bertillon, and Georges Bertillon, *La Vie et les œuvres du docteur L.-A. Bertillon* (Paris: Masson, 1883), 72. In an article published in the *Gazette hebdomadaire de médecine et de chirurgie* 2, no. 40 (1855): 715, he refers to discussions on this matter that took place in an international conference on statistics. He appears to have played a leading role in them.

125. Bertillon, Bertillon, and Bertillon, *La Vie et les œuvres*, 71–72.

126. Louis-Adolphe Bertillon, *Conclusions statistiques contre les détracteurs de la vaccine, précédées d'un Essai sur la méthode statistique appliquée à l'étude de l'homme* (Paris: Victor Masson, 1857), 17.

127. Bertillon, *Conclusions*, vi–vii (italics in the original).

128. Bertillon, Bertillon, and Bertillon, *La Vie et les œuvres*, 149–150.

129. The series of seven articles by Bertillon was published in *Le Moniteur des hôpitaux* 5, nos. 89–104 (July–August 1857).

130. Bertillon, *Conclusions*, 10.

131. *Conclusions*, 2–3.

132. *Conclusions*, 4–7.

133. D. Baxby, *Jenner's Smallpox Vaccine: The Riddle of Vaccinia Virus and Its Origin* (London: Heinemann, 1981), xiii.

134. Jean Baptiste Édouard Bousquet, "Vaccine. Extrait du rapport sur la vaccine pour l'année 1853," *Le Moniteur des hôpitaux* 3, no. 82 (1855): 653–56.

135. Armand Bayard, *L'Influence de la vaccine sur la population, ou de la gastro-entérite varioleuse avant et depuis la vaccine* (Paris: Victor Masson, 1855).

136. "Travaux originaux. De la prétendue substitution de la fièvre typhoïde à la variole, depuis l'introduction de la vaccine, par le docteur Barth," *Gazette hebdomadaire de médecine et de chirurgie* 1, no. 1 (1853): 4–6.

137. Bénédict Teissier, "De la vaccination et de son influence prétendue sur la production de la fièvre typhoïde," *Gazette hebdomadaire de médecine et de chirurgie* 1, no. 13 (1853): 188–91.

138. Bertillon, *Conclusions*, 47. See also Bertillon, Bertillon, and Bertillon, *La Vie et les œuvres*, 23, 73.

139. Dr. Leudet, "L'influence de la vaccine sur la population," *Gazette hebdomadaire de médecine et de chirurgie* 1, no. 26 (1854): 414.

140. Hector Carnot, *Essai de mortalité comparée avant et depuis l'introduction de la vaccine en France* (Autun: Dejussieu, 1849), 5.

141. There is a brief discussion of Bertillon's work on the vaccine and on mortality statistics in Schweber, *Disciplining Statistics*, 57.

142. He provides an indicative and abbreviated list of these publications in Bertillon, *Conclusions statistiques*, 50n1.

143. Louis-Adolphe Bertillon, review of *Eléments de statistique humaine, ou démographie comparée*, by Achille Guillard, *Gazette hebdomadaire de médecine et de chirurgie* 2, no. 39 (1855): 711.

144. Louis-Adolphe Bertillon, "Feuilleton. Congrès international de statistique," *Gazette hebdomadaire de médecine et de chirurgie* 2, no. 39 (1855): 698 (italics in the original).

145. "Feuilleton. Congrès international de statistique," 698.

146. Quetelet, *Lettres*, 343-44.

147. Baxby, *Jenner's Smallpox Vaccine*, 5, 63. See also N. J. Willis, "Edward Jenner and the Eradication of Smallpox," *Scottish Medical Journal* 42, no. 4 (1997): 118-21.

148. Bertillon, *Conclusions*, vi.

149. *Conclusions*, 44.

150. He also delivered a formal address on the matter to the Société de Statistique de Paris (*Journal de la Société de Statistique de Paris*, 3e année (1862): 29-30.

151. Bertillon, *Conclusions*, 83, 155.

152. Bertillon, Bertillon, and Bertillon, *La Vie et les œuvres*, 23.

153. Stigler, *The History of Statistics*, 201n27.

154. Quetelet, *Anthropométrie*, 31-32.

155. *Anthropométrie*, 32.

156. *Anthropométrie*, 183.

157. *Anthropométrie*, 68.

158. Quetelet, *Sur l'homme*, 2:41.

159. *Sur l'homme*, 2:48.

160. *Sur l'homme*, 2:53.

161. Quetelet, *Anthropométrie*, 68.

162. *Anthropométrie*, 16 (italics in the original).

163. *Anthropométrie*, 14 (italics in the original).

164. *Anthropométrie*, 45 (italics in the original).

165. *Anthropométrie*, 63.

166. *Anthropométrie*, 174.

167. *Anthropométrie*, 186.

168. *Anthropométrie*, 186.

169. *Anthropométrie*, 240.

170. *Anthropométrie*, 32.

171. *Anthropométrie*, 296.

172. Georges Canguilhem, *Le Normal et le pathologique* (Quadrige: Presses Universitaires de France, 1966), 99. At this point, Canguilhem was summarizing and taking as his own a critical analysis developed by Maurice Halbwachs, *La Théorie de l'homme moyen. Essai sur Quetelet et la statistique morale* (Paris: Alcan, 1913).

173. Canguilhem, *Le Normal et le pathologique*, 100.

174. *Le Normal et le pathologique*, 100. Ian Hacking makes a comparable point in the strongest of terms: "Quetelet had *made* mean stature, eye-colour, artistic faculty and disease into real quantities. Once he had done that (and it is never recorded that in 1844 he had constructed this entirely new kind of reality) deviation from the means was just natural deviation, deviation made by nature, and that could not be conceived of as error" (*The Taming of Chance*, 113).

175. Quetelet, *Anthropométrie*, 44.

176. Quetelet, *Sur l'homme*, 2:277.
177. Quetelet, *Anthropométrie*, 44.
178. Canguilhem, *Le Normal et le pathologique*, 99.
179. *Le Normal et le pathologique*, 99.
180. There is one use of it in *Anthropométrie*, 307, where he comments that the mother of a dwarf was of average height and showed "nothing abnormal."

CHAPTER FOUR

1. Libby Schweber observes that "all nineteen of the founding members were medical doctors, all were republicans, all were anticlerical, and all were positivists." Libby Schweber, *Disciplining Statistics: Demography and Vital Statistics in France and England, 1830-1885* (Durham, NC: Duke University Press, 2006), 61.
2. *Bulletins de la Société d'Anthropologie de Paris*, series 1, vol. 2, 1861: 32. The title of this journal will henceforth be abbreviated in our text as *BSAP*. Subsequent references to the general discussions of the proceedings of the Society will be given in the notes in the form "*BSAP*," followed by the series number, the volume number, the date, and the page number.
3. *BSAP*, series 1, vol. 2 (1861): 32-33.
4. On the influence of the Society across a range of countries, see Joy Dorothy Harvey, "Races Specified. Evolution Transformed: The Social Context of Scientific Debates Originating in the Société d'Anthropologie de Paris" (PhD diss., Harvard University, 1983), 79-83.
5. Claude Blanckaert underlines the importance of the Society and of these three figures in Claude Blanckaert, "Méthode des moyennes et notion de 'série suffisante' en anthropologie physique (1830-1880)," in *Moyenne, milieu, centre: Histoires et usages*, ed. Jacqueline Feldman, Gérard Lagneau, Benjamin Matalon (Paris: Editions de l'EHESS, 1991), 214. The Société d'Anthropologie de Paris was founded at Broca's instigation in 1859.
6. Joy Dorothy Harvey shows an understanding of this purpose when she observes in her examination of the Society that "the interactive quality of discussion and debate, controversy and complaint, brings to the surface the preconceptions of the scientists in a manner which a less dynamic process of idea exchange cannot provide." Harvey, "Races Specified," 2. For a condensed version of Harvey's thesis, see Joy Dorothy Harvey, "Evolutionism Transformed: Positivists and Materialists in the *Société d'Anthropologie de Paris* from Second Empire to Third Republic," in *The Wider Domain of Evolutionary Thought*, ed. D. Oldroyd and J. Langham (Dordrecht: Springer, 1983), 289-310.
7. A French translation of Camper's 1770 original appeared as *Dissertation sur les variétés naturelles qui caractérisent la physionomie des hommes des divers climats et des différents âges*, trans. H. J. Jansen (Paris: Jansen, 1791).
8. Broca referred to phrenology in general terms in a monograph extracted from the *BSAP* of 1879 entitled *Étude des variations craniométriques et de leur influence sur les moyennes. Détermination de la série suffisante* (Extraits des bulletins de la Société d'Anthropologie de Paris), séance du 18 décembre 1879, 7. The monograph was published in Paris by Hennuyer in 1880.
9. Paul Broca, "Sur le volume et la forme du cerveau suivant les individus et suivant les races," *BSAP*, series 1, vol. 2 (1861): 191-92. Broca himself engaged in the study of areas of the brain, notably the area that governed speech. On occasion, he presented some of that work to the Society. See for example *BSAP*, series 1, vol. 4 (1864): 200-2.
10. Paul Broca, "Étude des variations craniométriques et de leur influence sur les moyennes," *BSAP*, series 3, vol. 2 (1879): 756.
11. Groups of skulls were, however, sometimes studied as such by phrenologists. Nancy Stepan, *The Idea of Race in Science: Great Britain 1800-1960* (London: Macmillan, 1982), 24.

12. *BSAP*, series 3, vol. 2 (1879): 770.
13. *BSAP*, series 2, vol. 1 (1866): 460.
14. Louis-Adolphe Bertillon, "Sur les ossements des Eyzies," *BSAP*, series 2, vol. 3 (1868): 559.
15. *BSAP*, series 2, vol. 3 (1868): 568.
16. In the great debate of 1830, Étienne Geoffroy Saint-Hilaire had made much the same distinction to characterize the difference between his work and Cuvier's. Cuvier, he said, was engaged in "the search for differences," while he himself was engaged in "the search for analogies." Étienne Geoffroy Saint-Hilaire, *Principes de philosophie zoologique, discutés en mars 1830, au sein de l'Académie royale des sciences* (Paris: Pichon et Didier, Rousseau, 1830), 52.
17. *BSAP*, series 3, vol. 2 (1879): 756.
18. *BSAP*, series 3, vol. 2 (1879): 756.
19. Paul Topinard, "De la méthode en craniométrie," *BSAP*, series 2, vol. 8 (1873): 859.
20. *BSAP*, series 2, vol. 8 (1873): 854.
21. Paul Topinard, *L'Anthropologie*, 5th ed. (Paris: Reinwald, 1895), 216–17.
22. Paul Broca, *Mémoires d'anthropologie zoologique et biologique* (Paris: Reinwald, 1877), 500.
23. *BSAP*, series 2, vol. 8 (1873): 857.
24. *BSAP*, series 2, vol. 8 (1873): 857–58.
25. Broca formulated this question in Paul Broca, *Étude des variations craniométriques*, 16.
26. Paul Broca, "Sur les crânes basques de Saint-Jean-de-Luz," *BSAP*, series 2, vol. 3 (1867) 51.
27. *BSAP*, series 1, vol. 5 (1864): 708–9.
28. *BSAP*, series 1, vol. 5 (1864): 713. For some discussion of Davis's own view of race and his collecting of skeletons, see Paul Turnbull, "British Anthropological Thought in Colonial Practice: The Appropriation of Indigenous Australian Bodies," in *Foreign Bodies: Oceania and the Science of Race 1750–1940*, ed. Bronwen Douglas and Chris Ballard (Canberra: ANU ePress, 2008), 206, 219–223. Turnbull points out that by arguing that the Neanderthal skull was marked by pathology Davis believed he was making a strong point against Darwinian evolutionism.
29. Broca, *Étude des variations craniométriques*, 17–18.
30. *Étude des variations craniométriques*, 17–18.
31. *Étude des variations craniométriques*, 4–5.
32. *Étude des variations craniométriques*, 4–5.
33. Paul Broca, "Sur la détermination de l'âge moyen," *BSAP*, series 3, vol. 2, no. 2 (1879): 312.
34. *BSAP*, series 3, vol. 2 (1879): 771.
35. *BSAP*, series 3, vol. 2 (1879): 771.
36. *BSAP*, series 3, vol. 2 (1879): 310.
37. Broca, *Étude des variations craniométriques*, 16.
38. *BSAP*, series 3, vol. 2 (1879): 777.
39. Bronwen Douglas, after speaking of Broca's "harshly racialist" anthropology, makes the following observation with which we concur: "the blanket charge of racism is too blunt and anachronistic an instrument for most historians who want to discriminate precisely between discourses, ideologies, vocabularies and authors and to understand them in contemporary terms." Douglas, "Climate to Crania: Science and the Racialization of Human Difference," in *Foreign Bodies*, 66, 72. Nancy Stepan argues comparably that work done at that time ought not to be dismissed as "pseudoscientific." Stepan, *The Idea of Race in Science*, xvi.
40. *BSAP*, series 3, vol. 2 (1879): 493–94.
41. *BSAP*, series 3, vol. 2 (1879): 499.
42. *BSAP*, series 1, vol. 2 (1865): 261–62.
43. *BSAP*, series 1, vol. 2 (1865): 262.
44. A. de Gobineau, *Essai sur l'inégalité des races humaines*, vol. 1 (Paris: Firmin Didot, 1853), 47, 58.
45. Paul Broca, *Mémoires d'anthropologie*, 493.

46. *Mémoires d'anthropologie*, 515.
47. It was first published in the *Journal de la physiologie de l'homme et des animaux* in 1859 and 1860. Paul Weindling accurately observes that "Gobineau had virtually no impact on the bourgeois Paris anthropologists, who preferred . . . comparative anatomy to the vagaries of history." Paul Weindling, *Health, Race and German Politics between National Unification and Nazism, 1870–1945* (Cambridge: Cambridge University Press, 1989), 51–52.
48. Adolphe Quetelet, *Anthropométrie, ou Mesure des différentes facultés de l'homme* (Brussels: C. Muquardt, 1870), 263.
49. Quetelet, *Anthropométrie*, 314. Theodore Porter comments briefly on the significance of Quetelet's work with respect to the conflict between monogenism and polygenism. Theodore Porter, *The Rise of Statistical Thinking, 1820–1900* (Princeton, NJ: Princeton University Press, 1986), 108.
50. Adolphe Quetelet, "Sur les indiens O-Jib-Be-Wa's et les proportions de leur corps," *Bulletin de l'Académie Royale des Sciences et Belles-Lettres de Belgique*, 13, part 1 (1846): 70–76.
51. Adolphe Quetelet, "Sur les proportions de la race noire," *Bulletin de l'Académie Royale des Sciences et Belles-Lettres de Belgique*, 21, part 1 (1854): 96–100.
52. Quetelet, *Anthropométrie*, 322.
53. *Anthropométrie*, 322–3.
54. Broca, *Mémoires*, 503. An influential classification into five races had been made by the German Johann Friedrich Blumenbach in the last decade of the eighteenth century. Blumenbach set out a spectrum of "the five principal varieties of mankind" disposed around the white Caucasian, considered to be the primeval one. Johann Friedrich Blumenbach, *On the Natural Variety of Mankind, The Anthropological Treatises of Johann Friedrich Blumenbach*, trans. Thomas Bendyshe (London: Green, Longman, Roberts, and Green, 1865), 209, 264.
55. Quetelet, *Anthropométrie*, 323.
56. *Anthropométrie*, 323n1.
57. Broca, *Mémoires*, 500.
58. Quetelet, *Anthropométrie*, 58.
59. P. N. Gerdy, *Physiologie médicale, didactique et critique* (Paris: Roret, Béchet, 1830), lxviii–lxix.
60. *BSAP*, series 2, vol. 7 (1872): 29.
61. *BSAP*, series 1, vol. 2 (1861): 181.
62. Paul Broca, "Sur le plan horizontal de la tête et sur la méthode trigonométrique," *BSAP*, series 2, vol. 8 (1873): 49.
63. "Sur le plan horizontal de la tête," 49.
64. Broca, *Étude des variations craniométriques*, 4.
65. *BSAP*, series 1, vol. 1 (1859): 12.
66. *BSAP*, series 1, vol. 1 (1864): 65.
67. *BSAP*, series 1, vol. 1 (1864): 140.
68. *BSAP*, series 1, vol. 1 (1863): 594 (italics in the original).
69. *BSAP*, series 1, vol. 1 (1863): 767.
70. *BSAP*, series 2, vol. 1 (1866): 19.
71. *BSAP*, series 2, vol. 1 (1866): 20.
72. *BSAP*, series 1, vol. 1 (1864): 223–42.
73. *BSAP*, series 1, vol. 1 (1864): 478–79.
74. This exchange can be considered characteristic of Broca's control and diplomatic management of the Society. For some discussion of that question, see Elizabeth Williams, "Anthropological Institutions in Nineteenth-Century France," *Isis*, 76 (September 1985): 331–48.
75. *BSAP*, series 1, vol. 3 (1862): 282.
76. *BSAP*, series 1, vol. 3 (1862): 282.

77. *BSAP*, series 1, vol. 3 (1862): 282 (italics in the original).
78. *BSAP*, series 1, vol. 3 (1862): 283.
79. *BSAP*, series 1, vol. 3 (1862): 292.
80. *BSAP*, series 1, vol. 3 (1862): 295-96.
81. *BSAP*, series 1, vol. 1 (1859): 254.
82. *BSAP*, series 1, vol. 5 (1864): 572.
83. *BSAP*, series 2, vol. 1 (1866): 23.
84. *BSAP*, series 2, vol. 1 (1866): 23.
85. *BSAP*, series 1, vol. 2 (1860): 69-70.
86. *BSAP*, series 1, vol. 2 (1860): 70.
87. *BSAP*, series 2, vol. 1 (1866): 70-71.
88. *BSAP*, series 2, vol. 1 (1866): 71-72.
89. *BSAP*, series 2, vol. 1 (1866): 72.
90. *BSAP*, series 2, vol. 1 (1866): 73-74.
91. *BSAP*, series 1, vol. 1 (1861): 75.
92. *BSAP*, series 1, vol. 1 (1862): 139.
93. *BSAP*, series 1, vol. 1 (1862): 139.
94. *BSAP*, series 2, vol. 3 (1868): 25.
95. *BSAP*, series 2, vol. 4 (1869): 71-72.
96. *BSAP*, series 2, vol. 3 (1868): 26.
97. *BSAP*, series 2, vol. 9 (1874): 71.
98. *BSAP*, series 2, vol. 9 (1874): 96.
99. *BSAP*, series 1, vol. 2 (1865): 457.
100. *BSAP*, series 1, vol. 2 (1865): 457.
101. *BSAP*, series 1, vol. 2 (1865): 458.
102. Nancy Stepan gives a long list of Broca's instruments. Stepan, *The Idea of Race in Science*, 208n47.
103. *BSAP*, series 1, vol. 2 (1865): 657.
104. *BSAP*, series 1, vol. 1 (1864): 435-36.
105. Blumenbach, *On the Natural Variety of Mankind*, 237.
106. *BSAP*, series 1, vol. 1 (1861): 520.
107. *BSAP*, series 1, vol. 1 (1861): 517-18.
108. *BSAP*, series 1, vol. 1 (1861): 520.
109. *BSAP*, series 1, vol. 2 (1865): 20.
110. *BSAP*, series 1, vol. 2 (1865): 20.
111. *BSAP*, series 2, vol. 1 (1866): 385.
112. *BSAP*, series 2, vol. 1 (1866): 386.
113. *BSAP*, series 2, vol. 1 (1866): 386.
114. *BSAP*, series 2, vol. 1 (1866): 386.
115. *BSAP*, series 2, vol. 1 (1866): 387.
116. *BSAP*, series 2, vol. 1 (1866): 387.
117. *BSAP*, series 2, vol. 1 (1866): 387-88.
118. *BSAP*, series 2, vol. 1 (1866): 388.
119. *BSAP*, series 2, vol. 1 (1866): 388.
120. *BSAP*, series 2, vol. 1 (1866): 96.
121. *BSAP*, series 2, vol. 1 (1866): 14.
122. *BSAP*, series 2, vol. 1 (1866): 16.
123. *BSAP*, series 2, vol. 1 (1866): 18.
124. *BSAP*, series 2, vol. 1 (1866): 18.
125. *BSAP*, series 2, vol. 1 (1866): 19.
126. *BSAP*, series 2, vol. 1 (1866): 389.

127. *BSAP*, series 2, vol. 3 (1868): 734.
128. *BSAP*, series 2, vol. 1 (1866): 97.
129. *BSAP*, series 2, vol. 1 (1866): 406.
130. See, for example, Clémence Royer, "Remarques sur le transformisme," *BSAP*, series 2, vol. 5 (1870): 265–317.
131. Claude Blanckaert, "Préface," in Paul Broca, *Mémoires d'anthropologie*), xxv.
132. Nancy Stepan, *The Idea of Race in Science*, 94. For further discussion of the use of this notion by anthropologists, see George W. Stocking Jr., *Race, Culture, and Evolution: Essays in the History of Anthropology* (New York: Free Press, 1968), 42–68.
133. *BSAP*, series 3, vol. 2 (1879): 469.
134. *BSAP*, series 3, vol. 2 (1879): 469.
135. *BSAP*, series 1, vol. 1 (1859): 12.
136. *BSAP*, series 1, vol. 1 (1859): 11.
137. Broca, *Mémoires d'anthropologie*, 500 (italics in the original).
138. Topinard, *L'Anthropologie*, 226 (italics in the original).
139. *BSAP*, series 3, vol. 2 (1879): 757.
140. Broca, *Étude des variations craniométriques*, 4.
141. Broca, *Mémoires d'anthropologie*, 502.
142. Broca, *Mémoires d'anthropologie*, 501.
143. Lorraine Daston and Peter Galison, *Objectivity* (New York: Zone Books, 2007), 58.
144. *Objectivity*, 16.
145. *Objectivity*, 104.
146. *Objectivity*, 44.
147. *Objectivity*, 60.
148. *Objectivity*, 34, 123.

CHAPTER FIVE

1. Mauro Forno, "Scienziati e mass-media: Lombroso e gli studiosi positivisti nella stampa tra otto e novecento," in *Cesare Lombroso. Gli scienziati e la nuova Italia*, ed. Silvano Montaldo (Bologna: Il Mulino, 2010), 224–25.
2. Cesare Lombroso, *L'uomo delinquente in rapporto all'antropologia, alla giurisprudenza ed alle discipline carcerarie* (Turin: Bocca, 1889), 163, 168.
3. *L'uomo delinquente* (1889), 291.
4. *L'uomo delinquente* (1889), 163.
5. Cesare Lombroso, *L'uomo delinquente in rapporto all'antropologia, alla giurisprudenza ed alla psichiatria: Cause e rimedi* (Turin: Bocca, 1897), 617.
6. *L'uomo delinquente* (1897), 217.
7. Michel Foucault, *The Order of Things: An Archaeology of the Human Sciences* (New York: Pantheon Books, 1971), xv. Ian Hacking offers a breezily ironic characterization of Italian criminal anthropology: "The Italian countries took up statistics later but then engaged in the greatest parody of modern Europe ever conceived. [They] created a criminal anthropology of which Sherlock Holmes addicts have heard a distant echo." Ian Hacking, "Biopower and the Avalanche of Printed Numbers," *Humanities in Society* 5 (1982): 288. There is indeed much in Lombroso's writing that lends itself to ironic reading, but we are endeavoring to give an earnest account of it in the terms that were pertinent to him and to his contemporaries, including his outspoken critics.
8. Lombroso, *L'uomo delinquente* (1897), 609.
9. *Bulletins de la Société d'Anthropologie de Paris*, series 2, vol. 2, 1867: 348. The title of this journal will henceforth be abbreviated as *BSAP*.
10. *BSAP*, series 2, vol. 2 (1867): 348.

11. *BSAP*, series 2, vol. 2 (1867): 348.
12. *BSAP*, series 2, vol. 2 (1867): 349.
13. Lombroso, *L'uomo delinquente* (1897), 497.
14. *L'uomo delinquente* (1897), 497.
15. *BSAP*, series 2, vol. 2 (1867): 349.
16. *BSAP*, series 2, vol. 2 (1867): 352.
17. *BSAP*, series 2, vol. 2 (1867): 352.
18. *BSAP*, series 2, vol. 2 (1867): 351.
19. *BSAP*, series 2, vol. 2 (1867): 371-75.
20. Robert Nye, *Crime, Madness, and Politics in Modern France: The Medical Concept of National Decline* (Princeton, NJ: Princeton University Press, 1984), 62-63, points out that Hubert Lauvergne, a doctor at the convict station of Toulon during the July Monarchy, had attempted in 1841 to use phrenology to characterize criminals.
21. *BSAP*, series 2, vol. 2 (1867): 351.
22. *BSAP*, series 2, vol. 2 (1867): 351.
23. *BSAP*, series 2, vol. 2 (1867): 353.
24. *BSAP*, series 2, vol. 2 (1867): 378.
25. Adolphe Quetelet, "Sur la statistique morale et les principes qui doivent en former la base," *Mémoires de l'Académie Royale de Belgique* 21 (Brussels: Hayez, 1848): 21-34.
26. Adolphe Quetelet, "De l'influence du libre arbitre de l'homme sur les faits sociaux, et particulièrement sur le nombre des mariages," *Bulletin de la commission centrale de statistique* 3, [1846]: 136. https://archive.org/stream/bulletindelacom00statgoog - page/n11/mode/2up
27. Silvano Montaldo, "Premessa," in *Cesare Lombroso*, ed. Silvano Montaldo, 7.
28. Mauro Forno, "Scienziati e mass-media," 224-25.
29. Valeria Paola Babini, "Lo studio della mente: momenti di un dialogo tra Francia e Italia," in *Francia/Italia. Le filosofie dell'Ottocento*, ed. Renzo Ragghianti and Alessandro Savorelli (Pisa: Edizioni della Normale, 2007), 131.
30. Lombroso, *L'uomo delinquente* (1897), 503.
31. *L'uomo delinquente* (1897), 503.
32. *L'uomo delinquente* (1897), 504.
33. *L'uomo delinquente* (1897), 24.
34. See, for example, Paul Broca, "Recherches sur l'ethnologie de la France," *BSAP*, series 1, vol. 1 (1860): 12.
35. Lombroso, *L'uomo delinquente* (1897), 23.
36. *L'uomo delinquente* (1897), 26.
37. *L'uomo delinquente* (1897), 26-27.
38. *L'uomo delinquente* (1897), 513.
39. *L'uomo delinquente* (1897), 513.
40. *L'uomo delinquente* (1897), 514, 512.
41. Cesare Lombroso and Guiglelmo Ferrero, *La donna delinquente, la prostituta e la donna normale* (Torino: L. Roux, 1893), 261.
42. *La donna delinquente*, 262.
43. *La donna delinquente*, 262.
44. *La donna delinquente*, 261.
45. *La donna delinquente*, 261.
46. Lombroso, *L'uomo delinquente* (1897), 256.
47. *L'uomo delinquente* (1897), 628.
48. Léonce Manouvrier, "Les Crânes des suppliciés," *Archives de l'anthropologie criminelle et des sciences pénales. Médecine légale, judiciaire. Statistique criminelle. Legislation et droit* 1 (1886): 119-20. He discussed the place of criminal anthropology more briefly in Manouvrier, "La Place et

l'importance de la crâniologie anthropologique," *Matériaux pour l'histoire primitive et naturelle de l'homme. Revue mensuelle*, 3e série, tome 3 (January 1886): 17-20.
49. Manouvrier, "Les Crânes des suppliciés," 120.
50. Les Crânes des suppliciés," 120.
51. "Les Crânes des suppliciés," 120-21.
52. "Les Crânes des suppliciés," 121.
53. "Les Crânes des suppliciés," 125.
54. "Les Crânes des suppliciés," 126.
55. "Les Crânes des suppliciés," 128-29.
56. "Les Crânes des suppliciés," 129.
57. "Les Crânes des suppliciés," 130.
58. "Les Crânes des suppliciés," 130.
59. "Les Crânes des suppliciés," 136.
60. "Les Crânes des suppliciés," 136-37.
61. "Les Crânes des suppliciés," 137.
62. Lombroso and Ferrero, *La donna delinquente*, 262.
63. *Actes du deuxième congrès international d'anthropologie criminelle. Biologie et sociologie* (Paris, August 1889; repr. Lyon: Storck, 1890; Paris: Masson, 1890), 3. A summary version of this document was published as *Exposition universelle internationale de 1889. Direction générale de l'exploitation. Deuxième congrès international d'anthropologie criminelle. Session de Paris du 10 au 17 août 1889* (Paris: Imprimerie Nationale, 1889).
64. *Actes du deuxième congrès international d'anthropologie* (Storck and Masson), 28.
65. *Actes du deuxième congrès international*, 29.
66. *Actes du deuxième congrès international*, 29.
67. *Actes du deuxième congrès international*, 31.
68. *Actes du deuxième congrès international*, 31.
69. *Actes du deuxième congrès international*, 192.
70. *Actes du deuxième congrès international*, 32.
71. *Actes du deuxième congrès international*, 31.
72. *Actes du deuxième congrès international*), 33.
73. *Actes du deuxième congrès international*, 34.
74. *Actes du deuxième congrès international*, 33-34.
75. *Actes du deuxième congrès international*, 33.
76. *BSAP*, series 3, vol. 11 (1888): 645.
77. *BSAP*, series 3, vol. 11 (1888): 645.
78. *BSAP*, series 3, vol. 11 (1888): 646.
79. *BSAP*, series 3, vol. 11 (1888): 646-47.
80. Paul Topinard, "Criminologie et anthropologie," *Actes du deuxième congrès international*, 492.
81. "Criminologie et anthropologie," 492.
82. "Criminologie et anthropologie," 493.
83. "Criminologie et anthropologie," 493.
84. "Criminologie et anthropologie," 494.
85. *Exposition universelle de 1889 . . . Deuxième congrès international d'anthropologie criminelle*, 19.
86. R. Laschi, "Actes du II Congrès international d'Anthropologie criminelle," *Archivio di psichiatria, scienze penali ed antropologia criminale per servire allo studio dell'uomo alienato e delinquente* 12 (1891): 560.
87. Lombroso, *Pazzi ed anomali* (Città di Castell: Lapi, 1886), 4.
88. Cesare Lombroso, "I pazzi criminali," *Archivio di psichiatria, scienze penali ed antropologia criminale per servire allo studio dell'uomo alienato e delinquente* 9 (1888): 572.

89. Delia Frigessi, "Introduzione," in Cesare Lombroso, *Delitto, genio, follia*, ed. Delia Frigessi, Ferrucio Giacanelli, and Luisa Mangoni (Turin: Bollati Boringhieri, 1995), 359, mentions the idea of a "great social family" as a way of referring to all anomalous individuals fused together.

90. Renzo Villa, *Il deviante e i suoi segni: Lombroso e la nascita dell'antropologia criminale* (Milan: Franco Angeli, 1985), 267.

91. *Il deviante e i suoi segni*, 8. See also 104.

92. *Il deviante e i suoi segni*, 114–15.

93. *Il deviante e i suoi segni*, 114–15.

94. *Il deviante e i suoi segni*, 189.

95. For quotations of and specific responses to these criticisms, see Lombroso and Ferrero, *La donna delinquente*, 262, 287.

96. *La donna delinquente*, viii–ix.

97. *La donna delinquente*, viii.

98. *La donna delinquente*, 1.

99. *La donna delinquente*, 1–13.

100. *La donna delinquente*, 31–32.

101. *La donna delinquente*, 34.

102. *La donna delinquente*, 48.

103. *La donna delinquente*, 56.

104. *La donna delinquente*, 57.

105. *La donna delinquente*, 63.

106. *La donna delinquente*, 64–65.

107. *La donna delinquente*, 133.

108. All of this makes Lombroso a potential target for feminist deconstruction, although he has attracted relatively little attention in that regard. Lisa Downing is an exception. Her book *The Subject of Murder: Gender, Exceptionality, and the Modern Killer* (Chicago: University of Chicago Press, 2013) contains a strenuous critique of *La donna delinquente* in particular. See especially 63–67. She observes that "Lombroso's theory of the female criminal is riven with contradiction and paradox" (64).

109. Mary Gibson, *Born to Crime: Cesare Lombroso and the Origins of Biological Criminality* (Westport, CT: Praeger, 2002), 28.

110. *Born to Crime*, 29.

111. *Born to Crime*, 29.

112. *Born to Crime*, 29.

113. *Actes du deuxième congrès international*, 194.

114. *Actes du deuxième congrès international*, 196.

115. *Actes du deuxième congrès international*, 197.

116. *Pazzi ed anomali*, xii.

117. *Pazzi ed anomali*, xii, xv.

118. *Pazzi ed anomali*, 26.

119. *Pazzi ed anomali*, 11. In *L'uomo delinquente* (1897), 502, he refers to the popular practice of "giving importance to the physiognomy of criminals in deciding about their perversity."

120. *Pazzi ed anomali*, 27–34.

121. That argument was nicely articulated by Lombroso's colleague Virgilio Rossi in a review of Napoleone Colajanni's *L'alcolismo* published in "Sull'alcoolismo e le critiche di Colajanni," *Archivio di psichiatria, scienze penali ed antropologia criminale per servire allo studio dell'uomo alienato e delinquente* 7 (1886): 605–10, esp. 606.

122. *Pazzi ed anomali*, 11.

123. *Pazzi ed anomali*, 26.

124. *Pazzi ed anomali*, 26.

125. David Horn takes issue with Gibson on the matter of popular knowledge, arguing that "Lombroso frankly delighted in deploying traditional and popular culture to support his findings, and to undercut and isolate the 'metaphysical' claims made by classical jurists." See David G. Horn, *The Criminal Body: Lombroso and the Anatomy of Deviance* (New York: Routledge, 2003), 60.

CHAPTER SIX

1. Galton's early papers demonstrate that he began by exploring a number of visual technologies for the production of composite portraits, before settling on photography. These included the stereoscope (although this could combine only two images at a time), an "Iceland spar," and several devices of Galton's own invention. A discussion of the relative merits of these can be found in his article on "Composite Portraits Made by Combining Those of Many Different Persons into a Single Figure," *Nature* 18 (1878): 98–99. In *Generic Images* (London: Clowes, 1879), Galton describes using a magic lantern during public lectures to produce composite portraits in front of the eyes of the audience.

2. Galton described the technical process by which composite portraits were produced in an early article on the topic: "Suppose that there are eight portraits in the pack, and that under existing circumstances it would require an exposure of eighty seconds to give an exact photographic copy of any one of them," Galton explained. "We throw the image of each of the eight portraits in turn upon the same part of the sensitised plate for ten seconds. . . . The sensitised plate will now have had its total exposure of eighty seconds; it is then developed, and the print taken from it is the generalised picture of which I speak." "Composite Portraits," 97. Further accounts were given in "An Inquiry into the Physiognomy of Phthisis by the Method of 'Composite Portraiture,'" *Guy's Hospital Reports* 25 (1882): 475–93 and *Inquiries into Human Faculty and Its Development* (London: Macmillan, 1883).

3. Galton, "Composite Portraits," 97.

4. Francis Galton, "Notes on the Marlborough School Statistics," *Journal of the Anthropological Institute* 4 (1875): 133.

5. Galton, "Composite Portraits," 97.

6. Galton, *Generic Images*, 162.

7. Galton, *Natural Inheritance* (London and New York: Macmillan, 1889), 36.

8. *Natural Inheritance*, 36.

9. Galton, "Typical Laws of Heredity (1)," *Nature* 5 (1877): 492–93.

10. Galton, *Generic Images*, 163.

11. See, for example, *Journal de la Société de Statistique de Paris*, 10e année (1869): 38.

12. *Journal de la Société de Statistique de Paris*, 3e année (1862): 37.

13. *Journal de la Société de Statistique de Paris*, 6e année (1865): 3.

14. Wilhelm Lexis, "Sur la durée normale de la vie humaine et sur la théorie de la stabilité des rapports statistiques," *Annales de démographie internationale* 2, no. 5 (1878): 447.

15. Bertillon had presented two papers on the matter. The first, on "the different ways of measuring the duration of human life," was published in *Journal de la Société de Statistique de Paris*, 7e année (1866): 45–64. The second, on the "determination of mortality," was published in *Journal de la Société de Statistique de Paris*, 10e année (1869): 29–40. The second of these presentations applied the calculus of probabilities to the "probability of dying" during specified, relatively short periods.

16. Lexis, "Sur la durée normale de la vie humaine," 448.

17. Calculation of this sort was not entirely unprecedented. An anonymous article published in 1821 considered "probabilities in relation to the duration of life," and suggested, with no particular mathematical argument, that the average be calculated from the age of thirteen onward.

"Notions générales sur la population," *Recherches statistiques sur la ville de Paris et le département de la Seine. Recueil de tableaux dressés d'après les ordres de Monsieur le Comte de Chabrol, conseiller d'état, préfet du département* (Paris: Ballard, 1821), 19-26.

18. For some general discussion of Lexis's contribution to statistics, see Theodore Porter, *The Rise of Statistical Thinking, 1820-1900* (Princeton, NJ: Princeton University Press, 1986), 243-50.

19. Galton, *Natural Inheritance*, 6.

20. Lexis's work was taken up in France by Jacques Bertillon, a son of Louis-Adolphe and by that time himself a prominent statistician. In 1879, Bertillon made a teacherly presentation on the subject to the Société d'Anthropologie de Paris. Paul Broca, who had been a lifelong champion of statistical method in medicine and anthropology, resisted the innovation quite strenuously. See Jacques Bertillon, "Sur la vie moyenne et la vie normale," *Bulletins de la Société d'anthropologie de Paris*, series 3, vol. 2, no. 2 (1879): 468-69 and Paul Broca, "Détermination de la série suffisante," *Bulletins de la Société d'anthropologie de Paris*, series 3, vol. 2, no. 2 (1879): 763-64. See also Paul Broca, "Sur la détermination de l'âge moyen," *Bulletins de la Société d'anthropologie de Paris*, series 3, vol. 2, no. 2 (1879): 299.

21. Karl Pearson, *The Life, Letters and Labours of Francis Galton*, vol. 2 (Cambridge: Cambridge University Press, 1924), 298.

22. For a discussion of the geometry, see Alain Desrosières, *La Politique des grands nombres. Histoire de la raison statistique* (Paris: La Découverte, 2000), 145.

23. Francis Galton, "Typical Laws of Heredity," *Proceedings of the Royal Institution* 9 (1877): 297.

24. Galton, "Typical Laws of Heredity (3)," 532.

25. See Bertillon, "Sur la vie moyenne et la vie normale," 468-69.

26. Desrosières, *La Politique des grands nombres*, 86, comments that Galton and Karl Pearson brought together the questions about heredity arising from Darwin's work, normal distributions of attributes of the human species drawn from Quetelet, and techniques of adjustment drawn from the theory of errors of measurement.

27. For Desrosières, *La Politique des grands nombres*, 144, Galton's key innovation was that he no longer considered, as Quetelet had, that the regular distribution of human attributes was merely the net effect of a large number of aleatory causes. Galton was positing that, among those various causes, heredity was by far the most powerful.

28. For an early use of the phrase "mathematical law of deviation," see Galton, "Typical Laws of Heredity (1)," 493.

29. Galton, "Typical Laws of Heredity (1)," 493.

30. "Typical Laws of Heredity (2)," 512.

31. Galton, *Natural Inheritance*, 51.

32. *Natural Inheritance*, 55. On 57, he spoke again of "the Normal curve."

33. Galton, "Typical Laws of Heredity (1)," 493.

34. Galton, *Natural Inheritance*, 3.

35. Galton put it in narrowly mathematical terms by designating a median value as M, then expressing every value in the range under consideration as M plus or minus D (*Natural Inheritance*, 51-52). A calculation of averages was necessary to establish M, but the most significant statistical work would be done on D.

36. *Natural Inheritance*, 41.

37. *Natural Inheritance*, 41.

38. *Natural Inheritance*, 41.

39. Galton, "Typical Laws of Heredity (2)," 513.

40. "Typical Laws of Heredity (2)," 513.

41. "Typical Laws of Heredity (2)," 513.

42. "Typical Laws of Heredity (2)," 514.

43. Galton, *Essays in Eugenics* (London: Eugenics Education Society, 1909), 89.

44. Allan Sekula, "The Body and the Archive," *October* 39 (1986): 40.
45. Pearson, *The Life . . . of Francis Galton*, 1: 301. Cf. 2: 283: "Galton developed composite photography in his search for a method of ascertaining whether physiognomy is an index to mind."
46. *The Life . . . of Francis Galton*, 1: 301.
47. Quoted in *The Life . . . of Francis Galton*, 2: 283–84.
48. Galton, *Inquiries into Human Faculty*, 9.
49. Galton, "Request for Prints of Photographic Portraits," *Nature* 73 (1906): 534.
50. Pearson, *The Life . . . of Francis Galton*, 1: 301.
51. Galton, "Composite Portraiture," 135.
52. Francis Galton, "On the Application of Composite Portraiture to Anthropological Purposes," *Report of the British Association for the Advancement of Science* 51 (1881): 690.
53. Galton, *Generic Images*, 161.
54. *Generic Images*, 161.
55. Nancy Stepan, *The Idea of Race in Science: Great Britain 1800–1960* (London: Macmillan, 1982), 84.
56. Elizabeth Edwards, ed., *Anthropology and Photography, 1860–1920* (New Haven and London: Yale University Press and The Royal Anthropological Institute, 1992); Daniel Novak, *Realism, Photography and Nineteenth-Century Fiction* (Cambridge: Cambridge University Press, 2008).
57. Anne Maxwell, *Picture Imperfect: Photography and Eugenics, 1870–1940* (Eastbourne: Sussex Academic Press, 2008), 97.
58. *Picture Imperfect*, 97.
59. Jonathan Crary, *Techniques of the Observer: On Vision and Modernity in the 19th Century* (Cambridge, MA: MIT Press, 1990); Novak, *Realism, Photography and Nineteenth-Century Fiction*; Maxwell, *Picture Imperfect*.
60. David Green, "Veins of Resemblance: Photography and Eugenics," *Oxford Art Journal* 7, no. 2 (1984): 6.
61. "Veins of Resemblance," 3.
62. Galton, "An Inquiry into the Physiognomy of Phthisis," 478–79.
63. Lorraine Daston and Peter Galison, *Objectivity* (New York: Zone Books, 2007), 121.
64. *Objectivity*, 42.
65. As Daston and Galison recognize, these two approaches to scientific images should not be understood as historically distinct: "Epistemic virtues do not replace one another like a succession of kings. Rather, they accumulate into a repertoire of possible forms of knowing" (*Objectivity*, 113).
66. *Objectivity*, 121.
67. Galton, "Composite Portraits," 97.
68. Popular portraits of babies and toddlers taken during this period, when exposure times remained very long, included the shadows of their mothers in the background, covered by a dark sheet. While these images seem strange to the contemporary eye, they were extremely popular at the end of the nineteenth century, in a way that Galton's portraits never were.
69. Galton, *Essays in Eugenics*, 49.
70. See, for instance Marta Braun, *Picturing Time: The Work of Étienne-Jules Marey, 1830–1904* (Chicago: University of Chicago Press, 1994) and Phillip Prodger, *Time Stands Still: Muybridge and the Instantaneous Photography Movement* (Oxford and New York: Oxford University Press, 2003).
71. Galton, "An Inquiry into the Physiognomy of Phthisis," 482.
72. Francis Galton, *Hereditary Genius: An Inquiry into Its Laws and Consequences* (London: Macmillan, 1869), xx.
73. Pearson, *The Life . . . of Francis Galton*, 1: 217.
74. *The Life . . . of Francis Galton*, 1: 109.
75. *The Life . . . of Francis Galton*, 1: 79.
76. Galton, *Inquiries into Human Faculty*, 1.

77. Galton, *Essays in Eugenics*, 68–69.
78. *Essays in Eugenics*, 42.
79. *Essays in Eugenics*, 33.
80. *Essays in Eugenics*, 38.
81. *Essays in Eugenics*, 81.
82. Galton, *Inquiries into Human Faculty*, 9.
83. Galton, *Hereditary Genius*, xviii.
84. Galton, *Essays in Eugenics*, 66.
85. Galton, *Inquiries into Human Faculty*, 33.
86. Galton, *Essays in Eugenics*, 36–37.
87. Pearson, *The Life . . . of Francis Galton*, 1: 111.
88. Francis Galton, "Deterioration of the British Race," *Times*, June 18, 1909.
89. Galton, *Essays in Eugenics*, 21.
90. *Essays in Eugenics*, 47.
91. Karl Pearson, *National Life from the Standpoint of Science*, 2nd ed. (London: Black, 1905), 18.

92. Desrosières, *La Politique des grands nombres*, 130, 403, notes a historical irony in the fact that techniques that would later find widespread application in the practice of statistics should have been developed in the service of "a scientifico-political war machine."

93. Daniel Kevles, *In the Name of Eugenics: Genetics and the Uses of Human Heredity* (Berkeley: University of California Press, 1985), 3–19.

94. Stepan, *The Idea of Race in Science*, 112.
95. Galton, *Essays in Eugenics*, 43.
96. Francis Galton, "Eugenics: Its Definition, Scope, and Aims," *The American Journal of Sociology* (July 1904): 21.
97. "Eugenics: Its Definition, Scope, and Aims," 21.
98. "Eugenics: Its Definition, Scope, and Aims," 9.
99. "Eugenics: Its Definition, Scope, and Aims," 11.
100. Judges provided the key sample in Galton's *Hereditary Genius*.
101. Galton, *Essays in Eugenics*, 24.
102. *Essays in Eugenics*, 25.
103. Kevles, *In the Name of Eugenics*, 59.
104. *In the Name of Eugenics*, 64.
105. Galton, *Essays in Eugenics*, 77–78.
106. *Essays in Eugenics*, 75.

107. Pearson is still widely acknowledged as the foundation figure in theoretical statistics. *Biometrika* is still in press and recognized as the most prestigious journal in its field.

108. Galton, *Essays in Eugenics*, 79.
109. *Essays in Eugenics*, 79–80.

110. Henry Pickering Bowditch, "Are Composite Portraits Typical Pictures?," *McClure's Magazine* 3, no. 4 (September 1894): 331-42. Lombroso speaks of his own occasional use of the technique in his essay *L'Anthropologie criminelle et ses récents progrès* (Paris: Alcan, 1890), 31–32. The somewhat eccentric Italian criminologist Napoleone Colajanni considered that Galton's composite photography was theoretically sound, even though he himself did not apply it in his own work. Colajanni declared that composite photography had "victoriously and definitively answered all the objections brought against it." *La sociologia criminale* (Catania: Tropea, 1889) 1: 349–50.

111. Walter Rogers Furness, *Composite Photography Applied to the Portraits of Shakespeare* (Philadelphia: R. M. Lindsay, 1885).

112. Josh Ellenbogen, *Reasoned and Unreasoned Images: The Photography of Bertillon, Galton, and Marey* (University Park, PA: Penn State University Press, 2012), 77.

113. Sekula, "The Body and the Archive": 40, 52.
114. Galton, *Inquiries into Human Faculty*, 15.

115. *Inquiries into Human Faculty*, 63.
116. Havelock Ellis, *The Criminal* (London: Walter Scott, 1890), 15.
117. Ellis, *The Criminal*, 49.
118. Ellis, *The Criminal*, 49.
119. Francis Galton, "Criminal Anthropology," *Nature* 42 (1890): 75.
120. "Criminal Anthropology," 75.
121. Ellis, *The Criminal*, 50.
122. Ellenbogen, *Reasoned and Unreasoned Images*, 77.
123. Galton's review of *Signaletic Instructions* was published in *Nature*, October 15, 1896. Francis Galton, "Signaletic Instructions, including the Theory and Practice of Anthropometrical Identification," *Nature* 54 (1896): 569–70.
124. Sekula, "The Body and the Archive," 19. Desrosières, *La Politique des grands nombres*, 155, distinguishes between Alphonse Bertillon and Galton with admirable succinctness: "The difference is that the former has his eye on the individual whereas the latter attends closely to distribution."
125. Sekula, "The Body and the Archive," 40.
126. "The Body and the Archive," 54.
127. "The Body and the Archive," 27.
128. Ellenbogen, *Reasoned and Unreasoned Images*, 115.
129. In a curious epilogue to this history, as Ellenbogen notes, it was Galton who "helped introduce the individuation technology that later replaced Bertillonage—fingerprinting. To Bertillon's dismay, fingerprinting increasingly forced his system out of major penal systems in the years after 1910" (*Reasoned and Unreasoned Images*, 77).
130. Galton, *Essays in Eugenics*, 19–20.
131. Galton, "Composite Portraits," 97–98.
132. Francis Galton, "Photographic Composites," *Photographic News* 29, no. 1398 (1885): 243.
133. Galton, *Inquiries into Human Faculty*, 11.
134. Galton, "An Inquiry into the Physiognomy of Phthisis," 475.
135. "An Inquiry into the Physiognomy of Phthisis," 475.
136. "An Inquiry into the Physiognomy of Phthisis," 475.
137. "An Inquiry into the Physiognomy of Phthisis," 481.
138. "An Inquiry into the Physiognomy of Phthisis," 481.
139. "An Inquiry into the Physiognomy of Phthisis," 481.
140. "An Inquiry into the Physiognomy of Phthisis," 483.
141. "An Inquiry into the Physiognomy of Phthisis," 483.
142. "An Inquiry into the Physiognomy of Phthisis," 483.
143. Francis Galton, "The Anthropometric Laboratory," *Fortnightly Review* 31 (1882): 338. In America, notes Kevles, "thousands of people filled out their 'Record of Family Traits' and mailed the forms to the Eugenics Record Office" (*In the Name of Eugenics*, 38).
144. Maxwell, *Picture Imperfect*, 95.
145. Frans Lundgren, "The Politics of Participation: Francis Galton's Anthropometric Laboratory and the Making of Civic Selves," *The British Journal for the History of Science* 46, no. 3 (2013): 457.
146. "The Politics of Participation," 457.
147. Francis Galton, "Proposal to Apply for Anthropological Statistics from Schools," *Journal of the Anthropological Institute* 3 (1874): 308–11; Francis Galton, "On the Height and Weight of Boys Aged 14, in Town and Country Public Schools," *Journal of the Anthropological Institute* 5 (1876): 174.
148. Francis Galton, "On the Anthropometric Laboratory at the Late International Health Exhibition," *Journal of the Anthropological Institute* 14 (1885): 207.
149. "On the Anthropometric Laboratory," 338.

150. "On the Anthropometric Laboratory," 332.

151. Galton claimed the number of participants was 9,337, and that each was measured in 17 different ways.

152. Galton, "On the Anthropometric Laboratory," 206. Subsequent documents, however, reveal that the number of people who visited the permanent Anthropometric Laboratory in London in the first three years of its existence were only a third of this figure. The atmosphere of an International Exhibition clearly boosted its popularity.

153. Lundgren, "The Politics of Participation," 12.

154. "The Politics of Participation," 12.

155. "The Politics of Participation," 12.

156. Pearson, *The Life . . . of Francis Galton*, 1: 305.

157. Adolphe Quetelet, *Anthropométrie, ou Mesure des différentes facultés de l'homme* (Brussels: Muquardt, 1870), 19.

158. *Bulletins de la Société d'anthropologie de Paris*, series 1, vol. 2 (1861): 139.

159. Galton, *Hereditary Genius*, vi.

160. *Hereditary Genius*, vi.

161. *Hereditary Genius*, 1.

162. *Hereditary Genius*, 29-30.

163. *Hereditary Genius*, 31-32.

164. *Hereditary Genius*, 2.

165. Galton, *Inquiries into Human Faculty*, 4.

166. *Inquiries into Human Faculty*, 4.

167. *Inquiries into Human Faculty*, 19.

168. Galton, *Hereditary Genius*, v.

169. *Hereditary Genius*, 26.

170. Stepan, *The Idea of Race in Science*, 131.

171. They discuss the age range in Alfred Binet and Théodore Simon, "Le Développement de l'intelligence chez les enfants," *L'Année psychologique*, no. 14 (1907): 59-60.

172. Binet did refer on occasion to Galton's work as a psychologist. There is some discussion of Galton's view of ideation in Alfred Binet, *L'Étude expérimentale de l'intelligence* (Paris: Reinwald, 1903), 155, but eugenics does not rate a mention.

173. Charles Spearman, *The Nature of "Intelligence" and the Principles of Cognition* (1923; repr., New York: Arno, 1973), 7. He observed that by 1923 the literature devoted to these tests had become "so immense that a special bulletin had to be published periodically in order to announce the latest contributions" (8). For comment on the divergence of perspective between Spearman and Binet, see Nancy Stepan, *The Idea of Race in Science*, 132-33.

174. Binet and Simon, "Le Développement de l'intelligence chez les enfants," 1.

175. "Le Développement de l'intelligence chez les enfants," 64.

176. "Le Développement de l'intelligence chez les enfants," 64.

177. "Le Développement de l'intelligence chez les enfants," 64.

178. Alfred Binet, *Les Enfants anormaux. Guide pour l'admission des enfants anormaux dans les classes de perfectionnement* (Paris: Colin, 1907), 6-7.

179. *Les Enfants anormaux*, 6.

180. Binet and Simon, "Le Développement de l'intelligence chez les enfants," 47.

181. Binet, *Les Enfants anormaux*, 6, 6n1.

182. *Les Enfants anormaux*, 7.

183. *Les Enfants anormaux*, 7.

184. Binet and Simon, "Le Développement de l'intelligence chez les enfants," 85.

185. "Le Développement de l'intelligence chez les enfants," 75.

186. "Le Développement de l'intelligence chez les enfants," 77.

187. "Le Développement de l'intelligence chez les enfants," 2.

188. "Le Développement de l'intelligence chez les enfants," 30.
189. "Le Développement de l'intelligence chez les enfants," 30.
190. "Le Développement de l'intelligence chez les enfants," 15.

CHAPTER SEVEN

1. George Beard, *A Practical Treatise on Nervous Exhaustion (Neurasthenia): Its Symptoms, Nature, Sequences, Treatment* (New York: Treat, 1880); *American Nervousness: Its Causes and Consequences* (New York: Treat, 1881).
2. For an account of the spermatorrhoea epidemic, see Elizabeth Stephens, *Anatomy as Spectacle: Public Exhibitions of the Body from 1700 to the Present* (Liverpool: Liverpool University Press, 2011).
3. Warren Estelle Lloyd, *Psychology, Normal and Abnormal: A Study of the Processes of Nature from the Inner Aspect* (Los Angeles: Baumgart, 1908), 19.
4. Richard von Krafft-Ebing, *Psychopathia Sexualis, with Especial Reference to Contrary Sexual Instinct: A Medico-Legal Study*. 7th ed. (Philadelphia: F. A. Davis; London: Rebman, 1894), 1.
5. William Robie, *Rational Sex Ethics: A Physiological and Psychological Study of the Sex Lives of Normal Men and Women, with Suggestions for a Rational Sex Hygiene with Reference to Actual Case Histories* (Boston: Badger, 1916); Isabel Davenport, *Salvaging of American Girlhood: A Substitution of Normal Psychology for Superstition and Mysticism in the Education of Girls* (New York: Dutton, 1924).
6. Sylvanus Stall, *What a Young Boy Ought to Know* (Philadelphia: Vir, 1897); *What a Young Husband Ought to Know* (Philadelphia: Vir, 1899); *What a Man of Forty-Five Ought to Know* (Philadelphia: Vir, 1901); *What a Young Man Ought to Know* (Philadelphia: Vir, 1904).
7. Mary Wood-Allen, *What a Young Woman Ought to Know* (Philadelphia: Vir, 1897); *What a Young Girl Ought to Know* (Philadelphia: Vir, 1904).
8. Krafft-Ebing, *Psychopathia Sexualis*, vi.
9. Ivan Crozier provides a detailed account of the complex publication history of *Sexual Inversion* in his critical edition of the text. Ivan Crozier, "Introduction: Havelock Ellis, John Addington Symonds and the Construction of *Sexual Inversion*." *Sexual Inversion: A Critical Edition*. (London: Palgrave Macmillan, 2008), 1–95.
10. Robie, *Rational Sex Ethics*, frontispiece.
11. Jennifer Terry, *An American Obsession: Science, Medicine, and Homosexuality in Modern Society* (Chicago: University of Chicago Press, 1999), 134.
12. Stall, *What a Young Man Ought to Know*, 24.
13. Jessie Murray, "Preface," in Marie Carmichael Stopes, *Married Love: A New Contribution to the Solution of Sex Difficulties* (London: Fifield, 1918), 1.
14. June Rose, *Marie Stopes and the Sexual Revolution* (London and Boston: Faber and Faber, 1992), 119.
15. Renate Hauser, "Krafft-Ebing's Psychological Understanding of Sexual Behaviour," in *Sexual Knowledge, Sexual Science: The History of Attitudes to Sexuality*, ed. Roy Porter and Mikuláš Teich (Cambridge: Cambridge University Press, 1994), 211.
16. Krafft-Ebing, *Psychopathia Sexualis*, iv.
17. *Psychopathia Sexualis*, v.
18. Sander L. Gilman, "The Struggle of Psychiatry with Psychoanalysis: Who Won?," *Critical Inquiry* 13, no. 2 (1987): 296.
19. See Harry Oosterhuis, *Stepchildren of Nature: Krafft-Ebing, Psychiatry, and the Making of Sexual Identity* (Chicago and London: University of Chicago Press, 2000), 171, for the German context and, for the British context, Ivan Crozier's introduction to his critical edition of *Sexual Inversion*. While Sean Brady has questioned the role of legislation in shaping the production of

nineteenth-century homosexual subjectivities (*Masculinity and Male Homosexuality in Britain, 1861–1913* (London: Palgrave Macmillan, 2005), it undoubtedly played a pivotal role in Krafft-Ebing's research.

20. Lucy Bland and Laura Doan, eds., *Sexology in Culture: Labelling Bodies and Desires* (London: Polity, 1998), 2.

21. Oosterhuis, *Stepchildren of Nature*, 87.

22. *Stepchildren of Nature*, 105.

23. Krafft-Ebing, *Psychopathia Sexualis*, 12.

24. *Psychopathia Sexualis*, 13.

25. *Psychopathia Sexualis*, 13.

26. *Psychopathia Sexualis*, 142.

27. *Psychopathia Sexualis*, 60.

28. *Psychopathia Sexualis*, 153.

29. See Oosterhuis, *Stepchildren of Nature*, 47.

30. *Stepchildren of Nature*, 65.

31. Michel Foucault, *The History of Sexuality, Volume One: An Introduction* (New York: Pantheon Books, 1978), 67.

32. Foucault, *History of Sexuality*, 1:67.

33. Magnus Hirschfeld, *Sexual Anomalies: The Origin, Nature, and Treatment of Sexual Disorders* (New York: Emerson, 1948), 29.

34. Krafft-Ebing, *Psychopathia Sexualis*, 34.

35. T. S. Clouston, "The Developmental Aspects of Criminal Anthropology," *The Journal of the Anthropological Institute of Great Britain and Ireland* 23 (1894): 215.

36. Harry H. Laughlin, "What Eugenics Is All About," in *A Decade of Progress in Eugenics: Scientific Papers of the Third International Congress of Eugenics*, ed. H. F. Perkins (Baltimore: Williams and Wilkins, 1934), 12.

37. The *Eugenics Record Office Newsletter*, the official record of eugenics research undertaken through fieldwork or at Cold Spring Harbor, published weekly between 1921 and 1929, did not use the general term "abnormal" either, while "subnormal" appeared a total of three times in the decade of its publication. The term "sur-normal," to refer to the exceptionally intelligent or talented, also appeared several times. "Normal" was not used at all.

38. Krafft-Ebing, *Psychopathia Sexualis*, 56.

39. Quoted in Oosterhuis, *Stepchildren of Nature*, 115.

40. Michel Foucault, *Discipline and Punish: The Birth of the Prison*, trans. Alan Sheridan (London and New York: Penguin, 1991), 191.

41. Michel Foucault, *Discipline and Punish*, 191.

42. Lauren Berlant, "On the Case," *Critical Inquiry* 33, no. 4 (2007): 663–64.

43. "On the Case," 664.

44. Joy Damousi, Birgit Lang, and Katie Sutton, "Introduction: Case Studies and the Dissemination of Knowledge," *Case Studies and the Dissemination of Knowledge*, ed. Joy Damousi, Birgit Lang, and Katie Sutton (New York and Abingdon: Routledge, 2015), 1.

45. Katie Sutton, "Sexological Cases and the Prehistory of Transgender Identity Politics in Interwar Germany," *Case Studies and the Dissemination of Knowledge*, ed. Joy Damousi, Birgit Lang, and Katie Sutton (New York and Abingdon: Routledge, 2005), 85–103.

46. Ian Hacking, "The Looping Effects of Human Kinds," in *Causal Cognition: A Multidisciplinary Debate*, ed. Dan Sperber, David Premack, and Ann James (Oxford: Oxford University Press, 1995), 351–83.

47. Krafft-Ebing, *Psychopathia Sexualis*, 103.

48. Havelock Ellis, *Sexual Inversion, Studies in the Psychology of Sex*, vol. 2, 3rd ed. (Philadelphia: Davis, 1901), 70.

49. Hauser, "Krafft-Ebing's Psychological Understanding of Sexual Behavior," 211.
50. "Krafft-Ebing's Psychological Understanding of Sexual Behavior," 211.
51. Jonathan Ned Katz, *The Invention of Heterosexuality*, 2nd ed. (Chicago: University of Chicago Press, 2007), 81–82.
52. Sigmund Freud, *Three Contributions to the Sexual Theory*, trans. A. A. Brill. (New York: The Journal of Nervous and Mental Disease Publishing Company, 1910), 23.
53. Freud, *Three Contributions to the Sexual Theory*, 80.
54. Sigmund Freud, *The Psychopathology of Everyday Life*, trans. A. A. Brill (New York: T. Fisher Unwin, 1914), 23.
55. Freud, *Three Contributions to the Sexual Theory*, 23.
56. Oosterhuis, *Stepchildren of Nature*, 78.
57. Gilman, "The Struggle of Psychiatry with Psychoanalysis," 299.
58. "The Struggle of Psychiatry with Psychoanalysis," 294.
59. Rita Felski, "Introduction," in *Sexology in Culture: Labelling Bodies and Desires*, ed. Lucy Bland and Laura Doan (London: Polity, 1998), 1.
60. Freud, *Three Contributions to the Sexual Theory*, 4.
61. *Three Contributions to the Sexual Theory*, 4.
62. Ellis, *Sexual Inversion*, v.
63. Freud, *Three Contributions to the Sexual Theory*, 23.
64. A. A. Brill, "Translator's Introduction to Freud," in Sigmund Freud, *The Psychopathology of Everyday Life*, trans. A. A. Brill. (New York: T. Fisher Unwin, 1914), i.
65. Freud, *The Psychopathology of Everyday Life*, 1.
66. This is reported in the minutes of a meeting of the Wiener Psychoanalytische Vereinigung held on 11 November 1908, where the topic of discussion was Albert Moll's book *Das Sexualleben des Kindes*. Hermann Nunberg and Ernst Federn, eds., *Protokoll*, vol. 2 of *Protokolle Der Wiener Psychoanalytischen Vereinigung* (Giessen: Psychosozial-Verlag, 2008), 44. We are grateful to Katie Sutton for bringing this passage to our attention.
67. See Terry, *An American Obsession*, 120–21.
68. See Danielle Egan and Gail Hawkes, *Theorizing the Sexual Child in Modernity* (London: Palgrave Macmillan, 2010), 75–96.
69. Freud, *The Psychopathology of Everyday Life*, 3.
70. Katz, *The Invention of Heterosexuality*, 81 (italics in the original).
71. Robie, *Rational Sex Ethics*, 26.
72. C. W. Malchow, *The Sexual Life, Embracing the Natural Sexual Impulse, Normal Sexual Habits and Propagation, Together with the Sexual Physiology and Hygiene* (St. Louis: Mosby, 1907), 79.
73. Orson S. Fowler, *Sexual Science, Including Manhood, Womanhood, and Their Mutual Interrelations; Love, Its Laws, Power, etc.* (Philadelphia, Cincinnati, Chicago, and St. Louis: National Publishing Company, 1870), 14.
74. Lavinia Dock, *Hygiene and Morality: A Manual for Nurses and Others, Giving an Outline of the Medical, Social, and Legal Aspects of the Venereal Diseases* (New York: Putnam, 1910), 25.
75. For an account of these, see David Pivar, *Purity Crusade: Sexual Morality and Social Control, 1868–1900* (Westport, CT: Greenwood Press, 1973).
76. J. Haller and R. Haller. *The Physician and Sexuality in Victorian America* (Urbana, IL: University of Illinois Press. 1974), 263.
77. Robie, *Rational Sex Ethics*, 24.
78. Wood-Allen, *What a Young Girl Ought to Know* (Philadelphia: Vir, 1904) 21–22.
79. *What a Young Girl Ought to Know*, 21–22.
80. Frank Lydston, *Sex Hygiene for the Male and What to Say to the Boy* (Chicago: Riverton, 1912), 10.
81. Wood-Allen, *What a Young Girl Ought to Know*, 85.

82. John Kellogg, *Plain Facts for Old and Young: Embracing the Natural History of Hygiene of Organic Life* (Burlington, IA: Segner, 1890), 375.

83. *Plain Facts for Old and Young*, 328.

84. *Plain Facts for Old and Young*, 397.

85. Bernarr Macfadden, *The Virile Powers of Superb Manhood: How Developed, How Lost, How Regained* (New York: Physical Culture Publishing Company, 1900), 16–17.

86. *The Virile Powers of Superb Manhood*, 36.

87. *The Virile Powers of Superb Manhood*, 96.

88. The American Social Hygiene Association developed from earlier purity organizations, which were mostly private and philanthropic, and from the American Social Hygiene Division of the War Department, established during the First World War.

89. As the correspondence about this film in the archives of the US National Museum of Health and Medicine reveals, overwhelming public demand meant that the film almost immediately began to circulate among civilian audiences, as rotary clubs, local colleges, YMCAs and council groups around the country bombarded the Motion Picture Production Section with requests for copies to show their local audiences. Copies of the film were soon being sent around the country on very short-term loan (due to the high demand and the limited number of prints that could be produced during wartime), for the price of postage only, on the condition that the film would be screened free to the general public. Macfadden's journal *Physical Culture* was among the popular publications applauding the film and its message: "For the first time in the history of any government, radical, aggressive and thoroughly practical measures are being used to protect the army against what is regarded by many students as its greatest enemy." Carl Easton Williams, "Make the Army 'Fit to Fight': A Review of the War Department film by Edward H. Griffith, Exposing the Dangers of Wine, Women and Disease," *Physical Culture* 40, no. 2 (1918): 38.

90. C. W. Malchow, *The Sexual Life*, 80.

91. Marie Carmichael Stopes, *Married Love: A New Contribution to the Solution of Sex Difficulties* (London: Fifield, 1918), xii.

92. Stopes was a significant figure in British science, working first at Manchester University and then at University College London, where Karl Pearson and F. W. Weldon were also based. Stopes would later become a member of the British Eugenics Society.

93. Stopes, *Married Love*, 8.

94. Laura Doan, "Marie Stopes's Wonderful Rhythm Charts: Normalizing the Natural," *Journal of the History of Ideas* (forthcoming).

95. "Marie Stopes's Wonderful Rhythm Charts."

96. Robert Latou Dickinson and Lura Beam, *The Single Woman: A Medical Study in Sex Education* (Baltimore: Williams and Wilkins, 1934), xvi.

97. *The Single Woman*, xvi.

98. Robie, *Rational Sex Ethics*, 96–104.

99. Georges Canguilhem, *The Normal and the Pathological*, trans. Carolyn R. Fawcett (New York: Zone Books), 239.

100. Havelock Ellis, *The Task of Social Hygiene* (Boston: Houghton Mifflin, 1912), i.

101. Lillian Gilbreth, *The Home-Maker and Her Job* (New York and London: Appleton, 1927), 20.

102. *The Home-Maker and Her Job*, 104.

103. *The Home-Maker and Her Job*, 11–12.

CHAPTER EIGHT

1. The faces were modeled on composite photographs from a range of male and female college students. See Catherine Newman Howe, "Average Joes and Mean Girls: The Representation and

Transformation of the Average American, 1890–1945," (PhD diss., University of California, Santa Barbara, 2012), 59.

2. See Dudley Sargent, "The Physical Proportions of the Typical Man," *Scribner's Magazine* 2, no. 1 (1887): 3–17; Carolyn de la Peña, "Dudley Allen Sargent: Health Machines and the Energized Male Body," *Iron Game History* 8, no. 2 (October 2003): 3–19; Martha Verbrugge, *Able-Bodied Womanhood: Personal Health and Social Change in Nineteenth-Century Boston* (Oxford: Oxford University Press, 1988), 131–35.

3. Sargent, "The Physical Proportions of the Typical Man," 11.

4. Michel Foucault, *Discipline and Punish: The Birth of the Prison*, trans. Alan Sheridan (London and New York: Penguin, 1977), 144.

5. *Discipline and Punish*, 146.

6. *Discipline and Punish*, 144.

7. Howe, "Average Joes and Mean Girls," 78.

8. Anna Creadick argues, in *Perfectly Average: The Pursuit of Normality in Postwar America* (Amherst, MA: University of Massachusetts Press, 2010), that the word "normal" came into widespread use only in 1945, and by 1963 had begun to fall out of favor once more.

9. Tony Bennett, *The Birth of the Museum: History, Theory, Politics* (London: Routledge, 1995), 59–88.

10. Sargent, "The Physical Proportions of the Typical Man," 10.

11. "The Physical Proportions of the Typical Man," 6.

12. Sargent identified the swimmer Annette Kellerman as having the closest to an ideal female figure, while the bodybuilder George Sandow had the closest to an ideal male figure. Dudley Sargent, "Modern Woman Getting Nearer the Perfect Figure," *Sunday Magazine*, December 4, 1910, 4.

13. "Modern Woman Getting Nearer the Perfect Figure," 4.

14. "Far From Ideal: Actual Measurements Disappointing—the American Girl Is Not the Model of Grace She Is Supposed to Be," *Chicago Herald*, quoted in Howe, "Average Joes and Mean Girls," 81.

15. The first congress was held in London in 1912, in commemoration of Francis Galton, who had died the previous year. The third, and final, congress was again held at the American Museum of Natural History, in 1932.

16. Harry H. Laughlin, *The Second International Exhibition of Eugenics Held September 22 to October 22, 1921, in Connection with the Second International Congress of Eugenics in the American Museum of Natural History, New York: An Account of the Organization of the Exhibition, the Classification of the Exhibits, the List of Exhibitors, and a Catalog and Description of the Exhibits* (Baltimore: Williams and Wilkins, 1923), 12.

17. Laughlin, *The Second International Exhibition of Eugenics*, 12.

18. Although there is no information about who produced this statue, there is a good likelihood that it was made some decades earlier by Robert Tait McKenzie. McKenzie was the first medical director of physical education at McGill University, and from 1900 he began to produce "sports statues" based on the anthropometric data of college students he had collected over the previous years. Heather Prescott provides an account of these statues in her study of college student anthropometrics in the twentieth century (Prescott, "Using the Student Body: College and University Students as Research Subjects in the United States during the Twentieth Century," *Journal of the History of Medicine* 57, no. 1 (2002): 7).

19. See, for instance, Mary Coffey, "The American Adonis: A Natural History of the Average American (Man), 1921–1932," in *Popular Eugenics: National Efficiency and American Mass Culture in the 1930s*, ed. Susan Currell and Christina Cogdell (Athens: Ohio University Press, 2006), 185–216; Christina Codgell, *Eugenic Design: Streamlining America in the 1930s* (Philadelphia: University of Pennsylvania Press, 2004); Robert Rydell, *World of Fairs: The Century-of-Progress Expositions* (Chicago: University of Chicago Press, 1993).

20. Coffey, "The American Adonis," 198.

21. Charles Davenport, ed., *Scientific Papers of the Second International Congress of Eugenics* (Baltimore: Williams and Wilkins, 1923), 2. In his Presidential Address at the Third International Congress on Eugenics, at which "the Average Young American Male" was again on display, Charles Davenport further argued that immigration would introduce "a possible biological disharmony arising in the hybrid offspring of peoples widely unlike genetically; i.e., having marked structural, including neuronic, differences." Charles Davenport, "Presidential Address: The Development of Eugenics," in *A Decade of Progress in Eugenics: The Scientific Papers of the Third International Congress of Eugenics*, ed. Henry Farnham Perkins (Baltimore: Williams and Wilkins, 1934), 22.

22. Harry H. Laughlin, "What Eugenics Is All About," in *A Decade of Progress in Eugenics: Scientific Papers of the Third International Congress if Eugenics*, ed. H. F. Perkins (Baltimore: Williams and Wilkins, 1934), 13.

23. Edward Alden Jewell, "The Masterpiece and the Modeled Chart," *New York Times*, September 18, 1932, 9. Catherine Howe provides a detailed account of the contemporary reception to the 1932 exhibition of Davenport's statue in her doctoral thesis, "Average Joes and Mean Girls," note 1.

24. Harry H. Laughlin, *Eugenical Sterilization in the United States* (Chicago: Psychopathic Laboratory of the Municipal Court of Chicago, 1922), 446.

25. For a detailed account of the application of eugenics policies in the United States during the first half of the twentieth century, see Edwin Black, *War Against The Weak: Eugenics and America's Campaign to Create a Master Race* (New York: Thunder's Mouth, 2004).

26. "The Average Man Found by Science. He Is Shown to Be Superstitious, Ill Educated, Conventional and Possessing the Mind of a Boy of 14 Years," *New York Times*, 1 May 1927, 4. Hollingworth was at this time the president of the American Psychological Association. He was well known for his highly publicized study of the effects of caffeine on the body and the brain, which he undertook at the request of Coca Cola.

27. Sarah E. Igo, *The Averaged American: Surveys, Citizens, and the Making of a Mass Public* (Cambridge, MA: Harvard University Press, 2007), 11.

28. "The Average Man Found by Science," 4.

29. Robert S. Lynd and Helen Merrell Lynd, *Middletown: A Study in Modern American Culture* (San Diego, New York, London: Harcourt, Brace and Company, 1929), 5.

30. *Middletown*, 15.

31. *Middletown*, 22.

32. This was articulated in the eugenics debates at the Congress noted above, in which anthropometricians like Charles Davenport argued that eugenics was not driven by assumptions about racial superiority, but rather by scientific research on the importance of biological purity. In his Presidential Address at the Third Congress on Eugenics, Davenport argued that "the practical problem is not one of inferiority or superiority of races, but primarily of racial differences." Davenport, "Presidential Address," 23.

33. Igo, *The Averaged American*, 80.

34. "Some Rattling Good Stories," *Good Housekeeping*, June 1929, 204.

35. Igo, *The Averaged American*, 94.

36. Rydell, *World of Fairs*, 56.

37. The archive of the New York World's Fair recordings, held in the New York Public Library, is available online: "Interviews with the Typical American Family Contest Winners," New York Public Library, podcast audio, 1939–1940 New York World's Fair, https://itunes.apple.com/us/itunes-u/1939-40-new-york-worlds-fair/id430390294?mt=10.

38. Igo, *The Averaged American*, 64, 58.

39. Earnest Albert Hooton, *Young Man, You Are Normal* (New York: Putnam, 1945), 3.

40. Clark H. Heath, *What People Are: A Study of Normal Young Men*, The Grant Study, Department of Hygiene, Harvard University (Cambridge, MA: Harvard University Press, 1945), 12. For a

further account of the methodologies used in the Grant study, see also Creadick, *Perfectly Average*, 44–60.

41. Hooton, *Young Man, You Are Normal*, 2, 11. The selection criteria for participation in the study were "those who had done satisfactory work in secondary school and college, and whose health and college records indicated no physical or psychological abnormalities."

42. *Young Man, You Are Normal*, 7.

43. *Young Man, You Are Normal*, 3.

44. *Young Man, You Are Normal*, 8.

45. *Young Man, You Are Normal*, 4.

46. *Young Man, You Are Normal*, 4–5.

47. *Young Man, You Are Normal*, 3.

48. George Vaillant, *Adaptation to Life* (Cambridge, MA: Harvard University Press, 1977).

49. Earnest Albert Hooton, "What Is an American?," *American Journal of Physical Anthropology* 22, no. 1 (1936): 26.

50. Montague Francis Ashley-Montagu, "Apes, Men and Morons by Earnest Hooton: Review," *Science and Society* 2, no. 2 (1938): 283.

51. Sheldon provided two detailed reports on this work: William Herbert Sheldon, *The Varieties of Human Physique: An Introduction to Constitutional Psychology* (New York and London: Harper, 1940); William Herbert Sheldon, *Atlas of Men: A Guide for Somatotyping the Adult Male at All Ages* (New York and London: Harper, 1954).

52. After its initial publications in 1945, the Grant study was resumed in 1977 and updated findings published in Vaillant, *Adaptation to Life*. The study is still ongoing.

53. Dickinson used a method that combined physical examinations, anatomical illustrations (which he drew himself), psychological profiles, and personal histories. Dickinson's technique, as he explained, was to conduct "pelvic examinations" of his patients as he questioned them, in order to stimulate "free associations with the erotic life." These "associational elements are startled into expression." Robert Latou Dickinson and Lura Beam, *The Single Woman: A Medical Study in Sex Education* (Baltimore: Williams and Wilkins, 1934), 22.

54. See Cogdell, *Eugenic Design*, 196; Creadick, *Perfectly Average*, 19.

55. Although these statues are usually dated to 1943 (in studies by Creadick, Codgell, and in Julian B. Carter, *The Heart of Whiteness: Normal Sexuality and Race in America, 1880–1940* [Durham, NC: Duke University Press, 2007]), Dickinson's correspondence makes clear that they were produced at the start of 1945.

56. Dickinson to Richard Gill, 4 January, 1945, Robert Latou Dickinson papers, 1881–1972 (inclusive), 1883–1950 (bulk), B MS c72, Boston Medical Library, Francis A. Countway Library of Medicine, Boston, MA.

57. Dickinson to E. A. Hooton, 24 April, 1945, Dickinson papers.

58. Dickinson to Richard Gill, 4 January, 1945, Dickinson papers.

59. The Bureau of Home Economics records, from which Norma was modeled, emphasize that what they have produced in the published version of these data consisted of a normal distribution of these measurements, from which some data have been excluded in order to ensure the representationality of the results.

60. Although Dickinson was using additional data sets, the records in these collections collectively numbered in the low thousands, and would have modified only slightly the results of the millions of data sets collected by Davenport.

61. Dickinson to Richard Gill, 4 January, 1945, Robert Latou Dickinson papers. Dickinson added a note that suggested he has been actively collaborating with the anthropometricians collecting the data used to model Norma, continuing: "It is a study of mine of many years with anthropologists and others from the Bureau of Home Economics also."

62. Dickinson to E. A. Hooton, 24 April, 1945, Dickinson papers.

63. This argument developed one of the central claims of Hooton's 1936 article "What Is an American?"

64. Harry L. Shapiro, "Portrait of the American People," *Natural History Magazine* 54 (June 1945): 252–53.

65. This museum was the first permanent museum dedicated to health in the United States, established in 1940.

66. The museum began to manufacture and sell copies of these pieces immediately, manufacturing them in a range of materials, at different price points.

67. An account of the manufacture and exhibition of the Transparent Man can be found in Susan Currell and Christina Codgell, ed., *Popular Eugenics: National Efficiency and American Mass Culture in the 1930s* (Athens: Ohio University Press, 2006), 365–66.

68. The exhibition was recurated in 1943, and all the displays relating to Nazi eugenic programs were destroyed. A detailed account of this exhibition is provided in Robert Rydell, Christina Codgell, and Mark Largent, "The Nazi Eugenics Exhibition in the United States, 1934–1943," in *Popular Eugenics*, ed. Currell and Cogdell, 359–84.

69. "Health Museum's Sex Work Hailed," *Cleveland Plain Dealer*, July 10, 1945, 3.

70. Carter, *The Heart of Whiteness*, 31.

71. Creadick, *Perfectly Average*, 2.

72. Creadick, *Perfectly Average*, 12.

73. Josephine Robertson, "Are You Norma, Typical Woman?," *Cleveland Plain Dealer*, September 9, 1945, 1.

74. Josephine Robertson, "Are You Norma, That Rare Individual?," *Cleveland Plain Dealer*, September 9, 1945, 1, 8. Robertson's text is a direct summary of Shapiro's article, focusing on his discussion of the emergence of an American type.

75. Josephine Robertson, "Norma Is Appealing Model in Opinion of City's Artists," September 5, 1945, 1, 10.

76. "Gee Norma, We're Glad You're Home Again!" editorial cartoon, *Cleveland Plain Dealer*, September 13, 1945, 3.

77. Josephine Robertson, "Theatre Cashier, 23, Wins Title of 'Norma,' Besting 3,863 Entries," *Cleveland Plain Dealer*, September 23, 1945, 1.

78. "Theatre Cashier," 1.

79. "Theatre Cashier," 1.

80. "Theatre Cashier," 1.

81. Josephine Robertson, "Dr. Clausen Finds Norma Devout, but Still Glamorous," *Cleveland Plain Dealer*, September 23, 1945, 1.

82. Michel Foucault, *The History of Sexuality, Volume One: An Introduction* (New York: Pantheon Books, 1978), 144.

83. The large set of data on WWI demobilized soldiers did find extensive commercial application after the war. However, it was not originally collected for this purpose.

84. Ruth O'Brien and William C. Shelton, *Women's Measurements for Garment and Pattern Construction* (Washington: United States Government Printing Office, 1941), 1. The program served the secondary purpose of providing employment during the Depression. The Federal Work Projects Administration grant enabled a series of measuring implements and survey charts to be developed, and for a training program in their use to be established, thus providing work for large numbers of women as anthropometricians in urban and rural locations across the country.

85. *Women's Measurements*, 1.

86. *Women's Measurements*, 1.

87. *Women's Measurements*, 25.

88. *Women's Measurements*, 21.

89. *Women's Measurements*, 2.

90. *Women's Measurements*, 21.

91. Reports on Norma in the *Cleveland Plain Dealer* repeatedly claimed that the statue was produced from the averaged data of 15,000 American women, but it can be presumed that Dickinson had access to the final, tabulated data reproduced in *Women's Measurements for Garment and Pattern Construction*.

92. *Women's Measurements*, 28.

93. *Women's Measurements*, 28.

94. *Women's Measurements*, 44.

95. Earnest Albert Hooton, *A Survey in Seating* (Westport, CT: Greenwood Press, 1945).

96. Joyce L. Huff, "Freaklore: The Dissemination, Fragmentation, and Reinvention of the Legend of Daniel Lambert, King of Fat Men," in *Victorian Freaks: The Social Context of Freakery in Britain*, ed. Marlene Tromp (Columbus: Ohio State University Press, 2008), 45.

CHAPTER NINE

1. Howard A. Rusk, "Concerning Man's Basic Drive," review of *Sexual Behavior in the Human Male*, by Alfred Kinsey, *New York Times*, January 4, 1948, "Books" section, 1.

2. The relevant lyrics are: "According to the Kinsey report/every average man you know/much prefers to play his favorite sport/when the temperature is low/but when the thermometer goes way up/and the weather is sizzling hot/Mister Adam for his madam is not." Cole Porter, "Too Darn Hot," *Kiss Me, Kate* (New York: Off Broadway, 1948).

3. "Dr. Kinsey of Bloomington," *Time*, August 24, 1953, 55.

4. "Dr. Kinsey of Bloomington."

5. Lawrence Lariar, *Oh! Dr Kinsey! A Photographic Reaction to the Kinsey Report* (New York: Cartwrite, 1953).

6. These were published in William G. Cochran, Frederick Mosteller, and John W. Turkey, *Statistical Problems of the Kinsey Report on Sexual Behavior in the Human Male* (Washington, DC: American Statistical Association, 1954).

7. Alfred Kinsey, *Sexual Behavior in the Human Female* (Bloomington: Indiana University Press, 1953), 6.

8. *Sexual Behavior in the Human Female*, 6.

9. Harry Benjamin, "The Kinsey Report: Book Review and Roundup of Opinion," *American Journal of Psychotherapy* (July 1948): 400.

10. James H. Lade, review of *Sexual Behavior in the Human Male*, by Alfred Kinsey, *Health News* (April 1948): 16.

11. Sarah E. Igo, *The Averaged American: Surveys, Citizens, and the Making of a Mass Public* (Cambridge, MA: Harvard University Press, 2007), 223.

12. Kinsey, *Sexual Behavior in the Human Female*, 6. Katherine Davis, whose *Factors in the Sex Life of Twenty-Two Hundred Women* (New York: Harper, 1929) was referenced earlier, was a criminologist and superintendent at the New York Reformatory for Women from 1901–1918, and then head of the Bureau for Social Hygiene from 1918–1927. Davis, like Kinsey, restricted her study to average or ordinary women, all of whom were white and college educated. Dickinson, the creator of the composite statues of Norma and Normman, actually focused his study on the women who had been referred to him with gynecological problems, although he did use his physical examinations of these patients to model his sculptures of normal sexual anatomy.

13. Alfred Kinsey, *Sexual Behavior in the Human Male* (Philadelphia: Saunders, 1948), 12.

14. Kinsey, *Sexual Behavior in the Human Female*, 6. Kinsey restricted the data in his reports to those which had been collected from interviews with native white Americans. He interviewed a wider range of subjects, albeit in much lower numbers, including Europeans, African-Americans,

and prisoners; however, the data from these interviews were not used in the two reports. Kinsey was following the established statistical methods examined earlier, which found that combining "types" of data would render the results meaningless.

15. *Sexual Behavior in the Human Male*, 3.

16. In comparison, Katherine Davis undertook less than half the number of Kinsey's 5,000 interviews for *Factors in the Sex Life of Twenty-Two Hundred Women*. Robert Dickinson and Lura Beam, however, drew on the same number of interviews as did Kinsey for *The Single Woman: A Medical Study in Sex Education* (Baltimore: Williams & Wilkins, 1934).

17. As Brenda Weber notes in her account of regional newspaper reviews of *Sexual Behavior in the Human Female*, Akron's *Beacon Journal* reported the case of one local woman who returned her copy of the book because "she thought it would be 'full of hot stuff, but all it has is a lot of figures.'" Brenda Weber, "Talking Sex, Talking Kinsey," *Australian Feminist Studies* 25, no. 64 (2010): 192.

18. Statistical studies of sex had been undertaken, often with records collected in medical contexts like hospitals, from the second half of the nineteenth century. Mostly, however, these were figures on numbers of prostitutes or people with venereal diseases, designed as social reports on issues of public health. Kinsey was more interested in sexual behavior as a sphere of experience.

19. Howard Chiang, "Liberating Sex, Knowing Desire: *Scientia Sexualis* and Epistemic Turning Points in the History of Sexuality," *History of the Human Sciences* 23, no. 5 (2010): 53.

20. Ian Hacking, *The Taming of Chance* (Cambridge: Cambridge University Press, 1990), 160–69.

21. A significant difference between the two reports was that the first did not take emotional states or psychological conditions into account, and explicitly opposed its approach to those used in psychology and psychoanalysis. However, following trenchant criticism from the psychological disciplines, which argued that it was impossible to deduce anything meaningful from sexual behavior without taking emotions and psychology into account, the second volume included more data on and discussion of those areas. For many researchers in both quantitative and qualitative fields of research, that proved a great improvement and made the second volume of more consequence and scientific importance.

22. Donna Drucker, *The Classification of Sex: Alfred Kinsey and the Organization of Knowledge* (Pittsburgh: University of Pittsburgh Press, 2014), 5.

23. Herbert Hyman and Joseph Barmack, "Special Review: Sexual Behavior in the Human Female," *Psychological Bulletin* 51, no. 4 (1954): 418.

24. Lewis M. Terman, "Kinsey's 'Sexual Behavior in the Human Male': Some Comments and Criticisms," *Psychological Bulletin* 45, no. 5 (1948): 443.

25. Kinsey, *Sexual Behavior in the Human Male*, 5.

26. *Sexual Behavior in the Human Male*, 7.

27. Heike Bauer, "Sexology Backward: Hirschfeld, Kinsey and the Reshaping of Sex Research in the 1950s," in *Queer 50s: Rethinking Sexuality in the Postwar Years*, ed. Heike Bauer and Matt Cook (Hampshire: Palgrave Macmillan, 2012), 133–49.

28. "Sexology Backward," 139.

29. "Sexology Backward," 139.

30. Hyman and Barmack, "Special Review," 421.

31. Cochran, Mosteller, and Turkey, *Statistical Problems of the Kinsey Report*.

32. Terman, "Kinsey's 'Sexual Behavior in the Human Male,'" 443.

33. Hyman and Barmack, "Special Review," 421.

34. Donna Drucker, "'A Most Interesting Chapter in the History of Science': Intellectual Responses to Alfred Kinsey's *Sexual Behavior in the Human Male*," *History of the Human Sciences* 21, no. 1 (2012): 75–98.

35. "A Most Interesting Chapter," 82.

36. "A Most Interesting Chapter," 82.
37. Weber, "Talking Sex," 190. See also Igo, *The Averaged American*, 254.
38. Weber, "Talking Sex," 190.
39. Wardell Baxter Pomeroy, *Dr. Kinsey and the Institute for Sex Research* (New York: Harper and Row, 1972), 295.
40. Lionel Trilling, "The Kinsey Report, The Liberal Imagination," *The New York Review of Books* (1950), 223.
41. "The Kinsey Report, The Liberal Imagination," 223.
42. "The Kinsey Report, The Liberal Imagination," 223.
43. "The Kinsey Report, The Liberal Imagination," 223–42.
44. Weber, "Talking Sex," 191.
45. Igo, *The Averaged American*, 251.
46. Kinsey, *Sexual Behavior in the Human Female*, 7.
47. Jules Archer, "Are you Sexually Normal?," *Eye* (August 1950): 23.
48. M. Ernst, A. Stone, R. Benedict, and S. B. Wortis, *The Kinsey Report*, sound recording (New York: WMCA, 1948), quoted in Drucker, "A Most Interesting Chapter," 85.
49. M. F. Montagu, "Understanding Our Sexual Desires," in *About the Kinsey Report: Observations by 11 Experts on Sexual Behavior in the Human Male*, ed. D. P. Geddes and E. Curie (New York: New American Library, 1948), 64.
50. "Understanding our Sexual Desires," 64.
51. Michael Warner, *The Trouble with Normal: Sex, Politics, and the Ethics of Queer Life* (Cambridge, MA: Harvard University Press, 1999), 55.
52. Kinsey, *Sexual Behavior in the Human Male*, 623–36. It should be noted that Davis, writing thirty years before Kinsey, reported a similar incidence of homosexual activity among her female interview sample. Indeed, Davis's statistics for same-sex contact between women matched those reported in Kinsey's study: one in three women had carried a homosexual attraction "to the point of overt expression," she wrote (*Factors in the Sex Life*, 214).
53. Kinsey, *Sexual Behavior in the Human Male*, 646.
54. *Sexual Behavior in the Human Male*, 656.
55. *Sexual Behavior in the Human Male*, 7.
56. *Sexual Behavior in the Human Male*, 7.
57. *Sexual Behavior in the Human Male*, 58.
58. *Sexual Behavior in the Human Male*, 58.
59. *Sexual Behavior in the Human Male*, 7.
60. *Sexual Behavior in the Human Male*, 28. Kinsey elaborated: "while emphasizing the continuity of the gradations between exclusively heterosexual and exclusively homosexual histories, it has seemed desirable to develop some sort of classification which could be based on the relative amounts of heterosexual and homosexual experience or response in each history. . . . An individual may be assigned a position on this scale, for each period in his life. . . . A seven-point scale comes nearer to showing the many gradations that actually exist." *Sexual Behavior in the Human Male*, 639.
61. Trilling, "The Kinsey Report, The Liberal Imagination," 234.
62. Kinsey, *Sexual Behavior in the Human Male*, 329.
63. Review of *Sexual Behavior in the Human Male*, by Alfred Kinsey, *Journal of the American Medical Association* 136, no. 6 (Feb. 7, 1948): 430.
64. Rusk, "Concerning Man's Basic Drive."
65. Igo, *The Averaged American*, 229.
66. Kinsey, *Sexual Behavior in the Human Male*, 7.
67. *Sexual Behavior in the Human Male*, 37.

CONCLUSION

1. Canguilhem, *The Normal and the Pathological*, trans. Carolyn R. Fawcett (New York: Zone Books, 1989), 243.
2. Michel Foucault, *Discipline and Punish: The Birth of the Prison*, trans. Alan Sheridan (London and New York: Penguin, 1977), 144.
3. *Gazette médicale de Paris* 5, no. 19 (May 1837): 289–94.
4. *Le Moniteur des hôpitaux* 6 (1857): 41–47.
5. *Bulletins de la Société d'Anthropologie de Paris*, series 3, vol. 2 (1879): 101–6.
6. *Bulletins de la Société d'Anthropologie de Paris*, series 2, vol. 3 (1879): 187.
7. Anson Rabinach, *The Human Motor: Energy, Fatigue and the Origins of Modernity* (Berkeley: University of California Press, 1992).
8. Edna Yost, with Lillian Gilbreth, *Normal Lives for the Disabled* (New York: Macmillan, 1944).
9. Adam Curtis, *HyperNormalisation* (BBC iPlayer, October 2016); Alexei Yurchak, *Everything Was Forever, Until It Was No More: The Last Soviet Generation* (Princeton, NJ: Princeton University Press, 2013).

Bibliography

PRINTED SOURCES

Abrams, Philip. *The Origins of British Sociology, 1834-1914*. Chicago: University of Chicago Press, 1968.
Ackerknecht, Erwin. *Medicine at the Paris Hospital, 1794-1848*. Baltimore: Johns Hopkins University Press, 1967.
Actes du deuxième Congrès international d'anthropologie criminelle. Biologie et sociologie. Paris, August 1889. Reprinted, Lyon: Storck, 1890; Paris: Masson, 1890.
Actes du premier Congrès international d'anthropologie criminelle: Biologie et sociologie. Turin: Bocca Frères, 1886-87.
Adams, Mary Louise. *The Trouble with Normal: Postwar Youth and the Making of Heterosexuality*. Toronto: University of Toronto Press, 1997.
Annales d'hygiène publique et de médecine légale 1829-33.
Ahmed, Sara. *Queer Phenomenology: Orientations, Objects, Others*. Durham, NC: Duke University Press, 2006.
American Journal of the Medical Sciences, 1837-38.
American Social Hygiene Association. "1919. Poster campaign." Social Welfare History Archives. http://special.lib.umn.edu/swha/. Accessed October 31, 2009.
Anderson, Warwick. "The Case of the Archive." *Case Studies and the Dissemination of Knowledge*, edited by Joy Damousi, Birgit Lang, and Katie Sutton, 15-30. New York: Routledge, 2015.
Archer, Jules. "Are you Sexually Normal?" *Eye* (August 1950): 23.
Archivio di psichiatria, neuropatologia, antropologia criminale e medicina legale. 30 vols. Turin: Bocca, 1880-1909.
Archives de l'Académie des Sciences, Paris. Dossier 7. December 1762. Ms Procès-verbaux.
Armatte, Michel. "Modèles statistiques de l'homogénéité et de la stabilité d'une population au XIXe siècle." In *Les Ménages. Mélanges en honneur de Jacques Desabie*, edited by Philippe L'Hardy and Claude Thélot, 81-102. Paris: INSEE, 1989.
———. "La Moyenne à travers les traités de statistique du XIXe siècle." In *Moyenne, milieu, centre: Histoire et usages*, edited by Jacqueline Feldman, Gérard Lagneau, and Benjamin Matalon, 85-106. Paris: Editions de l'EHESS, 1991.
Armstrong, Tim. *Modernism, Technology and the Body: A Cultural Study*. Cambridge: Cambridge University Press, 1998.

BIBLIOGRAPHY

Ashley-Montagu, Montague Francis. "Apes, Men and Morons by Earnest Hooton: Review." *Science and Society* 2, no. 2 (1938): 282–86.

"The Average Man Found by Science. He Is Shown to Be Superstitious, Ill Educated, Conventional and Possessing the Mind of a Boy of 14 Years." *New York Times*, May 1, 1927, 4.

Balbi, Adrien. "Aperçu des principales classifications du genre humain." *Journal des travaux de la Société Française de Statistique Universelle* 2, no. 14 (August 1836): 67–68.

Babini, Valeria Paola. "Lo studio della mente: momenti di un dialogo tra Francia e Italia." In *Francia/Italia. Le filosofie dell'Ottocento*, edited by Renzo Rhagghianti and Alessandro Savorelli, 119–40. Pisa: Edizioni della Normale, 2007.

Baer, Abraham Adolf. *Der Verbrecher in anthropologischer Beziehung*. Leipzig: Thieme, 1893.

Banucci, Piero, ed. *Il museo di antropologia criminale "Cesare Lombroso" dell'Università di Torino*. Turin: Libreria Cortina, 2011.

Barnes, Harry Elmer, and Negley K. Teeters. *New Horizons in Criminology*. 2nd ed. New York: Prentice-Hall, 1951.

Barth, Dr. "Travaux originaux. De la prétendue substitution de la fièvre typhoïde à la variole, depuis l'introduction de la vaccine." *Gazette hebdomadaire de médecine et de chirurgie* 1, no. 1 (1853): 4–6.

Bauer, Heike. "Sexology Backward: Hirschfeld, Kinsey and the Reshaping of Sex Research in the 1950s." In *Queer 50s: Rethinking Sexuality in the Postwar Years*, edited by Heike Bauer and Matt Cook, 133–49. Hampshire: Palgrave Macmillan, 2012.

Baxby, D. *Jenner's Smallpox Vaccine: The Riddle of Vaccinia Virus and Its Origin*. London: Heinemann, 1981.

Bayard, Armand. *L'Influence de la vaccine sur la population, ou de la gastro-entérite varioleuse avant et depuis la vaccine*. Paris: Victor Masson, 1855.

Beard, George. *A Practical Treatise on Nervous Exhaustion (Neurasthenia): Its Symptoms, Nature, Sequences, Treatment*. New York: Treat, 1880.

———. *American Nervousness: Its Causes and Consequences*. New York: Treat, 1881.

Becker, Peter. "Lombroso come 'luogo della memoria' della criminologia." In *Cesare Lombroso: Gli scienziati e la nuova Italia*, edited by Silvano Montaldo, 33–51. Bologna: Il Mulino, 2010.

Beizer, Janet. *Ventriloquized Bodies: Narratives of Hysteria in Nineteenth Century France*. Ithaca, NY: Cornell University Press, 1994.

Belier, Alain. "Cesare Lombroso: sa vie, son œuvre et sa contribution à la naissance de la criminologie." PhD diss., Université de Paris, 1979.

Benjamin, Harry, "The Kinsey Report: Book Review and Roundup of Opinion," *American Journal of Psychotherapy* (July 1948): 400.

Bennett, Tony. *The Birth of the Museum: History, Theory, Politics*. London: Routledge, 1995.

Bérard, F. *Discours sur les améliorations progressives de la santé publique par l'influence de la civilisation* (Paris: Gabon, 1826).

Berger, Stefan. *The Search for Normality: National Identity and Historical Consciousness in Germany since 1800*. Providence, RI: Berghahn Books, 1997.

Berlant, Lauren. "On the Case," *Critical Inquiry* 33, no. 4 (2007): 663–72.

Berlant, Lauren, and Michael Warner. "Sex in Public." *Critical Inquiry* 24, no. 2 (Winter 1998): 547–66.

Bernard, Claude. *De la physiologie générale*. Paris: Hachette, 1872.

———. *Introduction à l'étude de la médecine expérimentale*. Paris: Baillière, 1865.

Bertillon, Alphonse. *Identification anthropométrique: Instructions signalétiques*. Paris: Melun, 1892.

Bertillon, Jacques. *Album de statistique graphique*. n.p. n.d. Copy available in the Bibliothèque Nationale de France.

———. *La Statistique humaine de la France (naissance, mariage, mort)*. Paris: Baillière, 1880.

———. "Sur la vie moyenne et la vie normale." *Bulletins de la Société d'Anthropologie de Paris*, series 3, vol. 2, no. 2 (1879): 468–82.

Bertillon, Jacques, Alphonse Bertillon, and Georges Bertillon. *La Vie et les œuvres du docteur L.-A. Bertillon*. Paris: Masson, 1883.

Bertillon, [Louis-Adolphe]. *Conclusions statistiques contre les détracteurs de la vaccine, précédées d'un Essai sur la méthode statistique appliquée à l'étude de l'homme*. Paris: Victor Masson, 1857.

———. "Moyenne." In *Dictionnaire encyclopédique des sciences médicales*, série 2, tome 10, edited by A. Dechambre and L. Lereboullet, 298–324. Paris: Masson and Asselin, 1876.

———. Review of Achille Guillard, *Éléments de statistique humaine, ou démographie comparée* (Paris: Guillaumin, 1855). *Gazette hebdomadaire de médecine et de chirurgie* 2, no. 39 (1855): 711.

———. "Sur les ossements des Eyzies." *Bulletins de la Société d'Anthropologie de Paris*, series 2, vol. 3 (1868): 554–74.

Berzero, Antonella, and Maria Carla Garbarino. *La Scienza in chiaro scuro: Lombroso e Mantegazza a Pavia tra Darwin e Freud*. Pavia: Pavia University Press, 2010.

Bichat, Xavier. *Recherches physiologiques sur la vie et la mort*. 3rd ed. Paris: Brosson, 1805.

Bienaymé, Irénée-Jules. "Considérations à l'appui de la découverte de Laplace sur la loi de probabilité dans la méthode des moindres carrés." *Comptes rendus des séances de l'Académie des Sciences* 37 (1853): 309–24.

———. *Mémoire sur la probabilité des erreurs d'après la méthode des moindres carrés*. Paris: Imprimerie impériale, 1858.

Binet, Alfred. *Les Enfants anormaux. Guide pour l'admission des enfants anormaux dans les classes de perfectionnement*. Paris: Colin, 1907.

———. *L'Étude expérimentale de l'intelligence*. Paris: Reinwald, 1903.

Binet, Alfred, and Théodore Simon. "Le Développement de l'intelligence chez les enfants." *L'Année psychologique*, no. 14 (1907): 1–94.

Black, Edwin. *War Against the Weak: Eugenics and America's Campaign to Create a Master Race*. New York: Thunder's Mouth, 2004.

Black, Nick. "Commentary: That Was Then, This Is Now." *International Journal of Epidemiology* 30 (2001): 1251.

Blanckaert, Claude. "Méthode des moyennes et notion de 'série suffisante' en anthropologie physique (1830–1880)." In *Moyenne, milieu, centre: Histoire et usages*, edited by Jacqueline Feldman, Gérard Lagneau, and Benjamin Matalon, 213–43. Paris: Editions de l'EHESS, 1991.

———. "Préface." In Paul Broca, *Mémoires d'anthropologie*, i–xliii. (Paris: Place, 1989).

Bland, Lucy, and Laura Doan, eds. *Sexology in Culture: Labelling Bodies and Desires*. London: Polity, 1998.

Bleuler, Eugen. *Der geborene Verbrecher: Eine kritische Studie*. Munich: J. F. Lehmann, 1896.

Block, Maurice. *Statistique de la France comparée avec les autres états de l'Europe*. Paris: Amyot, 1860.

———. *Traité théorique et pratique de la statistique*. Paris: Guillaumin, 1878.

Blumenbach, Johann Friedrich. *On the Natural Variety of Mankind, The Anthropological Treatises of Johann Friedrich Blumenbach*. Translated by Thomas Bendyshe. London: Green, Longman, Roberts, and Green, 1865.

Bollet, Alfred Jay. "Pierre Louis: The Numerical Method and the Foundation of Quantitative Medicine." *American Journal of the Medical Sciences* 266, no. 2 (1973): 93–101.

Bonnet, Charles. *Considérations sur les corps organisés*. Vol. 3. Amsterdam: Rey, 1762.

Bono, G. B. "Della capacità orbitale e cranica e dell'indice cefaloorbitale nei normali, nei pazzi, nei cretini e deliquenti." *Annali universali di medicina e chirurgia* 1, no. 256 (1881): 299–308.

Bouillaud, Jean-Baptiste. *Clinique médicale de l'Hôpital de la Charité, ou Exposition statistique des diverses maladies traitées à la clinique de cet hôpital*. Paris: Baillière, 1837.

Bouillaud, Jean-Baptiste. *Essai sur la philosophie médicale et sur les généralités de la clinique médicale, précédé d'un Résumé philosophique des principaux progrès de la médecine et suivi d'un Parallèle des résultats de la saignée coup sur coup avec ceux de l'ancienne méthode, dans le traitement des phlegma[s]ies aiguës*. Paris: Rouvier and Le Bouvier, 1836.

Bournet, Albert. "Chronique italienne." *Archives de l'anthropologie criminelle et des sciences pénales* 1 (1886): 69–76.
Bousquet, Jean Baptiste Édouard. "Vaccine. Extrait du rapport sur la vaccine pour l'année 1853." *Le Moniteur des hôpitaux* 3, no. 82 (August 1855): 653–56.
Bowditch, Henry Pickering. "Are Composite Photographs Typical Pictures?" *McClure's Magazine* 3, no. 4 (September 1894): 331–42.
Bowles, Gordon Townsend. *New Types of Old Americans at Harvard and at Eastern Women's Colleges*. With a foreword by Earnest A. Hooton. Cambridge, MA: Harvard University Press, 1932.
Brady, Sean. *Masculinity and Male Homosexuality in Britain, 1861–1913*. London: Palgrave Macmillan, 2005.
Braude, Hillel D. *Intuition in Medicine: A Philosophical Defense of Clinical Reasoning*. Chicago: University of Chicago Press, 2012.
Braun, Marta. *Picturing Time: The Work of Étienne-Jules Marey, 1830–1904*. Chicago: University of Chicago Press, 1994.
Braunstein, Jean-François. *Broussais et le matérialisme. Médecine et philosophie au XIXe siècle*. Paris: Méridiens-Klincksieck, 1986.
Brian, Eric. "Moyenne." In *Dictionnaire des concepts nomades en sciences humaines*, edited by Olivier Christin, 313–25. Paris: Métailié, 2010.
Brill, A. A. "Translator's Introduction to Freud." In Sigmund Freud, *The Psychopathology of Everyday Life*, i–vi. Translated by A. A. Brill. New York: T. Fisher Unwin, 1914.
Bristow, Joseph. "Symond's History, Ellis's Hereditary: *Sexual Inversion*." In *Sexology in Culture: Labelling Bodies and Desires*, edited by Lucy Bland and Laura Doan, 79–99. London: Polity Press, 1998.
Broca, Paul. "Détermination de la série suffisante." *Bulletins de la Société d'Anthropologie de Paris*, series 3, vol. 2, no. 2 (Paris: Victor Masson, 1879): 756–820.
———. "Du degré d'utilité de la statistique," *Le Moniteur des hôpitaux* 6 (January 1857): 41–47.
———. *Étude des variations craniométriques et de leur influence sur les moyennes: Détermination de la série suffisante*. [Extraits des bulletins de la Société d'Anthropologie de Paris, séance du 18 décembre 1879.] Paris: Hennuyer, 1880.
———. *Mémoires d'anthropologie zoologique et biologique*. Paris: Reinwald, 1877.
———. "Recherches sur l'ethnologie de la France," *Bulletins de la Société d'Anthropologie de Paris*, series 1, vol. 1 (Paris: Victor Masson, 1860): 1–56.
———. "Sur la détermination de l'âge moyen." *Bulletins de la Société d'Anthropologie de Paris*, series 3, vol. 2, no. 2 (1879): 298–317.
———. *Sur la détermination de l'âge moyen*. Paris: Hennuyer, 1879.
———. "Sur le plan horizontal de la tête et sur la méthode trigonométrique." *Bulletins de la Société d'Anthropologie de Paris*, series 2, vol. 8 (1873): 48–96.
———. "Sur le volume et la forme du cerveau suivant les individus et suivant les races." *Bulletins de la Société d'Anthropologie de Paris*, series 1, vol. 2 (1861): 139–204.
———. "Sur les crânes basques de Saint-Jean-de-Luz," *Bulletins de la Société d'Anthropologie de Paris*, series 2, vol. 3 (1868): 43–107.
Broussais, François-Joseph-Victor. *De l'irritation et de la folie, ouvrage dans lequel les rapports du physique et du moral sont établis sur les bases de la médecine physiologique*. Paris: Delaunay, 1828.
———. *Examen de la doctrine médicale généralement adoptée, et des systèmes modernes de nosologie*. Paris: Gabon, 1816.
———. *Histoire des phlegmasies ou inflammations chroniques, fondée sur de nouvelles observations de clinique et d'anatomie pathologique. Ouvrage présentant un tableau raisonné des variétés et des combinaisons diverses de ces maladies, avec leur différentes modes de traitement*. 3rd ed. Paris: Gabon, Crochard, 1822.
———. *Recherches sur la fièvre hectique. Considérée comme dépendante d'une lésion d'action des différens systèmes, sans vice organique*. Paris: Méquignon, 1803.

Buckle, Henry Thomas. *History of Civilisation in England*. 3rd ed. London: Appelton, 1861.
Buffon, Georges Louis Leclerc, Comte de. *Essai d'arithmétique morale*. Supp. 4 of *Histoire naturelle, générale et particulière*. Paris: Imprimerie royale, 1777.
———. *Œuvres completes*. Vol. 5. Paris: Rapet, 1818.
Bulferetti, Luigi. *Cesare Lombroso*. Turin: Utet, 1975.
Bulletins de l'Académie de médecine, 1837.
Bulletins de la Société d'Anthropologie de Paris, 1859–88.
Bulmer, Martin, Kevin Bales, and Kathryn Kish Sklar, eds. *The Social Survey in Historical Perspective, 1880–1940*. Cambridge: Cambridge University Press, 1987.
Cabanis, Pierre-Jean-Georges. *Du degré de certitude de la médecine*. Paris: F. Didot, 1798.
Callens, Stéphane. "Les Moyennes positivistes." In *Moyenne, milieu, centre: Histoire et usages*, edited by Jacqueline Feldman, Gérard Lagneau, and Benjamin Matalon, 169–92. Paris: Editions de l'EHESS, 1991.
Camper, Pierre. *Dissertation sur les variétés naturelles qui caractérisent la physionomie des hommes des divers climats et des différents âges*. Translated by H. J. Jansen. Paris: Jansen, 1791.
Canguilhem, Georges. *La Connaissance de la vie*. 2nd ed. Paris: Vrin, 2009.
———. *Études d'histoire et de philosophie des sciences*. Paris: Vrin, 1983.
———. *Idéologie et rationalité dans l'histoire des sciences de la vie: Nouvelles études d'histoire et de philosophie des sciences*. 2nd ed. Paris: Vrin, 2009.
———. "Monstrosity and the Monstrous." *Diogène* 40 (1962): 27–42.
———. *The Normal and the Pathological*. Translated by Carolyn R. Fawcett. New York: Zone Books, 1989.
———. *Le Normal et le pathologique*. Paris: Quadrige/PUF, 1966.
Carnot, Hector. *Essai de mortalité comparée avant et depuis l'introduction de la vaccine en France*. Autun: Dejussieu, 1849.
Carter, Julian B. *The Heart of Whiteness: Normal Sexuality and Race in America, 1880–1940*. Durham, NC: Duke University Press, 2007.
Challis, Debbie. *The Archeology of Race: The Eugenic Ideas of Francis Galton and Flinders Petrie*. London: Bloomsbury, 2013.
Chen, Tar Timothy. "A History of Statistical Thinking in Medicine." In *Advanced Medical Statistics*, edited by Ying Lu and Ji-Qian Fang, 3–20. Singapore: World Scientific, 2015.
Chiang, Howard. "Liberating Sex, Knowing Desire: *Scientia Sexualis* and Epistemic Turning Points in the History of Sexuality." *History of the Human Sciences* 23, no. 5 (2010): 42–69.
Civiale, Jean. "Des calculs arrêtés ou développés dans l'urètre (Extrait d'un mémoire lu à l'Académie royale de médecine)." *Gazette médicale de Paris* 1, no. 2 (1831): 65–67.
———. *Des Résultats de la lithotritie méthodiquement appliquée aux seuls cas qui la comportent*. Paris: Martinet, 1847.
———. *Nouvelles remarques historiques sur la lithotritie*. Paris: Renouard, 1843.
———. *Observations historiques sur la lithotritie*. Paris: Renouard, 1843.
———. "Quelques remarques sur la lithotritie." *Mémoires de l'Académie Royale de médecine* 6 (1835): 243–97.
———. "Quelques remarques sur la taille hypogastrique; lues à l'académie des sciences par M. le Dr Civiale." *Gazette médicale de Paris* 1, no. 2 (1831): 201–4.
———. *Résultats cliniques de la lithotritie, pendant les années 1860-1864*. Paris: Baillière, 1865.
———. *Traité de l'affection calculeuse ou Recherches sur la formation, les caractères physiques et chimiques, les causes, les signes et les effets pathologiques de la pierre et de la gravelle, suivies d'un Essai de statistique sur cette maladie*. Paris: Crochard, 1838.
———. *Traité pratique et historique de la lithotritie*. Paris: Baillière, 1847.
Clouston, T. "The Developmental Aspects of Criminal Anthropology." *The Journal of the Anthropological Institute of Great Britain and Ireland* 23 (1894): 215–25.
Cochran, William G., Frederick Mosteller, and John W. Turkey, *Statistical Problems of the Kinsey*

Report on Sexual Behavior in the Human Male. Washington, DC: American Statistical Association, 1954.
Codgell, Christina. *Eugenic Design: Streamlining America in the 1930s*. Philadelphia: University of Pennsylvania Press, 2004.
Coffey, Mary. "The American Adonis: A Natural History of the Average American (Man), 1921–1932." In *Popular Eugenics: National Efficiency and American Mass Culture in the 1930s*, edited by Susan Currell and Christina Cogdell, 185–216. Athens, OH: Ohio University Press, 2006.
Colajanni, Napoleone. *La sociologia criminale*. 2 vols. Catania: Tropea, 1889.
Colombo, Giorgio. *La scienza infelice: il Museo di antropologia criminale di Cesare Lombroso*. Turin: P. Boringhieri, 1975.
Comptes rendus hebdomadaires de l'Académie des Sciences, 1835–37.
Comte, Auguste. *Cours de philosophie positive*, edited by Michel Serres, François Dagognet, and Allal Sinaceur. Paris: Hermann, 1975.
———. *Système de politique positive*. 2 vols. Paris: Mathias, 1851.
Converse, Jean M. *Survey Research in the United States: Roots and Emergence, 1890–1960*. Berkeley: University of California Press, 1987.
Cournot, Antoine-Augustin. *Exposition de la théorie des chances et des probabilités*. Paris: Hachette, 1843.
Crary, Jonathan. *Techniques of the Observer: On Vision and Modernity in the 19th Century*. Cambridge, MA: MIT Press, 1990.
Creadick, Anna. *Perfectly Average: The Pursuit of Normality in Postwar America*. Amherst, MA: University of Massachusetts Press, 2010.
Crease, Robert P. *World in the Balance: The Historic Quest for an Absolute System of Measurement*. New York: W. W. Norton, 2011.
Crookshank, Edgar March. *History and Pathology of Vaccination*. London: Lewis, 1889.
Crozier, Ivan. "Introduction: Havelock Ellis, John Addington Symonds and the Construction of *Sexual Inversion*." In *Sexual Inversion: A Critical Edition*. London: Palgrave Macmillan, 2008, 1–95.
Cruveilhier, Jean. *Anatomie pathologique du corps humain*. Paris: J.-B. Baillière, 1829–42.
Cryle, Peter. "*Les Choses et les Mots*: Missing Words and Blurry Things in the History of Sexuality." *Sexualities* 12, no. 4 (2009): 439–52.
Cullen, Michael J. *The Statistical Movement in Early Modern Britain*. Hassocks, UK: Harvester, 1975.
Currell, Susan, and Christina Codgell. *Popular Eugenics: National Efficiency and American Mass Culture in the 1930s*. Athens, OH: Ohio University Press, 2006.
Cuvier, Georges. *Leçons d'anatomie comparée*. Paris: Crochard, Fantin, An XIV [1805].
———. *Principes de philosophie zoologique, discutés en mars 1830, au sein de l'Académie royale des sciences*. Paris: Pichon et Didier, Rousseau, 1830.
———. *Tableau élémentaire de l'histoire naturelle des animaux*. Paris: Baudoin, An 6 [1797].
Cvetkovich, Ann. *Mixed Feelings: Feminism, Mass Culture, and Victorian Sensationalism*. New Brunswick, NJ: Rutgers University Press, 1992.
Damousi, Joy, Birgit Lang, and Katie Sutton. "Introduction: Case Studies and the Dissemination of Knowledge." In *Case Studies and the Dissemination of Knowledge*, edited by Joy Damousi, Birgit Lang, and Katie Sutton, 1–12. New York: Routledge, 2015.
———, eds. *Case Studies and the Dissemination of Knowledge*. New York: Routledge, 2015.
Daston, Lorraine. *Classical Probability in the Enlightenment*. Princeton, NJ: Princeton University Press, 1988.
———, ed. *Biographies of Scientific Objects*. Chicago: University of Chicago Press, 2000.
Daston, Lorraine J. "Rational Individuals versus Laws of Society: From Probability to Statistics." In *Ideas in History*. Vol. 1 of *The Probabilistic Revolution*, edited by Lorenz Krüger, Lorraine J. Daston, and Michael Heidelberger, 295–304. Cambridge, MA: MIT Press, 1987.

Daston, Lorraine, and Fernando Vidal, eds. *The Moral Authority of Nature*. Chicago: University of Chicago Press, 2004.
Daston, Lorraine, and Peter Galison, *Objectivity*. New York: Zone Books, 2007.
Davenport, Charles, ed. *Scientific Papers of the Second International Congress of Eugenics*. Baltimore: Williams and Wilkins, 1923.
——. "Presidential Address: The Development of Eugenics." In *A Decade of Progress in Eugenics: The Scientific Papers of the Third International Congress of Eugenics*, edited by Henry Farnham Perkins, 17-22. Baltimore: Williams and Wilkins, 1934.
Davenport, Charles, and Albert G. Love. *Army Anthropology: Based on Observations Made on Draft Recruits, 1917-1918, and on Veterans at Demobilization, 1919*. Washington: War Department Government Printing Office, 1921.
——. *Defects Found in Drafted Men: Statistical Information Compiled from Draft Records*. Washington: War Department Government Printing Office, 1920.
Davenport, Isabel. *Salvaging of American Girlhood: A Substitution of Normal Psychology for Superstition and Mysticism in the Education of Girls*. New York: Dutton, 1924.
Davis, Katherine Bement. *Factors in the Sex Life of Twenty-Two Hundred Women*. New York: Harper, 1929.
Davis, Lennard. *Enforcing Normalcy: Disability, Deafness and the Body*. London: Verso, 1995.
Debierre, Charles-Marie. *Le Crâne des criminels*. Lyon: A. Storck, 1895.
Delaharpe, M. J. "Coup d'œil statistique sur la fièvre typhoïde à l'hôpital de Lausanne, de 1836 à 1850." *Gazette hebdomadaire de médecine et de chirurgie*, série 1, tome 1, no. 39 (1854): 631-32.
De la Peña, Carolyn. "Dudley Allen Sargent: Health Machines and the Energized Male Body." *Iron Game History* 8, no. 2 (2003): 3-19.
Desrosières, Alain. "Homogéneité ou hétérogénéité d'une population: de Quetelet à Lexis." In *Les Ménages. Mélanges en honneur de Jacques Desabie*, 67-80, edited by Philippe L'Hardy and Claude Thélot. Paris: INSEE, 1989.
——. "Masses, individus, moyennes: la statistique sociale au XIXe siècle." In *Moyenne, milieu, centre. Histoires et usages*, edited by Jacqueline Feldman, Gérard Lagneau, and Benjamin Matalon, 249-73. Paris: Editions de l'EHESS, 1991.
——. *The Politics of Large Numbers: A History of Statistical Reasoning*. Translated by Camille Naish. Cambridge, MA: Harvard University Press, 1998.
——. *La Politique des grands nombres. Histoire de la raison statistique*. Paris: La Découverte, 2000.
De Swaan, Abram. *The Management of Normality: Critical Essays in Health and Welfare*. London: Routledge, 1990.
Devereux, George. "Normal and Abnormal: The Key Problem of Psychiatric Anthropology." In *Some Uses of Anthropology, Theoretical and Applied*, edited by Joseph B. Casagrande and Thomas Gladwin. Washington: Anthropological Society of Washington, 1956.
Dickinson, Robert Latou. Papers, 1881-1972 (inclusive), 1883-1950 (bulk). B MS c72. Boston Medical Library. Francis A. Countway Library of Medicine, Boston, MA.
Dickinson, Robert Latou, and Lura Beam. *The Single Woman: A Medical Study in Sex Education*. Baltimore: Williams & Wilkins, 1934.
Dickinson, Robert Latou, and Abram Belskie. *Human Sex Anatomy: A Topographical Hand Atlas*. Baltimore: Williams & Wilkins, 1949.
Didier, Emmanuel. "Gabriel Tarde and Statistical Movement." In *The Social after Gabriel Tarde*, edited by Matei Candea, 163-76. New York: Routledge, 2010.
Doan, Laura. *Disturbing Practices: History, Sexuality, and Women's Experience of Modern War*. Chicago: University of Chicago Press, 2013.
——. Forthcoming. "Marie Stopes's Wonderful Rhythm Charts: Normalizing the Natural." *Journal of the History of Ideas*.

Dock, Lavinia. *Hygiene and Morality: A Manual for Nurses and Others, Giving an Outline of the Medical, Social, and Legal Aspects of the Venereal Diseases.* New York: Putnam, 1910.

Dolza, Delfina. *Essere figlie di Lombroso: due donne intellettuali tra '800 e '900.* Milan: F. Angeli, 1990.

Double, François-Joseph. "Philosophie médicale. Observations sur l'application du calcul à la thérapeutique." *Gazette médicale de Paris* 2, no. 5 (1837): 289–94.

Doubovitzki, Pierre. *Reproduction fidèle des discussions qui ont eu lieu sur la lithrotripsie [sic] et la taille, à l'Académie royale de Médecine en 1835, à l'occasion d'un rapport de M. Velpeau sur ces deux opérations.* Paris: Ducessois, 1835.

Douglas, Bronwen. "Climate to Crania: Science and the Racialization of Human Difference." In *Foreign Bodies: Oceania and the Science of Race 1750–1940*, edited by Bronwen Douglas and Chris Ballard, 33–96. Canberra: ANU ePress, 2008.

Downing, Lisa. "'Citizen-Paraphiliac': Normophilia and Biophilia in John Money's Sexology." In Lisa Downing, Iain Moreland, and Nikki Sullivan, *Fuckology: Critical Essays on John Money's Diagnostic Concepts*, 58–181. Chicago: University of Chicago Press, 2015.

———. *The Subject of Murder: Gender, Exceptionality, and the Modern Killer.* Chicago: University of Chicago Press, 2013.

Dreger, Alice Domurat. *One of Us: Conjoined Twins and the Future of Normal.* Cambridge, MA: Harvard University Press, 2004.

Driver, Edwin D. "Charles Buckman Goring." In *Pioneers in Criminology*, edited by Hermann Mannheim, 334–48. Chicago: Quadrangle Books, 1960.

"Dr. Kinsey of Bloomington," *Time*, August 24, 1953, 55.

Drouard, Alain. "Les Trois âges de la fondation française pour l'étude des problèmes humains." *Population* 6 (1983): 1017–37.

Drucker, Donna. "'A Most Interesting Chapter in the History of Science': Intellectual Responses to Alfred Kinsey's *Sexual Behavior in the Human Male*." *History of the Human Sciences* 21, no. 1 (2012): 75–98.

———. *The Classification of Sex: Alfred Kinsey and the Organization of Knowledge.* Pittsburgh: University of Pittsburgh Press, 2014.

Dubut de Laforest, Jean-Louis. *Morphine: Roman contemporain.* Paris: Dentu, 1891.

Dugdale, Richard. *"The Jukes": A Study of Crime, Pauperism, Disease and Heredity.* New York: G. P. Putnam, 1877.

Durkheim, Émile. "Crime et santé sociale." *Revue philosophique* 39 (1895): 518–23.

———. *Les Règles de la méthode sociologique.* 2nd ed. Paris: Alcan, 1901.

———. *Le Suicide.* Paris: Alcan, 1897.

Edwards, Elizabeth, ed. *Anthropology and Photography, 1860–1920.* New Haven and London: Yale University Press and the Royal Anthropological Institute, 1992.

Egan, Danielle, and Gail Hawkes. *Theorizing the Sexual Child in Modernity.* London: Palgrave Macmillan, 2010.

Ehrard, Jean. *L'Idée de nature en France dans la première moitié du XVIIIe siècle.* Paris: Albin Michel, 1994.

Eknoyan, Garabed. "Adolphe Quetelet (1799–1874)—The Average Man and Indices of Obesity." *NDT Nephrology Dialysis Transplantation* 23, no. 1 (2007): 47–51. Accessed June 4, 2013. doi:10.1093/ndt/gfm517.

———. "Emergence of Quantification in Clinical Investigation and the Quest for Certainty in Therapeutics: The Road from Hammurabi to Kefauver." *Advances in Chronic Kidney Disease* 12, no. 1 (2005): 88–95.

Ellenbogen, Josh. *Reasoned and Unreasoned Images: The Photography of Bertillon, Galton, and Marey.* University Park, PA: Penn State University Press, 2012.

Ellis, Havelock. *The Criminal.* London: Walter Scott, 1890.

———. *Sex in Relation to Society.* Vol. 6 of *Studies in the Psychology of Sex.* Philadelphia: Davis, 1927.
———. *Sexual Inversion.* Vol. 2 of *Studies in the Psychology of Sex.* 3rd ed. Philadelphia: Davis, 1901.
———. *The Task of Social Hygiene.* Boston: Houghton Mifflin, 1912.
Ernst, Waltraud, ed. *Histories of the Normal and the Abnormal: Social and Cultural Histories of Norms and Normativity.* New York: Routledge, 2006.
———. "The Normal and the Abnormal: Reflections on Norms and Normativity." In *Histories of the Normal and the Abnormal,* 1–39. London: Routledge, 2006.
Espinas, Alfred. "Gabriel Tarde: la criminalité comparée." *Revue philosophique* 24 (1887): 91.
Esquirol, Jean-Étienne Dominique. "Rapport statistique sur la maison royale de Charenton." *Annales d'hygiène publique et de médecine légale* 1, no. 1 (1829): 100–51.
Eugenics Record Office Newsletter, 1921–29.
Ewald, François. "Norms, Discipline and the Law," *Representations* 30 (1990): 138–61.
Exposition universelle internationale de 1889. Direction générale de l'exploitation. Deuxième congrès international d'anthropologie criminelle. Session de Paris du 10 au 17 août 1889. Paris: Imprimerie Nationale, 1889.
Feldman, Jacqueline, Gérard Lagneau, and Benjamin Matalon. "Introduction." In *Moyenne, milieu, centre. Histoires et usages,* edited by Jacqueline Feldman, Gérard Lagneau, and Benjamin Matalon, 9–28. Paris: Editions de l'École des Hautes Études en Sciences Sociales, 1991.
Feldman, Jacqueline, Gérard Lagneau, and Benjamin Matalon, eds. *Moyenne, milieu, centre. Histoires et usages.* Paris: Editions de l'École des Hautes Études en Sciences Sociales, 1991.
Felski, Rita. "Introduction." In *Sexology in Culture: Labelling Bodies and Desires,* edited by Lucy Bland and Laura Doan, 1–8. London: Polity, 1998.
Féré, Charles. *L'Instinct sexuel: Évolution et dissolution.* Paris: Alcan, 1899.
Ferrero, Guillaume. *Les Lois psychologiques du symbolisme.* Paris: Alcan, 1894.
Ferri, Enrico. *I nuovi orizzonti del diritto e della procedura penale.* 2nd ed. Bologna: Zanichelli, 1884.
———. *La scuola criminale positiva.* Naples: E. Detken, 1885.
———. *Sociologia criminale.* 3rd ed. Turin: Bocca, 1892.
Forno, Mauro. "Scienziati e mass-media: Lombroso e gli studiosi positivisti nella stampa tra otto e novecento." In *Cesare Lombroso: gli scienziati e la nuova Italia,* edited by Silvano Montaldo, 207–32. Bologna: Il Mulino, 2010.
Foucault, Michel. *Les Anormaux: Cours au Collège de France; 1974–1975.* Paris: Seuil/Gallimard, 1999.
———. *The Birth of the Clinic: An Archeology of Medical Perception.* Translated by A. M. Sheridan. London: Routledge, 1973.
———. *Discipline and Punish: The Birth of the Prison.* Translated by Alan Sheridan. London: Penguin, 1977.
———. *Dits et écrits: 1954–1988,* edited by Daniel Defert and François Ewald. 4 vols. Paris: Gallimard, 1994.
———. *Histoire de la sexualité I: La Volonté de savoir.* Paris: Gallimard, 1976.
———. *The History of Sexuality, Volume One: An Introduction.* New York: Pantheon Books, 1978.
———. "Introduction par Michel Foucault." In Vol. 3 of *Dits et écrits 1954–1988,* edited by Daniel Defert and François Ewald, 429–42. Paris: Gallimard, 1994.
———. Introduction to *On the Normal and the Pathological,* by Georges Canguilhem. Dordrecht: Reidel, 1978.
———. *Les Mots et les choses: une archéologie des sciences humaines.* Paris: Gallimard, 1966.
———. *Naissance de la clinique.* Paris: PUF, 1963.
———. "Nietzsche, Genealogy, History." In *Language, Counter-Memory, Practice: Selected Essays and Interviews,* edited by D. F. Bouchard, 139–64. Ithaca, NY: Cornell University Press, 1977.
———. *The Order of Things: An Archaeology of the Human Sciences.* New York: Pantheon Books, 1971.

———. "A Preface to Transgression." In *Language, Counter-Memory, Practice*, edited by Donald Bouchard, 29–52. Ithaca, NY: Cornell University Press, 1977.
———. "La Vie: l'expérience et la science." In Vol. 4 of *Dits et écrits 1954–1988*, edited by Daniel Defert and François Ewald, 763–76. Paris: Gallimard, 1994.
Fourier, Joseph. *Théorie analytique de la chaleur*. Paris: Didot, 1822.
Fowler, Orson S. *Sexual Science, Including Manhood, Womanhood, and Their Mutual Interrelations; Love, Its Laws, Power, etc*. Philadelphia, Cincinnati, Chicago, and St. Louis: National Publishing Company, 1870.
Freeman, Elizabeth, ed. "Queer Temporalities." Special issue, *GLQ* 13, nos. 2–3 (2007).
———. *Time Binds: Queer Temporalities, Queer Histories* (Durham, NC: Duke University Press, 2010).
Freud, Sigmund. *A General Introduction to Psychoanalysis*. Translated by A. A. Brill. New York: Boni and Liveright, 1920.
———. *The Psychopathology of Everyday Life*. Translated by A. A. Brill. New York: T. Fisher Unwin, 1914.
———. *Three Contributions to the Sexual Theory*. Translated by A. A. Brill. New York: The Journal of Nervous and Mental Disease Publishing Company, 1910.
Frigessi, Delia. "Introduzione." In *Cesare Lombroso, Delitto, genio, follia*, edited by Delia Frigessi, Ferrucio Giacanelli, and Luisa Mangoni, 333–75. Turin: Bollati Boringhieri, 1995.
Frigessi, Delia, Ferrucio Giacanelli, and Luisa Mangoni, eds. *Cesare Lombroso, Delitto, genio, follia*, Turin: Bollati Boringhieri, 1995.
Furness, Walter Rogers. *Composite Photography Applied to the Portraits of Shakespeare*. Philadelphia: R. M. Lindsay, 1885.
Fuster, Joseph Jean Nicolas. "De l'application du calcul à la thérapeutique." *Gazette médical de Paris* 1, no. 3 (Janvier 1832): 25–27.
Galton, Francis. "The Anthropometric Laboratory." *Fortnightly Review* 31 (1882): 338.
———. "Composite Portraits Made by Combining Those of Many Different Persons into a Single Figure." *Nature* 18 (1878): 97–100.
———. "Criminal Anthropology." *Nature* 42 (1890): 75–76.
———. "Deterioration of the British Race." *Times*. June 18, 1909.
———. *Essays in Eugenics*. London: Eugenics Education Society, 1909.
———. "Eugenics: Its Definition, Scope, and Aims." *The American Journal of Sociology* (July 1904): 1–25.
———. *Generic Images*. London: Clowes, 1879.
———. *Hereditary Genius: An Inquiry into its Laws and Consequences*. London: Macmillan, 1869.
———. *Inquiries into Human Faculty and Its Development*. London: Macmillan, 1883.
———. *Inquiries into Human Faculty and Its Development*. 2nd ed. London: Dent, 1907.
———. "An Inquiry into the Physiognomy of Phthisis by the Method of 'Composite Portraiture.'" *Guy's Hospital Reports* 25 (1882): 475–93.
———. *Life History Album*. London: Macmillan, 1884.
———. *Natural Inheritance*. London and New York: Macmillan, 1889.
———. "Notes on the Marlborough School Statistics." *Journal of the Anthropological Institute* 4 (1875): 130–35.
———. "On the Anthropometric Laboratory at the Late International Health Exhibition." *Journal of the Anthropological Institute* 14 (1885): 205–21.
———. "On the Application of Composite Portraiture to Anthropological Purposes." *Report of the British Association for the Advancement of Science* 51 (1881): 690–91.
———. "On the Height and Weight of Boys Aged 14, in Town and Country Public Schools." *Journal of the Anthropological Institute* 5 (1876): 174–81.
———. "Photographic Composites." *Photographic News* 29, no. 1398 (1885): 243.
———. "Proposal to Apply for Anthropological Statistics from Schools." *Journal of the Anthropological Institute* 3 (1874): 308–11.

———. *Record of Family Faculties, Consisting of Tabular Forms and Directions for Entering Data, with an Explanatory Preface.* London: Macmillan, 1884.

———. "Request for Prints of Photographic Portraits." *Nature* 73 (1906): 534.

———. "Signaletic Instructions, Including the Theory and Practice of Anthropometrical Identification." *Nature* 54 (1896): 569–70.

———. "Typical Laws of Heredity." *Nature* 5 (1877): 492–533.

———. "Why Do We Measure Mankind?" *Lippincott's Monthly Magazine* 45 (1890): 236–41.

Garland-Thomson, Rosemarie, ed. *Freakery: Cultural Spectacles of the Extraordinary Body.* New York: New York University Press, 1996.

Garland-Thomson, Rosemarie. *Extraordinary Bodies: Figuring Physical Disability in American Culture and Literature.* New York: Columbia University Press, 1997.

Garnier, Jean-Guillaume, and Adolphe Quetelet, eds. *Correspondance mathématique et physique.* Gand: Vandekerckove, 1 and 2 (1825–26).

Garnier, Pierre. *Hygiène de la génération: Anomalies sexuelles apparentes et cachées, avec 230 observations.* Paris: Garnier, 1889.

———. *Hygiène de la génération: La Stérilité humaine et l'hermaphrodisme.* Paris: Garnier, 1883.

Garraud, René. "Rapports du droit pénal et de la sociologie criminelle." *Archives de l'anthropologie criminelle et des sciences pénales* 1 (1886): 9–23.

Gauss, Carl Friedrich. *Theoria motus corporum coelestium in sectionibus conicis solem ambientium.* Hamburg: Perthes and Besser, 1809.

Gavarret, Jules. *De l'application de la statistique à la médecine. Réponse à l'examen critique auquel M. le docteur Valleix a soumis, dans le numéro de mai 1840 des Archives générales de médecine, l'ouvrage de M. Gavarret.* N.p.: Locquin, 1840.

———. *Principes généraux de statistique médicale, ou Développement des règles qui doivent présider à son emploi.* Paris: Bechet jeune and Labé, 1840.

Gazette hebdomadaire de médecine et de chirurgie, 1830–55.

Gazette médicale de Paris, 1830–1860.

"Gee Norma, We're Glad You're Home Again!" Editorial cartoon. *Cleveland Plain Dealer,* September 13, 1945, 3.

Gemelli, Agostino. *Le dottrine moderni della delinquenza: critica delle dottrine criminali positiviste.* Florence: Fiorentina, 1908.

Geoffroy Saint-Hilaire, Étienne. *Mémoire sur un enfant quadrupède, né à Paris et vivant, monstruosité déterminée par le nom générique d'iléadelphe.* Lyon: Boursy fils, 1831.

———. *Philosophie anatomique,* vol. I. *Pièces osseuses des organes respiratoires.* Paris: Baillière, 1818.

———. *Philosophie anatomique,* vol. II. *Des monstruosités humaines.* Paris: chez l'auteur, 1822.

———. *Principes de philosophie zoologique, discutés en mars 1830, au sein de l'Académie royale des sciences.* Paris: Pichon and Didier, Rousseau, 1830.

Geoffroy Saint-Hilaire, Isidore. *Histoire générale et particulière des anomalies de l'organisation chez l'homme et les animaux, ouvrage comprenant des recherches sur les caractères, la classification, l'influence physiologique et pathologique, les rapports généraux, les lois et les causes des monstruosités, des variétés et vices de conformation, ou Traité de tératologie.* 4 vols. Paris: Baillière, 1832–37.

———. *Vie, travaux et doctrine scientifique d'Étienne Geoffroy Saint-Hilaire, par son fils.* Paris: Bertrand, 1847.

Gerdy, P. N. *Physiologie médicale, didactique et critique.* Paris: Roret, Béchet, 1830.

Gibson, Mary. *Born to Crime: Cesare Lombroso and the Origins of Biological Criminality.* Westport, CT: Praeger, 2002.

———. *Cesare Lombroso, il determinismo biologico e la delinquenza minorile.* Naples: Edizioni Scientifiche Italiane, 2001.

———. "La crimonologia prima e dopo Lombroso." In *Cesare Lombroso: Gli scienziati e la nuova Italia,* edited by Silvano Montaldo, 15–32. Bologna: Il Mulino, 2010.

Gilbreth, Lillian. *The Home-Maker and Her Job*. New York and London: Appleton, 1927.
Gilman, Sander L. "The Struggle of Psychiatry with Psychoanalysis: Who Won?" *Critical Inquiry* 13, no. 2 (1987): 293–313.
Gilman, Sander L., Helen King, Roy Porter, George Rousseau, and Elaine Showalter. *Hysteria beyond Freud*. Berkeley: University of California Press, 1993.
Gioja, Melchiorre. *Esame d'un'opinione intorno all'indole, estensione e vantaggi delle statistiche*. Milan: Presso Gli Editori, 1826.
Gioja, Melchior. *Philosophie de la statistique*. Milan: Gabon, 1826.
Gobineau, A. de. *Essai sur l'inégalité des races humaines*, vol. 1. Paris: Firmin Didot, 1853.
Goldberg, Jonathan, and Madhavi Menon. "Queering History." *PMLA* 120, no. 5 (2005): 1608–17.
Goring, Charles. *The English Convict: A Statistical Study*. Montclair, NJ: Patterson Smith, 1972.
Gould, Stephen Jay. *The Mismeasure of Man*. 2nd ed. New York: Norton, 1996.
Green, David. "Veins of Resemblance: Photography and Eugenics." *Oxford Art Journal* 7, no. 2 (1984): 3–16.
Guerry, André-Michel. *Essai sur la statistique morale de la France*. Paris: Crochard, 1833. Reprint, Paris: Hachette, 1971.
———. "Motifs des crimes capitaux." *Annales d'hygiène publique et de médecine légale* 1, no. 8 (1832): 341–46.
———. *Statistique morale de l'Angleterre comparée avec la statistique morale de la France*. Paris: Baillière, 1864.
Guillard, Achille. *Eléments de statistique humaine, ou démographie comparée*. Paris: Guillaumin, 1855.
Guyot, Jules. *Bréviaire de l'amour expérimental: méditations sur le mariage selon la physiologie du genre humain*. Paris: Marpon/Flammarion, 1882.
Habermas, Jürgen. *Between Facts and Norms: Contributions to a Discourse Theory of Law and Democracy*. Cambridge: Polity, 1996.
Hacking, Ian. "Biopower and the Avalanche of Printed Numbers." *Humanities in Society* 5 (1982): 279–95.
———. *The Emergence of Probability: A Philosophical Study of Early Ideas about Probability, Induction and Statistical Inference*. Cambridge: Cambridge University Press, 1975.
———. "The Looping Effects of Human Kinds." In *Causal Cognition: A Multidisciplinary Debate*, edited by Dan Sperber, David Premack, and Ann James, 351–83. Oxford: Oxford University Press, 1995.
———. *The Taming of Chance*. Cambridge: Cambridge University Press, 1990.
———. "Was there a Probabilistic Revolution 1800–1930?" In *Ideas in History*. Vol. 1 of *The Probabilistic Revolution*, edited by Lorenz Krüger, Lorraine J. Daston, and Michael Heidelberger. Cambridge, MA: MIT Press, 1987.
Halberstam, J. Jack. *Gaga Feminism: Sex, Gender, and the End of Normal*. Boston: Beacon Press, 2012.
———. *In a Queer Time and Place: Transgender Bodies, Subcultural Lives*. New York: New York University Press, 2005.
Halbwachs, Maurice. *La Théorie de l'homme moyen: Essai sur Quetelet et la statistique morale*. Paris: Alcan, 1913.
Haller, J., and R. Haller. *The Physician and Sexuality in Victorian America*. Urbana: University of Illinois Press, 1974.
Halperin, David. *One Hundred Years of Homosexuality, and Other Essays on Greek Love*. New York: Routledge, 1990.
———. *Saint Foucault: Towards a Gay Hagiography*. New York: Oxford University Press, 1995.
Hankins, Frank. "Adolphe Quetelet as Statistician." PhD diss., Columbia University, 1908. Accessed 4 June 2013. http://www.archive.org/details/adolphequeteleta00hankuoft
———. *Adolphe Quetelet as Statistician*. New York: Columbia University, 1908.

Harrowitz, Nancy A. *Antisemitism, Misogyny and the Logic of Cultural Difference: Cesare Lombroso and Matilde Serao*. Lincoln: University of Nebraska Press, 1994.

Harvey, Joy Dorothy. "Evolutionism Transformed: Positivists and Materialists in the Société d'Anthropologie de Paris from Second Empire to Third Republic." In *The Wider Domain of Evolutionary Thought*, edited by D. Oldroyd and J. Langham, 289–310. Dordrecht: Springer, 1983.

———. "Races Specified. Evolution Transformed: The Social Context of Scientific Debates Originating in the Société d'Anthropologie de Paris." PhD diss., Harvard University, 1983.

Hastings, Donald W. *Impotence and Frigidity*. London: Churchill, 1963.

Hauser, Renate. "Krafft-Ebing's Psychological Understanding of Sexual Behaviour." In *Sexual Knowledge, Sexual Science: The History of Attitudes to Sexuality*, edited by Roy Porter and Mikuláš Teich, 210–30. Cambridge: Cambridge University Press, 1994.

Hawkins, Francis Bisset. *Elements of Medical Statistics; Containing the Substance of the Gulstonian Lectures Delivered at the Royal College of Physicians: with Numerous Additions, Illustrative of the Comparative Salubrity, Longevity, Mortality and Prevalence of Diseases in the Principal Countries and Cities of the Civilized World*. London: Longman, 1829.

"Health Museum's Sex Work Hailed." *Cleveland Plain Dealer*, July 10, 1945, 3.

Heath, Clark H. *What People Are: A Study of Normal Young Men*. Cambridge, MA: Harvard University Press, 1945.

Henslow, John Stevens. *The Principles of Descriptive and Physiological Botany*. London: Longman, 1835.

Herschel, John. *A Preliminary Discourse on the Study of Natural Philosophy*. Philadelphia, PA: Carey and Lea, 1831.

Herschel, John F. W. Review of *Lettres à S.A.R. le duc régnant de Saxe-Cobourg et Gotha sur la théorie des probabilités appliquée aux sciences morales et statistiques*, by Adolphe Quetelet. *Edinburgh Review*, no. 185 (1850): 1–57.

Heyes, Cressida. *Self-Transformations: Foucault, Ethics, and Normalized Bodies*. Oxford: Oxford University Press, 2007.

Hirschfeld, Magnus. *Sexual Anomalies: The Origin, Nature, and Treatment of Sexual Disorders*. New York: Emerson, 1948.

Hooton, Earnest Albert. *A Survey in Seating*. Westport CT: Greenwood Press, 1945.

Hooton, Earnest Albert. "What Is an American?" *American Journal of Physical Anthropology* 22, no. 1 (1936): 1–26.

———. *Young Man, You Are Normal*. New York: Putnam, 1945.

Horn, David G. *The Criminal Body: Lombroso and the Anatomy of Deviance*. New York: Routledge, 2003.

Howe, Catherine Newman. "Average Joes and Mean Girls: The Representation and Transformation of the Average American, 1890–1945." PhD diss., University of California, Santa Barbara, 2012.

Huff, Joyce L. "Freaklore: The Dissemination, Fragmentation, and Reinvention of the Legend of Daniel Lambert, King of Fat Men." In *Victorian Freaks: The Social Context of Freakery in Britain*, edited by Marlene Tromp, 37–59. Columbus: Ohio State University Press, 2008.

Hyman, Herbert, and Joseph Barmack. "Special Review: Sexual Behavior in the Human Female." *Psychological Bulletin* 51, no. 4 (1954): 418–32.

Igo, Sarah E. *The Averaged American: Surveys, Citizens, and the Making of a Mass Public*. Cambridge, MA: Harvard University Press, 2007.

Jaarsma, Ada Susanne. *Troubling the Normal: Contemporary Encounters with Kierkegaard*. Ann Arbor, MI: UMI Dissertation Services, 2005.

Jagose, Annamarie. *Orgasmology*. Durham, NC: Duke University Press, 2012.

Jagose, Annamarie. "The Trouble with Antinormativity." *differences: A Journal of Feminist Cultural Studies* 26, no. 1 (2015): 26–47.

Jewell, Edward Alden. "The Masterpiece and the Modeled Chart." *New York Times*, September 18, 1932, 9.
Joly, Henri. *La France criminelle*. Paris: Cerf, 1889.
Jones, Greta. *Sexual Hygiene in Twentieth Century Britain*. London: Croom Helm, 1986.
Jorland, Gérard. "La Sous-détermination des théories médicales par les statistiques: le cas Semmelweis." In *Body Counts: Medical Quantification in Historical and Sociological Perspective*, edited by Gérard Jorland, Annick Opinel, and George Weisz, 205–25. Montreal: McGill-Queen's University Press, 2005.
Journal de la Société de Statistique de Paris, 1860–74.
Journal de la Société française de statistique, 1860–75.
Journal of the American Medical Association. Review of *Sexual Behavior in the Human Male*, by Alfred Kinsey. 136, no. 6 (Feb. 7, 1948): 430.
Julin, Armand. "The History and Development of Statistics in Belgium." In *The History of Statistics, Their Development and Progress in Many Countries: Memoirs to Commemorate the Seventy-Fifth Anniversary of the American Statistical Association*, 125–75. New York: Macmillan, 1918.
Kang, Zheng. "La société de statistique de Paris au XIXe siècle: un lieu de savoir social." *Journal de la Société française de statistique* 134, no. 3 (1993): 49–61.
———. "Lieu de savoir social. La Société de Statistique de Paris au XIXe siècle, 1860–1914." PhD diss., EHESS, Paris, 1989.
Katz, Jonathan Ned. *The Invention of Heterosexuality*. 2nd ed. Chicago: University of Chicago Press, 2007.
Kellogg, John. *Plain Facts for Old and Young: Embracing the Natural History of Hygiene of Organic Life*. Burlington, IA: Segner, 1890.
Kevles, Daniel. *In the Name of Eugenics: Genetics and the Uses of Human Heredity*. Berkeley: University of California Press, 1985.
Kinsey, Alfred. *Sexual Behavior in the Human Male*. Philadelphia: Saunders, 1948.
Kinsey, Alfred. *Sexual Behavior in the Human Female*. Bloomington: Indiana University Press, 1953.
Kirby, Vicki. "Transgression: Normativity's Self-Inversion." *differences: a Journal of Feminist Cultural Studies* 26, no 1 (2015): 96–116.
Knapp, Friedrich. Vol.18 of *Bericht über die Schriften Quetelets zur Sozialstatistik und Anthropologie*: *Jahrbücher für Nazionalökonomie und Statistik*. Jena: Fischer, 1902.
Koch, Julius Ludwig August. *Die Frage nach dem geborenen Verbrecher*. Ravensburg: Otto Maier, 1894.
Kozar, Andrew J. *The Sport Sculpture of R. Tait McKenzie*. Champaign, IL: Human Kinetics Books, 1992.
Krafft-Ebing, Richard von. *Psychopathia Sexualis, with Especial Reference to Contrary Sexual Instinct: A Medico-Legal Study*. 7th ed. Philadelphia: F. A. Davis; London: Rebman, 1894.
Krauth, Stefan. *Die Hirnforschung und der gefährliche Mensch: Uber die Gefahren einer Neuauflage der biologischen Kriminologie*. Munster: Westfälisches Dampfboot, 2008.
Krüger, Lorenz, Lorraine J. Daston, and Michael Heidelberger, eds. *The Probabilistic Revolution. Volume 1: Ideas in History* (Cambridge, MA: MIT Press, 1987).
Kurella, Hans. *Cesare Lombroso als Mensch und Forscher*. Wiesbaden: Bergmann, 1910.
———. *Cesare Lombroso: A Modern Man of Science*. New York: Rebman, 1911.
———. *Cesare Lombroso und die Naturgeschichte des Verbrechens*. Hamburg: Verlagsanstalt und Druckerei Aktien-Gesellschaft, 1892.
———. *Naturgeschichte des Verbrechens: Grundzüge der kriminellen Anthropologie und Criminalpsychologie*. Stuttgart: Enke, 1893.
La Berge, Ann F. "Medical Statistics at the Paris School: What Was at Stake?" In *Body Counts: Medical Quantification in Historical and Sociological Perspective*, edited by Gérard Jorland, Annick Opinel, and George Weisz, 89–108. Montreal: McGill-Queen's University Press, 2005.

Lacassagne, Alexandre, René Garraud, and Henry Coutagne, eds. *Archives de l'anthropologie criminelle et des sciences pénales*. Paris: Masson, 1886.
Lacroix, Silvestre-François. *Traité élémentaire du calcul des probabilités*. Paris: Courcier, 1816.
Lade, James H. Review of *Sexual Behavior in the Human Male*, by Alfred Kinsey. *Health News* (April 1948): 16.
Lamarck, Jean-Baptiste de Monet. *Philosophie zoologique, ou Exposition des considérations relatives à l'histoire naturelle des animaux*. London: Macmillan, 1914. Accessed February 10, 2013. http://gallica.bnf.fr/ark:/12148/bpt6k5675762f.
Lancaster, Roger N. *The Trouble with Nature: Sex and Science in Popular Culture*. Berkeley: University of California Press, 2003.
Laplace, Pierre-Simon. *Essai philosophique sur les probabilités*. Paris: Bachelier, 1840.
———. *Théorie analytique des probabilités*, vol. 1. Paris: Gabay, 1995. Reprint of *Œuvres de Laplace*, vol. 7. Paris: Imprimerie Royale, 1847.
Lariar, Lawrence. *Oh! Dr. Kinsey! A Photographic Reaction to the Kinsey Report*. New York: Cartwrite, 1953.
Laschi, R. "Actes du II Congrès international d'Anthropologie criminelle." *Archivio di psichiatria, scienze penali ed antropologia criminale per servire allo studio dell'uomo alienato e delinquente* 12 (1891): 559–62.
Latour, Bruno. *Science in Action: How to Follow Scientists and Engineers through Society*. Cambridge, MA: Harvard University Press, 1988.
Laughlin, Harry H. *Eugenical Sterilization in the United States*. Chicago: Psychopathic Laboratory of the Municipal Court of Chicago, 1922.
———. *The Second International Exhibition of Eugenics held September 22 to October 22, 1921, in Connection with the Second International Congress of Eugenics in the American Museum of Natural History, New York: An Account of the Organization of the Exhibition, the Classification of the Exhibits, the List of Exhibitors, and a Catalog and Description of the Exhibits*. Baltimore: Williams and Wilkins, 1923.
———. "What Eugenics Is All About." In *A Decade of Progress in Eugenics: Scientific Papers of the Third International Congress of Eugenics*, edited by H. F. Perkins. Baltimore: Williams and Wilkins, 1934.
Laugier, H. "L'Homme normal." In *Encyclopédie française* 4. Paris: Société de Gestion de l'Encyclopédie Française, 1937.
Lauvergne, Hubert. *Les Forçats considérés sous le rapport physiologique, moral et intellectuel*. Paris: Baillière, 1841.
La Vergata, Antonello. "Lombroso e la degenerazione." In *Cesare Lombroso: Gli scienziati e la nuova Italia*, edited by Silvano Montaldo, 53–93. Bologna: Il Mulino, 2010.
Legendre, Adrien Marie. *Nouvelles méthodes pour la détermination des orbites des comètes*. Paris: Didot, 1805.
Legrand, Stéphane. *Les Normes chez Foucault*. Paris: PUF, 2007.
Le Moniteur des hôpitaux 1852–57.
Leonard, Eileen B. *Women, Crime, and Society: A Critique of Theoretical Criminology*. New York: Longman, 1982.
Leudet, Dr. "L'Influence de la vaccine sur la population." *Gazette hebdomadaire de médecine et de chirurgie* 1, no. 26 (March 1854): 414.
Leuret, François. "De la fréquence du pouls chez les aliénés considérée dans ses rapports avec les saisons, la température atmosphérique, les phases de la lune, l'âge, etc.; réfutation de l'opinion admise sur le décroissement de la fréquence du pouls chez les vieillards; note sur la pesanteur spécifique du cerveau des aliénés." *Gazette médicale de Paris* 1, no. 3 (1832): 621–22.
Levine, Philippa, and Alison Bashford, eds. *The Oxford Handbook of the History of Eugenics*. Oxford Handbooks Online, 2012.

Lexis, Wilhelm. "Sur la durée normale de la vie humaine et sur la théorie de la stabilité des rapports statistiques." *Annales de démographie internationale* 2, no. 5 (1878): 447-60.
Leys, Ruth. "The Turn to Affect: A Critique." *Critical Inquiry* 37, no. 3 (2011): 434-72.
L'Hardy, Philippe, and Claude Thélot, eds. *Les Ménages. Mélanges en honneur de J. Desabie*. Paris: INSEE, 1989.
Lloyd, Warren Estelle. *Psychology, Normal and Abnormal: A Study of the Processes of Nature from the Inner Aspect*. Los Angeles: Baumgart, 1908.
Lock, Margaret. "Accounting for Disease and Distress: Morals of the Normal and Abnormal." In *The Handbook of Social Studies in Health and Medicine*, edited by Gary L. Albrecht, Ray Fitzpatrick, and Susan C. Scrimshaw, 259-76. London: Sage, 2000.
Lombroso, Cesare. *L'Anthropologie criminelle et ses récents progrès*. Paris: Alcan, 1890.
———. *Delitto di libidine*. Turin: Bocca, 1886.
———. *Delitto, genio, follia*. edited by Delia Frigessi, Ferrucio Giacanelli, and Luisa Mangoni. Turin: Bollati Boringhieri, 1995.
———. "I pazzi criminali." *Archivio di psichiatria, scienze penali ed antropologia criminale per servire allo studio dell'uomo alienato e delinquente* 9 (1888): 572.
———. *Pazzi ed anomali*. Città di Castell: Lapi, 1886.
———. *La perizia psichiatrico-legale, coi metodi per eseguirla e la casuistica penale, classificata antropologicamente*. Turin: Bocca, 1905.
———. "Pro mea schola!" *Archivio di psichiatria* 5 (1884): 92-104.
———. *Sulla medicina legale del cadavere secondo gli ultimi studi di Germania ed Italia*. Turin: Bocca, 1877.
———. *Tre tribuni: Studiati da un alienista*. Turin: Bocca, 1887.
———. *L'uomo bianco e l'uomo di colore: letture su l'origine e la varietà delle razze umane*. 2nd ed. Turin: Bocca, 1892.
———. *L'uomo delinquente in rapporto all'antropologia, alla giurisprudenza ed alle discipline carcerarie*. Turin: Bocca, 1889.
———. *L'uomo delinquente in rapporto all'antropologia, alla giurisprudenza ed alla psichiatria: Cause e rimedi*. Turin: Bocca, 1897.
Lombroso, Cesare, and Guglielmo Ferrero. *La donna delinquente, la prostituta e la donna normale*. Turin: L. Roux, 1893.
———. *La donna delinquente, la prostituta e la donna normale*. Turin: Bocca, 1903.
———. *La Femme criminelle et la prostituée*. Translated by Louise Meille. Paris: Alcan, 1896.
———. *Criminal Woman, the Prostitute, and the Normal Woman*. Translated by Nicole Hahn Rafter and Mary Gibson. Durham, NC: Duke University Press, 2004.
Lombroso, Gina. *L'Anima della donna*. Bologna: Zanichelli, 1926.
Lombroso, Paola, and Gina Lombroso. *Cesare Lombroso: Appunti sulla vita; Le opere*. Turin: Bocca, 1906.
London Medical Gazette, 1837.
Lottin, Joseph. *Quetelet, statisticien et sociologue*. Louvain: Institut Supérieur de Philosophie, 1912.
Louis, Pierre Charles Alexandre. *Examen de l'examen de M. Broussais, relativement à la phtisie [sic] et à l'affection typhoïde*. Paris: Baillière, 1834.
———. *Pathological Researches on Phthisis*. Translated by Charles Cowan. Boston: Hilliard, Gray, 1836.
———. *Recherches anatomo-pathologiques sur la phthisie*. 2nd ed. Paris: Baillière, 1843.
Lucas, Aimé. *Les Jeunes filles folles ou égarées: Nouvel aperçu historique, statistique et administratif sur la prostitution, les prostituées et l'administration qui les surveille et les régit*. Paris: Terry, 1843.
Lundgren, Frans. "The Politics of Participation: Francis Galton's Anthropometric Laboratory and the Making of Civic Selves." *The British Journal for the History of Science* 46, no. 3 (2013): 445-66.

Lydston, Frank. *Sex Hygiene for the Male and What to Say to the Boy*. Chicago: Riverton, 1912.

———. *The Diseases of Society: The Vice and Crime Problem*. Philadelphia: J. B. Lippincott, 1906.

Lynd, Robert S., and Helen Merrell Lynd. *Middletown: A Study in Modern American Culture*. San Diego, New York, London: Harcourt Brace, 1929.

Macfadden, Bernarr. *The Virile Powers of Superb Manhood: How Developed, How Lost, How Regained*. New York: Physical Culture Publishing Company, 1900.

Macherey, Pierre. "The Natural History of Norms." In *Michel Foucault: Philosopher*, edited by Timothy Armstrong, 176–91. New York: Harvester Wheatsheaf, 1992.

Magnan, Valentin. *Des anomalies, des aberrations et des perversions sexuelles*. Paris: Delahaye et Lecrosnier, 1885.

Mailly, Edouard. "Essai sur la vie et les ouvrages de Quetelet." *Annuaire de l'Académie royale des sciences, des lettres et des beaux-arts de Belgique* 41 (1875): 109–297.

Mailly, Edouard. *Essai sur la vie et les ouvrages de L.-A.-J. Quetelet*. Brussels: Hayez, 1875.

Malchow, C. W. *The Sexual Life, Embracing the Natural Sexual Impulse, Normal Sexual Habits and Propagation, Together with the Sexual Physiology and Hygiene*. St. Louis: Mosby, 1907.

Manouvrier, Léonce. "Les Crânes des suppliciés." *Archives de l'anthropologie criminelle et des sciences pénales. Médecine légale, judiciaire. Statistique criminelle. Legislation et droit* 1 (1886): 119–41.

———. "La Place et l'importance de la crâniologie anthropologique." *Matériaux pour l'histoire primitive et naturelle de l'homme* 3, no. 3 (Janvier 1886): 4–20.

———. *Travaux du Dr L. Manouvrier de 1880 à 1889 (laboratoire d'anthropologie à l'École des hautes études, Paris)*. Paris: Librairies imprimeries réunies, n.d.

Mantegazza, Paolo. "Di alcune recenti proposte di riforma della craniologia." *AAE* 23 (1893): 45–55.

———. *Fisiologia del dolore*. Firenze: Paggi, 1880.

Markowitsch, Hans J., and Werner Siefer. *Tatort Gehirn: Auf der Suche nach dem Ursprung des Verbrechens*. Frankfurt am Main: Campus, 2007.

Martin, Biddy. "Extraordinary Homosexuals and the Fear of Being Ordinary." *differences* 6 (Summer-Fall 1994): 100–26.

Martin, Thierry. "Gabriel Tarde et la statistique criminelle." *Mathématiques et sciences humaines* 193 (2011): 27–35.

Matthews, J. Rosser. "Commentary: The Paris Academy of Science Report on Jean Civiale's Statistical Research and the Nineteenth-Century Background to Evidence-Based Medicine." *International Journal of Epidemiology* 30 (2001): 1249–50.

Matthews, J. Rosser. *Quantification and the Quest for Medical Certainty*. Princeton, NJ: Princeton University Press, 1995.

Maxwell, Anne. *Picture Imperfect: Photography and Eugenics, 1870–1940*. Eastbourne: Sussex Academic Press, 2008.

Mazzarello, Paolo. *Il genio e l'alienista: la visita di Lombroso a Tolstoj*. Naples: Bibliopolis, 1998.

McKenzie, Robert Tait. "The Search for Physical Perfection." In *University Lectures: Delivered by the Members of the Faculty in the Free Public Lecture Course 1914–1915*, 409–19. Philadelphia: University of Pennsylvania, 1915.

McRuer, Robert. *Crip Theory: Cultural Signs of Queerness and Disability*. New York: New York University Press, 2006.

McWhorter, Ladelle. *Bodies and Pleasures: Foucault and the Politics of Sexual Normalization*. Bloomington: Indiana University Press, 1999.

Messedaglia, Angelo. *La Statistica: i suoi metodi e la sua competenza*. Rome: Elzeviriana, 1879.

———. "Statistica morale dell'Inghilterra comparata alla statistica morale della Francia di M. A. Guerry: relazione critica." *Atti dell'I.R. Istituto Veneto di Scienze, Lettere ed Arti* 3, no. 10 (1864–65): 1068–85; 1135–68.

Micale, Mark. *Approaching Hysteria: Disease and its Interpretations*. Princeton, NJ: Princeton University Press, 1995.
Milet, Jean. *Gabriel Tarde et la philosophie de l'histoire*. Paris: Vrin, 1970.
Mill, John Stuart. *A System of Logic, Being a Connected View of the Principles of Evidence and the Methods of Scientific Investigation*, 4th ed. 1843. Vols. 6 and 7. Reprint, London: Parker, 1856.
Montagu, M. F. "Understanding Our Sexual Desires." In *About the Kinsey Report: Observations by 11 Experts on "Sexual Behavior in the Human Male,"* edited by D. P. Geddes and E. Curie, 59–69. New York: New American Library, 1948.
Montaldo, Silvano. "Premessa." In *Cesare Lombroso. Gli scienziati e la nuova Italia*, edited by Silvano Montaldo, 7-12. Bologna: Il Mulino, 2010.
Montaldo, Silvano, and Paolo Tappero. *Cesare Lombroso cento anni dopo*. Turin: Utet, 2009.
———. *Il museo di antropologia criminale "Cesare Lombroso."* Turin: Utet, 2009.
Montaldo, Silvano, ed. *Cesare Lombroso. Gli scienziati e la nuova Italia*. Bologna: Il Mulino, 2010.
Moran, Jeffrey. *Teaching Sex: The Shaping of Adolescence in the Twentieth Century*. Cambridge, MA: Harvard University Press, 2000.
Moreau, César. "Procès-verbal." *Journal des travaux de la Société Française de Statistique Universelle*, new series, 1, no. 6 (1835): 321-24.
Moreau de Jonnès, Alexandre. *Eléments de statistique, comprenant les principes généraux de cette science, et un aperçu historique de ses progrès*. Paris: Guillaumin, 1847.
Morel, Bénédict Auguste. *Traité des dégénérescences physiques, intellectuelles et morales de l'espèce humaine et des causes qui produisent ces variétés maladives*. Paris: Baillière, 1857.
Morell, Jack, and Arnold Thackray. *Gentlemen of Science: Early Years of the British Association for the Advancement of Science*. Oxford: Oxford University Press, 1981.
———, eds. *Gentlemen of Science: Early Correspondence of the British Association for the Advancement of Science*. London: Offices of the Royal Society, 1984.
Mosse, George L. *Toward the Final Solution: A History of European Racism*. New York: Ferting, 1978.
Mucchielli, Laurent. "Criminology, Hygienism, and Eugenics in France, 1870-1914: The Medical Debates on the Elimination of 'Incorrigible' Criminals." In *Criminals and Their Scientists: The History of Criminology in International Perspective*, edited by Peter Becker and Richard F. Wetzell, 207-29. Cambridge: Cambridge University Press, 2006.
Murphy, Terence D. "Medical Knowledge and Statistical Methods in Early-Nineteenth-Century France," *Medical History* 25 (1981), 311-13.
Murray, Jessie. "Preface." In Marie Carmichael Stopes, *Married Love: A New Contribution to the Solution of Sexual Difficulties*, 1-4. London: Fifield, 1918.
Naecke, P. *Verbrechen und Wahnsinn beim Weibe: Mit Ausblicken auf die Criminal-Anthropologie überhaupt*. Vienna: Braunmüller, 1894.
"Notions générales sur la population." *Recherches statistiques sur la ville de Paris et le département de la Seine. Recueil de tableaux dressés d'après les ordres de Monsieur le Comte de Chabrol, conseiller d'état, préfet du département* (Paris: Ballard, 1821): 8-78.
Novak, Daniel. *Realism, Photography and Nineteenth-Century Fiction*. Cambridge: Cambridge University Press, 2008.
Nunberg, Hermann, and Ernst Federn, eds., *Protokolle Der Wiener Psychoanalytischen Vereinigung*. 4 vols. Giessen: Psychosozial-Verlag, 2008.
Nye, Robert A. *Crime, Madness, and Politics in Modern France: The Medical Concept of National Decline*. Princeton, NJ: Princeton University Press, 1984.
———. "Heredity or Milieu: The Foundations of Modern European Criminological Theory." *ISIS* 67, no. 3 (1976): 334-55.
———. *Masculinity and Male Codes of Honor in Modern France*. New York: Oxford University Press, 1993.

O'Brien, Ruth, and William C. Shelton. *Women's Measurements for Garment and Pattern Construction*. Washington: United States Government Printing Office, 1941.

Oettingen, Alexander von. *Die Moralstatistik, inductiver Nachweis der Gesetzmässigkeit sittlicher Lebensbewegung in Organismus der Menschheit*. Erlangen: A. Deichert, 1868.

———. *Die Moralstatistik und die Christliche Sittenlehre*. Erlangen: Deichert, 1873.

Olin-Lauritzen, Sonja, and Lars-Christer Hydén. *Medical Technologies and the Life World: The Social Construction of Normality*. Abingdon: Routledge, 2007.

Oosterhuis, Harry. *Stepchildren of Nature: Krafft-Ebing, Psychiatry, and the Making of Sexual Identity*. Chicago and London: University of Chicago Press, 2000.

Palacio, Jean de. *Configurations décadentes*. Leuven: Peeters, 2007.

Pancaldi, Giuliano. *Darwin in Italy: Science across Cultural Frontiers*. Bloomington: Indiana University Press, 1991.

Parens, Erik, ed. *Surgically Shaping Children: Technology, Ethics, and the Pursuit of Normality*. Baltimore: Johns Hopkins University Press, 2006.

Pasquino, Pasquale. "Theatrum Politicum. The Genealogy of Capital—Police and the State of Prosperity." *Ideology and Consciousness* 4 (1978): 41–54.

Pearson, Karl. "Abraham De Moivre." *Nature* 117 (April 1926): 551.

———. *The Grammar of Science*. London: Scott, 1892.

———. *The Life, Letters and Labours of Francis Galton*. 3 vols. Cambridge: Cambridge University Press, 1914–30.

———. *National Life from the Standpoint of Science*. 2nd ed. London: Black, 1905.

Penta, Pasquale. *I pervertimenti sessuali*. Napoli: Luigi Piero, 1893.

Perrier, Edmond. *La Philosophie zoologique avant Darwin*. Paris: Alcan, 1884.

Picard, Jean-Paul. "Feuilleton. Congrès international de statistique." *Gazette hebdomadaire de médecine et de chirurgie* 2, no. 39 (1855): 697–700.

Pick, Daniel. *Faces of Degeneration: A European Disorder, c.1848–c.1918*. Cambridge: Cambridge University Press, 1989.

Pinel, Philippe. *La Médecine clinique rendue plus précise et plus exacte*. Paris: Brosson, 1804.

———. *Nosographie philosophique*. Paris: Maradan, 1797.

———. *Traité médico-philosophique sur l'aliénation mentale*. 2nd ed. Paris, 1809.

Pivar, David. *Purity Crusade: Sexual Morality and Social Control, 1868–1900*. Westport, CT: Greenwood Press, 1973.

Poisson, Dulong, Larrey, and Double. "Reprints and Reflections: Statistical Research on Conditions Caused by Calculi by Doctor Civiale." *International Journal of Epidemiology* 30 (2001): 1246–49.

Poisson, Siméon-Denis. "Analyse mathématique. Note sur la loi des grands nombres, par M. Poisson." *Comptes rendus hebdomadaires des séances de l'Académie des sciences* 2 (1836): 377–80.

———. "Calcul des probabilités—Recherches sur la probabilité des jugements, principalement en matière criminelle, par M. Poisson." *Comptes rendus hebdomadaires des séances de l'Académie des sciences* 1 (1835): 473–94.

———. *Recherches sur la probabilité des jugements en matière criminelle et en matière civile, précédées des Règles générales du calcul des probabilités*. Paris: Gabay, 2003. First published 1837 by Bachelier.

Pomeroy, Wardell Baxter, *Dr. Kinsey and the Institute for Sex Research*. New York: Harper and Row, 1972.

Poovey, Mary. *A History of the Modern Scientific Fact: Problems of Knowledge in the Sciences of Wealth and Society*. Chicago: University of Chicago Press, 1998.

Porter, Cole. "Too Darn Hot." *Kiss Me, Kate*. New York: Off Broadway, 1948.

Porter, Roy, and Mikuláš Teich, eds. *Sexual Knowledge, Sexual Science: The History of Attitudes to Sexuality*. Cambridge: Cambridge University Press, 1994.

Porter, Theodore. *Karl Pearson: The Scientific Life in a Statistical Age.* Princeton, NJ: Princeton University Press, 2004.

———. *The Rise of Statistical Thinking, 1820–1900.* Princeton, NJ: Princeton University Press, 1986.

———. "A Statistical Survey of Gases: Maxwell's Social Physics." *Historical Studies in the Physical Sciences* 12 (1981): 77–116.

Prescott, Heather Munro. "Using the Student Body: College and University Students as Research Subjects in the United States during the Twentieth Century." *Journal of the History of Medicine* 57, no. 1 (2002): 3–38.

Prodger, Phillip. *Time Stands Still: Muybridge and the Instantaneous Photography Movement.* Oxford and New York: Oxford University Press, 2003.

"Projet de classification pour l'établissement de la statistique générale d'un département." *Journal des travaux de la Société Française de Statistique Universelle,* new series, 2, no. 19 (1837): 401.

Pryce, Anthony. "Let's Talk About Sexual Behavior in the Human Male: Kinsey and the Invention of (Post)Modern Sexualities." *Sexuality and Culture* 10, no. 1 (2006): 63–93.

Pucheran, Jacques. "Considérations anatomiques sur les formes de la tête osseuse dans les races humaines." PhD diss., Paris, 1841.

Quetelet, Adolphe. *Anthropométrie, ou Mesure des différentes facultés de l'homme.* Brussels: Muquardt, 1870.

———. *Congrès international de statistique: Sessions de Bruxelles (1853), Paris (1855), Vienne (1857), Londres (1860), Berlin (1863), Florence (1867), La Haye (1869) et St-Petersbourg (1872).* Brussels: Hayez, 1873.

———. "De l'influence du libre arbitre de l'homme sur les faits sociaux, et particulièrement sur le nombre des mariages." *Bulletin de la commission centrale de statistique* 3 [1846]: 135–55.

———. "Hygiène morale." *Annales d'hygiène publique et de médecine légale* 1, no. 9 (1833): 308–36.

———. *Instructions populaires sur le calcul des probabilités.* Brussels: Hayez, 1828.

———. *Lettres à S.A.R. le duc régnant de Saxe-Cobourg et Gotha sur la théorie des probabilités appliquée aux sciences morales et politiques.* Brussels: Hayez, 1845.

———. *Statistique et astronomie.* Bruxelles: Hayez, n.d.

———. *Statistique internationale de l'Europe: Plan adopté par les délégués officiels des différents Etats, dans la septième session du congrès international tenu à La Haye en 1869.* Brussels: Hayez, 1869.

———. "Sur l'appréciation des documents statistiques et en particulier sur l'appréciation des moyennes." *Bulletin de la Commission centrale de statistique* [de Belgique] 2 (1845): 205–86.

———. "Sur la statistique morale et les principes qui doivent en former la base." *Mémoires de l'Académie Royale de Belgique* 21. Brussels: Hayez, 1848.

———. "Sur les indiens O-Jib-Be-Wa's et les proportions de leur corps." *Bulletin de l'Académie Royale des Sciences et Belles-Lettres de Belgique,* 13, part 1 (1846): 70–76.

———. "Sur les proportions de la race noire." *Bulletin de l'Académie Royale des Sciences et Belles-Lettres de Belgique,* 21, part 1 (1854): 96–100.

———. *A Treatise on Man and the Development of His Faculties.* Edinburgh: William and Robert Chambers, 1842.

———. *Sur l'homme et le développement de ses facultés ou Essai de physique sociale.* 2 vols. Paris: Bachelier, 1835.

———. *Théorie des probabilités.* Bruxelles: Jamar, 1853.

Quetelet, Adolphe, and Édouard Smits. *Recherches sur la reproduction et la mortalité de l'homme aux différents âges, et sur la population de la Belgique.* Brussels: Hauman, 1832.

———. "Statistique du choléra. Recherches sur la reproduction et la mortalité de l'homme aux différens âges, et sur la population de la Belgique." *Gazette médicale de Paris* 1, no. 3 (1832): 480–82.

Quetelet, Adolphe, ed. *Correspondance mathématique et physique.* 6 vols. Gand: Vandekerckove, 1825–30.

Rabinach, Anson. *The Human Motor: Energy, Fatigue and the Origins of Modernity.* Berkeley: University of California Press, 1992.

Rabinow, Paul. *French Modern: Norms and Forms of the Social Environment.* Cambridge, MA: MIT Press, 1989.

Raybeaud, Louis. "Recherches statistiques et historiques sur le mouvement et le progrès des races humaines." *Journal des travaux de la Société Française de Statistique Universelle* 2, no. 21 (1837): 551-55.

Rayer, Pierre-François-Olive. *Cours de médecine comparée.* Paris: Baillière, 1863.

———. "Sur la statistique médicale." *Bulletin de l'Académie royale de médecine.* Paris: Firmin-Didot, 1836.

———. *Sur la statistique médicale.* Paris: Firmin-Didot, 1855.

"Recherches mathématiques sur la population française aux 18e et 19e siècles." *Journal des travaux de la Société Française de Statistique Universelle* 1, no. 3 (1835): 356-59.

Reiffenberg, Baron de. "Bibliography of the Ancient Statistics of Belgium." *Nouveaux mémoires de l'Académie royale des sciences et belles lettres de Bruxelles* 7 (1832).

Reiss, Elizabeth. *Bodies in Doubt: An American History of Intersex.* Baltimore: Johns Hopkins University Press, 2009.

Reiss, Steven. *The Normal Personality: A New Way of Thinking about People.* Cambridge: Cambridge University Press, 2008.

Réveillé-Parise, Joseph-Henri. "Miscellanées, no. VII," *Gazette médicale de Paris* 4, nos. 12-13 (1836): 145-50, 177-82.

Journal of the American Medical Association. Review of *Sexual Behavior in the Human Male*, by Alfred Kinsey. 136, no. 6 (Feb. 7, 1948): 430.

Risueño d'Amador, [Benigno]. *Quels avantages la médecine pratique a-t-elle retirés de l'étude des constitutions médicales et des épidémies? Question proposée par l'Académie Royale de Médecine de Paris, pour le concours du prix Moreau de la Sarthe.* Montpellier: Picot, 1829.

———. *Mémoire sur le calcul des probabilités appliqué à la médecine. Lu à l'Académie Royale de Médecine dans sa séance du 25 avril 1837.* Paris: Baillière, 1837.

Robertson, Josephine. "Are You Norma, That Rare Individual?" *Cleveland Plain Dealer*, September 9, 1945, 1, 8.

———. "Are You Norma, Typical Woman?" *Cleveland Plain Dealer*, September 9, 1945, 1.

———. "Dr. Clausen Finds Norma Devout, but Still Glamorous." *Cleveland Plain Dealer*, September 23, 1945, 1.

———. "Norma Is Appealing Model in Opinion of City's Artists." September 15, 1945, 1, 10.

———. "Theatre Cashier, 23, Wins Title of 'Norma,' Besting 3,863 Entries." *Cleveland Plain Dealer*, September 23, 1945, 1.

Robie, William. *Rational Sex Ethics: A Physiological and Psychological Study of the Sex Lives of Normal Men and Women, with Suggestions for a Rational Sex Hygiene with Reference to Actual Case Histories.* Boston: Badger, 1916.

Rose, June. *Marie Stopes and the Sexual Revolution.* London and Boston: Faber and Faber, 1992.

Rose, Nikolas. *Governing the Soul: The Shaping of the Private Self.* 2nd ed. London: Free Association Books, 1999.

Rose, Steven, Leon J. Kamin, and R. C. Lewontin. *Not in Our Genes: Biology, Ideology and Human Nature.* New York: Pantheon Books, 1985.

Rossi, Virgilio. "Sull'alcoolismo e le critiche di Colajanni." *Archivio di psichiatria, scienze penali ed antropologia criminale per servire allo studio dell'uomo alienato e delinquente* 7 (1886): 605-10.

Roubaud, Félix [Dr. Rauland, pseud.]. *Le Livre des époux. Guide pour la guérison de l'impuissance, de la stérilité et de toutes les maladies des organes génitaux.* Paris: Chez les principaux libraires, 1859.

———. *Traité de l'impuissance et de la stérilité chez l'homme et la femme, comprenant l'exposition des moyens recommandés pour y remédier.* 3rd ed. Paris: Baillière, 1876.

Royer, Clémence. "Remarques sur le transformisme." *Bulletins de la Société d'Anthropologie de Paris*, series 2, vol. 5 (1870): 265-317.

Rümelin, Gustave. *Problèmes d'économie politique et de statistique.* Translated by A. de Riedmatten. Paris: Guillaumin, 1896.

Rusk, Howard A. "Concerning Man's Basic Drive." Review of *Sexual Behavior in the Human Male*, by Alfred Kinsey. *New York Times*, January 4, 1948. "Books" section, 1.

Rydell, Robert, Christina Codgell, and Mark Largent. "The Nazi Eugenics Exhibition in the United States, 1934-1943." In *Popular Eugenics: National Efficiency and American Mass Culture in the 1930s*, edited by Susan Currell and Christina Cogdell, 359-84. Athens: Ohio University Press, 2006.

Rydell, Robert. *World of Fairs: The Century-of-Progress Expositions.* Chicago: University of Chicago Press, 1993.

Sargent, Dudley. "Modern Woman Getting Nearer the Perfect Figure." *Sunday Magazine*, December 4, 1910, 4.

———. "The Physical Proportions of the Typical Man." *Scribner's Magazine* 2, no. 1 (1887): 3-17.

Say, Jean-Baptiste. *De l'objet et de l'utilité des statistiques.* Paris: Rignoux, 1827.

———. *Traité d'économie politique, ou Simple exposition de la manière dont se forment, se distribuent et se consomment les richesses.* Paris: Renouard, 1814.

Schweber, Libby. *Disciplining Statistics: Demography and Vital Statistics in France and England, 1830-1885.* Durham, NC: Duke University Press, 2006.

Sedgwick, Eve Kosofsky. *Touching Feeling: Affect, Pedagogy, Performativity.* Durham, NC: Duke University Press, 2003.

Sekula, Allan. "The Body and the Archive." *October* 39 (1986): 3-64.

Shapiro, Barbara J. *Probability and Certainty in Seventeenth-Century England.* Princeton, NJ: Princeton University Press, 1983.

Shapiro, Harry L. "Portrait of the American People." *Natural History Magazine* 54 (June 1945): 246-55.

Sheldon, William Herbert. *Atlas of Men: A Guide for Somatotyping the Adult Male at All Ages.* New York and London: Harper, 1954.

———. *The Varieties of Human Physique: An Introduction to Constitutional Psychology.* New York and London: Harper, 1940.

Shepherdson, Charles. *Vital Signs: Nature, Culture, Psychoanalysis.* New York: Routledge, 2000.

Sheynin, O. B. "On the History of Medical Statistics." *Archive for History of Exact Sciences* 26 (1982): 241-86.

Shove, Elizabeth. *Comfort, Cleanliness and Convenience: The Social Organization of Normality.* Oxford: Berg, 2003.

"Some Rattling Good Stories." *Good Housekeeping.* June 1929: 204.

Sigerist, Henry Ernst. *Introduction à la médecine.* Translated by Maurice Tenine. Paris: Payot, 1932.

Souberbielle, Joseph. *Renseignements adressés à l'Académie des sciences, sur quelques points de la statistique des affections calculeuses, présentée par M. Civiale, dans la séance du 26 août 1833.* Paris: Béthune, 1833.

Spade, Dean. *Normal Life: Administrative Violence, Critical Trans Politics, and the Limits of Law.* Brooklyn, NY: South End Press, 2011.

Spearman, Charles. *The Nature of "Intelligence" and the Principles of Cognition.* 1923. Reprinted, New York: Arno, 1973.

Stall, Sylvanus. *What a Man of Forty-Five Ought to Know.* Philadelphia: Vir, 1901.

———. *What a Young Boy Ought to Know.* Philadelphia: Vir, 1897.

———. *What a Young Husband Ought to Know.* Philadelphia: Vir, 1899.

———. *What a Young Man Ought to Know.* Philadelphia: Vir, 1904.

Stanley, Liz. *Sex Surveyed 1949-1994: From Mass-Observation's "Little Kinsey" to the National Survey and the Hite Reports.* London: Taylor and Francis, 1995.

Stekel, Wilhelm. *Disorders of the Instincts and the Emotions: The Parapathic Disorders; Frigidity in Woman.* New York: Boni and Liveright, 1926.

Stepan, Nancy. *The Idea of Race in Science: Great Britain 1800–1960.* London: Macmillan, 1982.

Stephens, Elizabeth. *Anatomy as Spectacle: Public Exhibitions of the Body from 1700 to the Present.* Liverpool: Liverpool University Press, 2011.

Stigler, Stephen. *The History of Statistics: The Measurement of Uncertainty before 1900.* Cambridge, MA: Belknap Press of Harvard University Press, 1986.

———. *Statistics on the Table: The History of Statistical Concepts and Methods.* Cambridge, MA: Harvard University Press, 1999.

Stocking Jr., George W. *Race, Culture, and Evolution: Essays in the History of Anthropology.* New York: Free Press, 1968.

Stopes, Marie Carmichael. *Married Love: A New Contribution to the Solution of Sex Difficulties.* London: Fifield, 1918.

Sutton, Katie. "Sexological Cases and the Prehistory of Transgender Identity Politics in Interwar Germany." *Case Studies and the Dissemination of Knowledge*, edited by Joy Damousi, Birgit Lang, and Katie Sutton, 85–103. New York and Abingdon: Routledge, 2005.

Tammeo, Giuseppe. *La prostituzione, saggio di statistica morale.* Turin: Roux, 1890.

———. "Prolusione al corso di Statistica." *Annali di Statistica* 2, no. 7 (1879): 1–24.

Tarde, Gabriel. *La Criminalité comparée.* Paris: Alcan, 1886.

———. "Criminalité et santé sociale." *Revue philosophique* 39 (1895):148–62.

———. "Problèmes de criminalité." *Revue philosophique* 21 (1886): 1–25.

———. "La Statistique criminelle du dernier demi-siècle." *Revue philosophique* 15 (1883): 49–82.

———. *Il tipo criminale: una critica al delinquente-nato di Cesare Lombroso*, edited by Sabina Curti. With an introduction by Sabina Curti. Verona: Ombre corte, 2010.

———. "Le Type criminel," *Revue philosophique* 19 (1885): 593–627.

Taylor, Ian, Paul Walton, and Jock Young. *The New Criminology: For a Social Theory of Deviance.* London: Routledge, 1973.

Teissier, Bénédict. "De la vaccination et de son influence prétendue sur la production de la fièvre typhoïde." *Gazette hebdomadaire de médecine et de chirurgie* 1, no. 13 (1853): 188–91.

Terman, Lewis M. "Kinsey's 'Sexual Behavior in the Human Male': Some Comments and Criticisms." *Psychological Bulletin* 45, no. 5 (1948): 443–59.

Terry, Jennifer. *An American Obsession: Science, Medicine, and Homosexuality in Modern Society.* Chicago: University of Chicago Press, 1999.

Topinard, Paul. *L'Anthropologie.* 5th ed. Paris: Reinwald, 1895.

———. "Criminologie et anthropologie." *Actes du deuxième congrès international d'anthropologie criminelle. Biologie et sociologie*, 3492–93. Paris: 1889. Reprinted, Lyon: Storck, 1890; Paris: Masson, 1890.

———. "De la méthode en craniométrie." *Bulletins de la Société d'Anthropologie de Paris*, series 2, vol. 8 (1873): 851–69.

———. *Éléments d'anthropologie générale.* Paris: Delahaye & Lecrosnier, 1885.

Traub, Valerie. "The New Unhistoricism in Queer Studies." *PMLA* 128, no. 1 (2013): 21–39.

Trilling, Lionel. "The Kinsey Report, The Liberal Imagination." *The New York Review of Books*, 1950, 223–42.

Tröhler, Ulrich. "Commentary: 'Medical Art' versus 'Medical Science': J. Civiale's Statistical Research on Conditions Caused by Calculi at the Paris Academy of Sciences in 1835." *International Journal of Epidemiology* 30 (2001): 1252–53.

———. "Quantification in British Medicine and Surgery 1750–1830, with Special Reference to Its Introduction into Therapeutics." PhD diss. University of London, 1978.

———. "Quantifying Experience and Beating Biases: A New Culture in Eighteenth-Century British Clinical Medicine." In *Body Counts: Medical Quantification in Historical and Sociological Perspective*, edited by Gérard Jorland, Annick Opinel, and George Weisz. Montreal: McGill-Queen's University Press, 2005.

Turnbull, Paul. "British Anthropological Thought in Colonial Practice: The Appropriation of

Indigenous Australian Bodies." In *Foreign Bodies: Oceania and the Science of Race 1750–1940*, edited by Bronwen Douglas and Chris Ballard, 205–28. Canberra: ANU ePress, 2008.

Turner, Stephen. *Explaining the Normative*. Cambridge: Polity, 2010.

Vaillant, George. *Adaptation to Life*. Cambridge, MA: Harvard University Press, 1977.

Vandenbroucke, Jan P. "Commentary: Treatment of Bladder Stones and Probabilistic Reasoning in Medicine: An 1835 Account and Its Lessons for the Present." *International Journal of Epidemiology* 30 (2001): 1253–58.

Vastel, Dr. "Rapport statistique sur la maison d'aliénés du Bon-Sauveur de Caen, pendant les années 1829 et 1830." *Annales d'hygiène publique et de médecine légale* 1, no. 8 (1832): 223–49.

Velo Dalbrenta, Daniele. *La scienza inquieta: saggio sull'antropologia criminale di Cesare Lombroso*. Padua: CEDAM, 2004.

Velpeau, Alfred, Paul Delmas, Joseph Souberbielle, Jacques Rochoux, and Jean Civiale. *Rapport et discussions à l'Académie royale de médecine sur la taille et la lithotritie, suivis de Lettres sur le même sujet par Messieurs Delmas, Souberbielle, Rochoux, Civiale, Velpeau*. Paris: Baillière, 1835.

Verbrugge, Martha. *Able-Bodied Womanhood: Personal Health and Social Change in Nineteenth-Century Boston*. Oxford: Oxford University Press, 1988.

Verplaetse, Jan. *Localizing the Moral Sense: Neuroscience and the Search for the Cerebral Seat of Morality, 1800–1930*. New York: Springer, 2009.

Viazzi, Pio. *Sui reati sessuali. Noti ed appunti di psicologia e giurisprudenza*. With a foreword by Enrico Morselli. Turin: Bocca, 1896.

Villa, Renzo. *Il deviante e i suoi segni: Lombroso e la nascita dell'antropologia criminale*. Milan: Franco Angeli, 1985.

———. "Letture recenti di Lombroso." *Studi Storici* 8, no. 2 (1977): 243–52.

Villermé, Louis René. "De la mortalité dans les divers quartiers de la ville de Paris." *Annales d'hygiène publique et de médecine légale* 1, no. 3 (1830): 294–331.

———. *Des prisons telles qu'elles sont et telles qu'elles devraient être: par rapport à l'hygiène, à la morale et à la morale politique*. Paris: Hachette, 1971.

———. "Mémoire sur la distribution de la population française." *Annales d'hygiène publique et de médecine légale* 1, no. 17 (1837): 245–80.

———. "Mémoire sur la mortalité dans les prisons." *Annales d'hygiène publique et de médecine légale* 1 (January 1829): 1–100.

———. "Mémoire sur la taille de l'homme en France." *Annales d'hygiène publique et de médecine légale* 1 (April 1829): 351–95.

———. *Recherches des causes de la richesse et de la misère des peuples civilisés, ou Application des principes de l'économie politique et des calculs de la statistique du [sic] gouvernement de l'État*. Paris: Renouard, n.d.

———. "Recherches sur la reproduction et la mortalité de l'homme aux différents âges, et sur la population de la Belgique, par MM. A Quetelet et Ed. Smits." *Annales d'hygiène publique et de médecine légale* 1, no. 8 (1832): 459–66.

———. "Statistique du choléra. *Recherches sur la reproduction et la mortalité de l'homme aux différens âges, et sur la population de la Belgique*, par MM. A Quetelet et Ed. Smits; rapport fait à l'Académie de médecine par M. Villermé," *Gazette médicale de Paris* 3, no. 68 (1832): 480–82.

———. "Sur la durée moyenne des maladies aux différens âges." *Annales d'hygiène publique et de médecine légale* 1, no. 2 (1829): 241–66.

Vogt, Carl. *Lectures on Man*. London: Anthropological Society of London, 1864.

Vold, George B. *Theoretical Criminology*. New York: Oxford University Press, 1958.

Wagner, Adolf. *Statistisch-anthropologische Untersuchung der Gesetzmässigkeit in den scheinbar willkürlichen menschlichen Handlungen*. Hamburg: Boyes and Geisler, 1864.

Walkowitz, Judith. *City of Dreadful Delight: Narratives of Sexual Danger in Late-Victorian London*. Chicago: University of Chicago Press, 1992.

Warman, Caroline. "From Pre-Normal to Abnormal: The Emergence of a Concept in Late Eighteenth-Century France," *Psychology and Sexuality* 1, no 3 (2010): 200–13.

Warner, J. H. *The Therapeutic Perspective: Medical Practice, Knowledge and Identity in America, 1820–1885*. Cambridge, MA: Harvard University Press, 1986.

Warner, Michael. *Fear of a Queer Planet*. Minneapolis: University of Minnesota Press, 1993.

———. *The Trouble with Normal: Sex, Politics and the Ethics of Queer Life*. Cambridge, MA: Harvard University Press, 1999.

Weber, Brenda. "Talking Sex, Talking Kinsey." *Australian Feminist Studies* 25, no. 64 (2010): 189–98.

Weeks, Jeffrey. *Sex, Politics and Society: The Regulation of Sexuality since 1800*. London & New York: Longman, 1989.

Weindling, Paul. *Health, Race and German Politics between National Unification and Nazism, 1870–1945*. Cambridge: Cambridge University Press, 1989.

Weisz, George. "From Clinical Counting to Evidence-Based Medicine." In *Body Counts: Medical Quantification in Historical and Sociological Perspective*, edited by Gérard Jorland, Annick Opinel, and George Weisz, 377–93. Montreal: McGill-Queen's University Press, 2005.

———. *The Medical Mandarins: The French Academy of Medicine in the Nineteenth and Early Twentieth Centuries*. Oxford: Oxford University Press, 1995.

Wiegman, Robyn, and Elizabeth A. Wilson. "Introduction: Antinormativity's Queer Conventions." "Queer Theory without Antinormativity." Special issue, *differences: a Journal of Feminist Cultural Studies* 26, no 1 (2015): 1–25.

Williams, Carl Easton. "Make the Army 'Fit to Fight': A Review of the War Department Film by Edward H. Griffith, Exposing the Dangers of Wine, Women and Disease." *Physical Culture* 40, no. 2 (1918): 38–39, 68–69.

Williams, Elizabeth. "Anthropological Institutions in Nineteenth-Century France." *Isis* 76 (September 1985): 331–48.

Willis, N. J. "Edward Jenner and the Eradication of Smallpox." *Scottish Medical Journal* 42, no. 4 (1997): 118–21.

Winterson, Jeannette. *Why Be Happy When You Can Be Normal?* London: Jonathan Cape, 2011.

Wolfgang, Marvin. "Cesare Lombroso, 1835–1909." In *Pioneers of Criminology*, edited by H. Mannheim, 232–91. 2nd ed. Montclair, NJ: Patterson Smith, 1972.

Wood-Allen, Mary. *What a Young Girl Ought to Know*. Philadelphia: Vir, 1904.

———. *What a Young Woman Ought to Know*. Philadelphia: Vir, 1897.

Yost, Edna, with Lilian Gilbreth. *Normal Lives for the Disabled*. New York: Macmillan, 1944.

Yurchak, Alexei. *Everything Was Forever, Until It Was No More: The Last Soviet Generation*. Princeton, NJ: Princeton University Press, 2013.

Ziino, Giuseppe. *Studi di medicina legale e varii*. Messina: Giorgio, 1907.

AUDIOVISUAL MATERIALS

Curtis, Adam. *HyperNormalisation*. BBC iPlayer, October 2016.

Ernst, M., A. Stone, R. Benedict, and S. B. Wortis. The Kinsey Report. Sound recording. New York: WMCA, 1948.

"Honolulu Trip." *The Jack Benny Show* season 4, episode 1. Columbia Broadcasting System. September 13, 1953 [original airing].

"Interviews with the Typical American Family Contest Winners." New York Public Library, podcast audio, 1939–40 New York World's Fair. https://itunes.apple.com/us/itunes-u/1939-40-new-york-worlds-fair/id430390294?mt=10. Accessed February 20, 2015.

Index

abnormality: abnormal children in French schooling, 252; category questioned by Binet, 252–53; constitutive of the normal in Freud, 277; not used by Lombroso, 182; not used by Quetelet, 140; rarely used in eugenics, 396n37; rejected by I. Geoffroy, 42; rejected by Kinsey, 345–48; term used by Krafft-Ebing, 268–71
Académie de Médecine, 64–99, 133, 230
Ackerknecht, Erwin, 369n5, 369n9, 373n130
American Statistical Association, 334, 339
analogy: in comparative anatomy, 29–31, 39; Étienne Geoffroy vs. Cuvier, 39, 64, 367n66; in Lombroso, 203–4
anatomy: comparative, 2, 27; composition, 29, 31; organization, 28–29, 31, 42, 46, 195; transcendental, 33–34
anomalies: in comparative anatomy, 34, 36; in criminal skulls, 193; determined through calculation, 153; in I. Geoffroy, 41–44; hereditary, 142–43; proliferation in criminal anthropology, 200–202
anthropology: conflict between French physical and Italian criminal, 184–90, 195–204; criminal, 18, 180–211, 239; cultural, 304; "negative anthropology" in Lombroso, 204; physical, 144–79
anthropometric charts, 295
anthropometric laboratory, 245–46, 296, 314, 394n151, 394n152

anthropometry: Broca's measurement techniques, 168–69, 194; disappointment for Lombroso, 193–94; distribution included in measurement, 223; expansion into popular culture, 310; gender difference, 138, 155; human proportions, 134–35, 158; measurement of conscripts, 127, 168, 175; participatory, 244–46; photography as method, 212–13; ready-made clothing, 325, 402n84; weighing brains, 155–56, 165
antinormativity, 6–7
Aristotle, 28
Armatte, Michel, 125, 377n63, 378n92
astronomy, 70
atavism: Lombroso criticized by Topinard, 201–2; Lombroso's theory, 190–93, 238–39
average: age of the population, 214; in American popular discourse, 303, 308; average man (*l'homme moyen*) applied to medicine, 109–11; calculation of, in anthropological series, 152; in composite portraits, 212; criticism of the average man concept, 126–27; health, 87; ongoing place in statistics, 123; patient, 2, 17, 86–87, 92, 105, 110–11; pictorial, 225; in Quetelet, 80, 114–17; two kinds of, 123–24, 127–28, 175, 214; used to define the type, 85
"Average Young American Male" statue, 299–301, 314–15

Babini, Valeria, 189–90
Barmack, Joseph, 339, 341
Bauer, Heike, 340, 404n27
Baxby, D., 132
Bayard, Dr., 129–30, 132
Beam, Lura, 288, 314
Beard, George, 261, 282
Belskie, Abram, 297, 299, 313–14, 317
Benedict, Ruth, 341
Benjamin, Harry, 403n9
Bennett, Tony, 297
Bérard, F., 95, 374n160
Berlant, Lauren, 272–73
Bernoulli, Jakob, 67–69, 72–73, 80
Bertillon, Alphonse, 240–41, 247
Bertillon, Jacques, 175, 217, 390n20
Bertillon, Louis-Adolphe: defense of vaccination, 128–29; on the instability of color in races, 162; on mortality statistics, 389n15; on Quetelet, 127; using statistics, 130–33
Bichat, Xavier, 49–54
binary thinking: in Broussais's physiology, 56–59; central in French anthropology, 198–203; in general, 15–16; homosexual vs. heterosexual, 345; in Krafft-Ebing, 269–70; less prevalent in anatomy than in physiology, 27, 34; normal vs. abnormal, 12, 42; not central in Italian criminal anthropology, 182, 199; not characteristic of sexology, 19
Binet, Alfred, 250–54, 394nn171–72
biometrics: following Galton, 18, 228, 236, 255; in Stopes, 287
biopolitics, 14
Black, Edwin, 400n25
Blanckaert, Claude, 174, 381n5
Bland, Lucy, 267, 276
Block, Maurice, 377n58
Blumenbach, Johann Friedrich, 169, 383n54
Bollet, Alfred Jay, 369n10
Bonnet, Charles, 46–47
Bouillaud, Jean-Baptiste, 90–93, 97, 105, 131, 373n145
Bowditch, Henry Pickering, 237, 392n110
Brady, Sean, 395–96n19
Braude, Hillel D., 369n1, 369n3
Braun, Marta, 391n70
Brian, Eric, 378n92
Brill, A. A., 277

British Association for the Advancement of Science, Anthropometric and Racial Committee, 225
Broca, Paul: autopsy of a criminal, 184–85; constituting series, 152; critique of Gobineau, 156–57; critique of phrenology, 146; defense of craniometry, 166–68; determining anomalies through calculation, 153; excavated skulls, 147; humans not a single race, 154; intellectual capacity, 248; mortality statistics, 390n20
Broussais, François-Joseph-Victor: development of physiological terms, 52–57; polemic against Louis, 88; use of "normal," 26, 52
Buffon, Georges-Louis, 47–48, 71

Cabanis, P. J. G., 57, 74–79, 83
calculus of probabilities: classical theory of, 66–71; in games of chance, 84; general, 16, 67–68, 70–73, 79; in medical practice, 83–85, 91; in Pinel, 78; in population statistics, 117–18, 125–26
Callens, Stéphane, 102
Camper, Petrus, 145, 169, 381n7
Canguilhem, Georges: etymology of "anomaly," 43; focus on medical writing, 37; normalization, 2, 288, 353; normativity, 1, 23–25, 27, 37, 140; quantitative thinking in medicine, 16, 44–45, 48, 59–62, 365n12; on Quetelet, 139–40; relative neglect of anatomy, 27, 38–39; vitalism, 23, 59; words and things, 24–27, 52
capital punishment, 185–86
Carnot, Hector, 130–33
Carter, Julian B., 5–6, 14, 318
case study, 268, 272, 279, 290, 314, 319, 336
Charcot, Jean-Martin, 275
Chiang, Howard, 336, 404n19
chrononormativity, 359
Civiale, Jean, 103–4, 375n15
Clouston, Thomas, 271
Cochran, William G., 403n6
Codgell, Christina, 399n19, 402nn67–68
Coffey, Mary, 300, 399n19
Colajanni, Napoleone, 388n121, 392n110
color: classifying average colors, 162; "colored savages," 191; "five-color" view of race, 159; general, 144; unreliable way of differentiating races, 162–63

composite photography. *See* photography: composite
composite statues. *See* statues, composite
composition. *See* anatomy
Condorcet, Nicolas de, 73, 82
consumer culture, 14-15, 20, 291, 302, 305, 307-8, 323-24, 326-27, 351, 354-55
counting. *See* numerical method
Cournot, Antoine-Augustin, 119, 126
Cousin, Victor, 56
craniometry: craniometry and race, 159-61, 164-68; integrity of the skull, 168-69; in Lombroso, 193-95; physical anthropology, 145-63, 177; skulls of criminals, 196-97, 239; stability of the skull in heredity, 143
Crary, Jonathan, 391n59
Creadick, Anna, 318, 337, 399n8
criminal anthropology. *See* anthropology: criminal
critical disability studies, 2
Crozier, Ivan, 395n9, 395n19
Currell, Susan, 402n67
curve: binomial, 118-22, 124, 218; normal, 4, 12-13, 119, 218-19, 254; significance of, 139-40
Cuvier, Georges, 39-41, 85

Damousi, Joy, 273
Darwin, Charles, 174, 217, 224, 229, 231, 255
Daston, Lorraine: classical theory of probabilities, 66-71; moral statistics, 113, 371n48; Quetelet's average man, 115, 372n85, 372n91, 377n71; scientific observation, 178, 226, 240, 391n65
data: call for collection by Galton, 245; data society, 15, 247, 353; Kinsey's treatment of, 335-39; organizing data in Louis's practice, 88; retrieval of, in Galton, 19, 220, 247
Davenport, Charles, 233, 297, 299-300, 302, 314, 400n21
Davenport, Isabel, 264
Davenport, Jane, 297, 300-301
Davis, Katherine, 265, 335, 403n12, 404n16, 405n52
Davis, Lennard, 5
deformity: found in criminals, 239; and heredity, 142-43; in natural history, 28; place of deformed individuals in a series, 151; as statistical exception in Quetelet, 116, 134, 137
degeneration: of Americans, 300, 304; of the British race in Galton, 228, 232, 262; in Krafft-Ebing, 267, 273-77; in Lombroso, 200, 204, 262
Delaharpe, M. J., 375n6
De la Peña, Carolyn, 399n2
Desrosières, Alain: Galton and A. Bertillon, 393n124; Galton and Pearson, 390nn26-27, 392n92; Gaussian law, 378n86; geometry, 390n22; history of statistical reasoning, 64, 67, 71; large numbers, 371n40; Moreau de Jonnès, 377n56; Quetelet, 115, 121-22, 376n35; Risueño, 83; statistics and the medical profession, 95, 369n2, 373n104
deviation: anomaly misconceived as statistical deviation, 44-49; "deviants" in Lombroso, 204; illness as deviation in Quetelet, 110; law of deviation in Galton, 217, 254
Dickinson, Robert Latou, 288, 297, 299, 313-14, 317-18, 320, 335, 401n55, 401n60
distribution. *See* errors; variations
Doan, Laura, 9-11, 267, 276, 287-88
Dock, Lavinia, 280-81
Douglas, Bronwen, 382n39
Downing, Lisa, 388n108
Dreger, Alice Domurat, 5
Drucker, Donna, 338, 341

Edwards, Elizabeth, 391n56
Eknoyan, Garabed, 369n10, 376n28
Ellenbogen, Josh, 237, 393n129
Ellis, Havelock, 239-40, 265, 273, 277, 289
epilepsy, 192-93, 203
Ernst, Waltraud, 363n9
errors, measurement and distribution of, 72, 119-24
eugenics: American and German, 233, 301, 317-18; British, 234-36; congresses, 300-302, 316, 399n15, 400n32; and evolution, 229, 255; exhibition, 297, 299-300; importance of intelligence, 250, 252; legislation, 302; policies in Galton, 18, 228-29, 250; positive eugenics, 231-32, 300-301; racial extinction, 232; supported by statistics, 230; the "unfit," 231-32; and violence, 13

Eugenics Record Office, 233, 393n143, 396n37

exhibition: international, 213, 245–46; of statues, 20, 294, 296–301, 314–15, 317, 324, 329, 333, 358, 394n152, 399nn16–17, 400n23, 402nn67–68

expérience (experience/experiment): general, 78, 82, 197; role of counting in clinical experience, 88, 90, 107–8

extreme individuals and statistics: exclusion of races, 216, 305–6; extremes in the study of series, 151–52; methodical exclusion, 154, 178, 214–15, 252; ready-made clothing, 325–27; two forms of exclusion, 153

Feldman, Jacqueline, 375n5, 377n72, 378n105
Felski, Rita, 276
fitness: children, 285; military, 284; people in general, 323, 358; the "unfit" in eugenics, 231–32
fixity: anatomical, 63; in craniometric method, 168–69; dried skulls preferable because fixed, 167–68; fixed characters in physical anthropology, 159; general property of skulls, 160–61; imposed by the use of numbers, 82, 84; in the medical sciences, 61, 75; "permanence" of race, 172; species not fixed, 173
Forno, Mauro, 189
Foucault, Michel: clinical statistics, 76–77, 96–98; genealogy, 10; individual normalization, 273, 288, 295, 298, 351, 353; institutional normalization, 4, 7–9, 14, 232, 353; words and things, 25
Fourier, Joseph, 365n9
Fowler, Orson, 280
Freeman, Elizabeth, 9
frequentist thinking: approach to crime, 188–89; definition of type, 175–76; general, 52, 60, 62
Freud, Sigmund, 274–79
Furness, Walter Rogers, 392n111

Galison, Peter, 178, 226, 240, 391n65
Gall, Franz, 187, 196, 240
Galton, Francis, 212–57; composite photography, 212–13, 221–27, 229, 236–40; criminal anthropology, 239; normal curve, 13
Garland-Thomson, Rosemarie, 5

Gavarret, Jules, 103
Gebhard, Bruno, 317, 320, 323
gender: anthropometric differences, 138, 155, 205, 215–16; criminal women different from criminal men, 205; normalized in intelligence testing, 253–54; women less sensitive and more mendacious, 206
genius: in Galton, 247; in Lombroso, 201
Geoffroy Saint-Hilaire, Étienne, 29–38; debate with Cuvier, 39–41, 382n16; natural history, 28; qualitative understanding of normality, 63
Geoffroy Saint-Hilaire, Isidore, 41–49; abnormal a false category, 12; constructing teratology, 35
geometry, "normal" a term in, 2, 25, 169
Gerdy, P. N., 160
Gibson, Mary, 207
Gilbreth, Lillian, 289–90, 357
Gilman, Sander, 266, 276
Gobineau, Arthur de, 156–58, 233
Goldberg, Jonathan, 364n31
Grant study of normal young men, 14, 20, 309–13, 328, 353
Gratiolet, Pierre, 155, 165–66
Green, David, 226
Guerry, André-Michel, 106
Guillard, Achille, 131, 380n143

Habermas, Jürgen, 363n9
Hacking, Ian: average man, 80, 115, 120, 380n174; Italian criminal anthropology, 385n7; looping effects, 273; medical statistics, 4, 13, 97, 337, 351; normal state, 2, 365n9; not progressive history, 66–67; probability, 72, 373n130
Halberstam, J. Jack, 9
Halbwachs, Maurice, 380n172
Hallé, Jean-Noël, 53
Haller, J., and R. Haller, 281
Halperin, David, 6–7, 11, 337–38
Harvey, Joy Dorothy, 381n4, 381n6
Hauser, Renate, 274
Hawkins, Francis Bisset, 95
Heath, Clark W., 309–12
Henslow, John Stevens, 37
heredity: criminal heredity, 187, 238; distribution in Galton, 217–18, 230; generational and ancestral heredity, 220; hereditary intellectual ability, 248–50; hereditary

normality, 142–44, 153, 179; opposed to atavism, 198–202; racial heredity, 143, 153–56, 170–72; reversion to mediocrity, 219, 224, 230–31; self-improvement, 295
Herschel, John F. W., 101, 106
heteronormativity. *See* normativity
heterosexuality, 5, 14, 345, 405n60
Hirschfeld, Magnus, 271, 340, 404n27
historical linguistics and race, 163–64, 171
Hollingworth, Harry, 303–4, 308, 400n26
homme moyen. *See* average
homonormativity, 359
homosexuality, 14, 336, 340, 343–45, 405n60
Hooton, Earnest A., 309, 312, 314, 316, 326
Horn, David G., 389n125
hospital: nontreatment of sexually transmitted diseases, 281; phthisis patients, 242; statistics, 96–97, 113
Howe, Catherine Newman, 296, 398n1
Huff, Joyce L., 327
hybridity, 164, 172–73
hygiene: exhibition, 317; racial, 229; social, 143
Hyman, Herbert, 339, 341
hypernormalization, 359

Igo, Sarah: acceptance of statistics, 342; average Americans, 303, 306; Kinsey's use of "normal," 349; surveys and participatory citizenship, 14, 297, 308, 356
individualization/individuation: through case studies, 272–73, 288, 290, 295; in criminal records, 241, 393n124; degrees of imperfection, 125; images in scientific observation, 226; individual equilibrium, 353; measured against the normal, 298, 302, 316–17, 319, 323–24, 326–27, 354; of patients, 100, 105, 108–11, 141; within a population, 8–9, 13–15; self-management, 357; through self-measurement, 255, 296
industrialization, 261, 304–5, 307, 327
insanity: accompanying sexual disease, 266; and crime, 185–86, 203–4; and degeneration, 232
intelligence: convention required for measurement, 251; distinguished from backwardness in schooling, 253; importance to eugenics, 250; lack of established measurement, 249–51; statistical study of hereditary "genius" in Galton, 248–49; testing, 247–54

Jagose, Annamarie, 7, 14
Jewell, Edward Alden, 400n23
Jorland, Gérard, 369–70n10

Katz, Jonathan Ned, 274, 279
Kellogg, John, 283, 286
Kevles, Daniel, 234–35, 393n143
Kinsey, Alfred: biology, 335; importance of statistics, 20; quantitative study of sex, 335–38, 341; questioning the heterosexual-homosexual binary, 345; reception of his work, 333–34, 347; statistical methods, 335–42; surveys, 14
Krafft-Ebing, Richard von, 264–67, 272, 275

La Berge, Ann F., 369–70n10
Lacassagne, Alexandre, 195, 197
Lade, James H., 403n10
Lagneau, Gérard, 375n5, 377n72, 378n105
Lang, Birgit, 273
Laplace, Pierre-Simon, 70–71, 73, 78, 80, 91
Largent, Mark, 402n68
large numbers: law of, 72, 86; problematic in clinical medicine, 90, 96–98, 108
Laughlin, Harry, 271, 302
Lécuyer, Bernard-Pierre, 95
Leudet, Dr., 130
Lexis, Wilhelm, 214–18, 229
Linnaeus, 28, 163, 174, 178
Lloyd, Warren, 263
Lock, Margaret, 14
Lombroso, Cesare: "failures" of method, 200, 207; mix of genres, 18, 181, 189–90; taking the normal for granted, 180–81, 210; use of proverbs in science, 206–10; wide readership, 180
Louis, Pierre Charles Alexandre, 84, 88–90, 102
Love, Albert G., 300
Lundgren, Frans, 244, 246
Lydston, Frank, 282
Lynd, Robert S. and Helen Merrell, 304–6

Macfadden, Bernarr, 283–84, 286
Malchow, C. W., 280, 285–86
Manouvrier, Léonce, 195–200, 207–8
Mantegazza, Paolo, 150

438 INDEX

Marey, Jules-Étienne, 227–28
masturbation, 184, 288, 291, 336, 343, 348, 356
Matalon, Benjamin, 375n5, 377n72, 378n105
Matthews, J. Rosser, 65–66, 83, 88, 370n19, 372n76, 372n83, 372n93, 373n130
Maxwell, Anne, 225, 244
McKenzie, Robert Tait, 399n18
McRuer, Robert, 1
mean. *See* average
median, 175, 215, 219
medical gaze, 4, 75–76, 89, 147–51
medical probability, 79, 98
mediocrity, 11, 115, 219, 231, 250, 316
Menon, Madhavi, 364n31
Middle America, 302, 304–9, 311
Mill, John Stuart, 378n95
Moll, Albert, 397n66
monstrosity, 33–36, 38, 41, 81, 117
Montagu, M. F., 344
Montaldo, Silvano, 189
Moreau de Jonnès, Alexandre, 113
mortality, 214–16, 389n15
Mosteller, Frederick, 403n6
Murray, Jessie, 265
Muybridge, Eadweard, 227–28

natural state, 51–54
nervous disease, 261–62
"new normal," 358
Norma, 297, 313–20, 328, 353; search for, 320–24
normal curve. *See* curve
normalization: eugenics policy, 231–32; Foucault's history of, 232, 247; gender in intelligence testing, 253–54; hypernormalization, 358–59; violent under Nazism, 233
normal school, 13, 25
normal state: in anatomical classification, 32–33; emergence in medical discourse, 25; in physiology, 16, 26, 49, 59–60; reference for identifying anomalies, 34, 45–46; stage of anatomical development, 31, 41; in teratology, 38–39, 48–49; used by Quetelet to refer to medicine, 108–11, 119, 133–34
normativity, 5, 269, 359. *See also* Canguilhem, Georges
"normcore," 358
Normman, 297, 313–20, 328, 353

Novak, Daniel, 391n56
number. *See* numerical method
numerical method, in medicine: attacked in the Académie de Médecine, 83–87; criticized by Quetelet, 101; damaging the reputation of statistics, 131–32; distinguished from established medical statistics, 94–95; "inflexibility" of numbers, 82; practiced and defended by Louis, 84, 87–90, 93
Nye, Robert, 386n20

O'Brien, Ruth, 402n84
Oosterhuis, Harry, 267
ordinary: Americans, 304–6, 334–35, 341; average, 219; crimes, 199–200; natural, 51–52; normal, 2, 7; questioned by Galton, 213; rules, 35, 37, 46; usual, 38, 45–49, 55, 197, 239
organization. *See* anatomy
outliers. *See* extreme individuals and statistics

Parent-Duchâtelet, A. J. B., 95
pathology, physiology of, 53–54
Pearson, Karl, 216, 221–22, 229, 231–33, 236, 392n107
perfection: in Bichat, 50; in Broussais, 54, 57–58, 61; concept in natural history, 28–29, 32; in Cuvier, 40–41; in Dickinson, 315–16; in Quetelet, 138; proper to a given race, 156
Périer, Jean-André, 142–43
Perrier, Edmond, 367n71
perversion, 20, 270, 274–77, 344
photography: in anthropology, 220, 225; composite, 212–13, 221–27, 229, 236–44, 389nn1–2, 398n1; of criminal types, 238; identity, 240; in scientific observation, 226
phrenology, 145–47, 187, 196, 240, 381n8
physiognomy, 165–67, 178, 221–22, 242–43
physiology, 26–27, 49–60
Pinel, Philippe, 77–78, 90
Pivar, David, 281
Poisson, Siméon-Denis, 67, 71–73, 76, 79, 82, 84, 112, 117
Pomeroy, Wardell, 338, 341
Poovey, Mary, 5
population health, 17
Porter, Theodore, 95, 118, 369n10, 375n1, 378n91, 383n49, 390n18

positive school (*scuola positiva*), 180–211
Prescott, Heather, 399n18
prisons, 183, 222–23, 239, 241
probabilistic thinking. *See* calculus of probabilities
Prodger, Philip, 391n70
Pruner-Bey, 147–49, 162–63, 187
psychiatry, 263, 266–67, 272
psychoanalysis, 263, 266, 274–78
psychology, 262–63
public health campaign, 280–86, 317
punch card machine, 247, 326, 336, 338–39, 355–56

Quatrefages, Armand de, 170–71
queer studies, 2, 5, 6–7, 11
Quetelet, Adolphe: anthropometry, 158; criticism of numerical method, 101–5, 107; defining the type by calculated average, 85, 137; distribution of errors, 119–20; moral statistics, 113–14, 121, 247–48; population statistics, 17, 96–97, 114–17, 377n60; tendency to crime measured statistically, 188–89, 192; unity of the human species, 136, 158

race: anthropometric studies of, 18; betterment of the race in Galton, 18, 233; central place in 1860s anthropology, 12, 17–18; criminals a race, 190–92; in Parisian physical anthropology, 143, 153–56; questioned by some anthropologists, 169–70; race studies, 2; racial hierarchy, 150, 155–56, 190–91; uncertain definition of, in anthropology, 170–73
range of variations. *See* variations
Rayer, Pierre-François-Olive, 93–94, 106, 374n156
ready-made clothes, 314, 325, 354–55
Reiss, Elizabeth, 5
Réveillé-Parise, Joseph-Henri, 78–79
reversion. *See* heredity
Risueño d'Amador, B. J. I., 82–93
Robertson, Josephine, 320–23
Robie, William, 264–65, 280
Rose, June, 395n14
Royal Institution, 216
Royer, Clémence, 174, 385n130
Ruggles, Theodora, 298–99
Rümelin, Gustave, 376n27
Rydell, Robert, 399n19, 402n68

Sanson, André, 170–73
Sargent, Dudley Allen, 294–96, 298, 317–18, 399n2, 399n12
"savages," 181, 191, 196–97, 250, 304
schools, 245, 250–54
Schweber, Libby, 112, 380n141, 381n1
scuola positiva. *See* positive school
Sekula, Allan, 221, 237, 240–41
self-improvement, 20, 294–96, 302, 318, 324, 327, 358
self-management, 263, 296, 357
self-measurement, 322, 324
sex advice literature, 264–65, 279–88
sexology, 263, 278, 318
sexual behavior, 264
sexual hygiene, 281–84, 291
sexual inversion, 265, 268, 273, 277
sexual pathology, 266–70, 274
Shapiro, Harry L., 316–18, 320–21
Sheldon, William Herbert, 312–13
Shelton, William C., 402n84
Sheynin, O. B., 369–70n10, 373n105
Simon, Théodore, 250–54, 394n171
Société d'Anthropologie de Paris, 142–79, 184–90, 201, 217, 381nn4–5
Société de Statistique de Paris, 123, 214, 380n150
Société Française de Statistique Universelle, 112–13
somatotypes, 313
Spade, Dean, 5
Spearman, Charles, 251
spermatorrhoea, 261, 395n2
Stall, Sylvanus, 264–65
standardization, 232, 253, 314, 325–27, 353–56
statistics: governmental, 101, 143–44; judicial, 67, 72–74, 82, 84, 112–13; kinds of, 112–14; medical, 63–65, 75–76, 82, 94–96, 101–5; moral, 73, 111–14, 120–21, 371n39; resistance to clinical statistics, 63, 74–76, 80–84, 87, 95, 98–99, 128; used to attack and defend vaccination, 129–30; use of "normal" in, 213–20
statues, composite, 294–98; "Average Young American Male," 297, 299; "Typical American," 294
Stepan, Nancy, 174, 224, 234, 250, 381n11, 382n39, 384n102
Stigler, Stephen M., 122–23, 133, 370n15
Stocking, George W., Jr., 385n132

Stopes, Marie, 265–66, 279, 286–88, 398n92
survey, 14, 297, 308–9, 356
Sutton, Katie, 273, 397n66

Teissier, Bénédict, 379n137
teratology, 41–48
Terman, Lewis, 340–41
time management, 289, 356–57
Topinard, Paul, 149–50, 195, 201
Transparent Man, 317, 402n67
Traub, Valerie, 364n34
Trilling, Lionel, 341–42, 348
Tröhler, Ulrich, 369n7, 373n130
Turkey, John W., 403n6
Turnbull, Paul, 382n28
Turner, Stephen, 363n9
type: anatomical in Cuvier, 41; anthropological, 136, 144; botanical, 37; criminal, 12, 180–211, 238–39, 241; ethnic, 127, 143, 153, 176–77; evolution of, in Galton, 219, 224–25; ideal type in Quetelet, 80, 85, 115, 124–26, 138, 175; perceived by scientific observation, 178; racial, 172, 176, 230–31, 233; statistically defined in I. Geoffroy, 45; type of the species, 53, 136, 158; unquestioning usage of the term, 86, 174–79; visible in composite portraits, 212–13, 223–24, 243
"Typical American Male and Female" statues, 294, 296
"Typical Families" competition, 307

urbanization, 261, 283–84, 290, 304–5

vaccination, 129–32
Vaillant, George, 401n48, 401n52
variability in patients, 84, 94, 109
variations, range of: Galton on Quetelet, 213; in "normal anatomy," 92; normality as a range, 199–200; patterns of distribution, 114–15, 216; in Quetelet, 116–17; variations in physical anthropology, 146, 149, 161
Verbrugge, Martha, 399n2
Villa, Renzo, 203
Villermé, Louis-René, 95–97

Warman, Caroline, 25, 37, 51–52
Warner, J. H., 369n10
Warner, Michael, 6–7, 337, 344
Weber, Brenda, 341–42, 404n17
Weindling, Paul, 383n47
Weisz, George, 64, 373n105, 373n135
Weldon, Walter Frank, 233, 235–36
whiteness, 318
Williams, Carl Easton, 398n89
Williams, Elizabeth, 383n74
Willis, N. J., 380n147
Winterson, Jeannette, 6
Wood-Allen, Mary, 264, 282
World's Fair, 294, 296, 298, 307, 314, 317

Yost, Edna, 357

Lightning Source UK Ltd.
Milton Keynes UK
UKOW06f2248211217
314835UK00003B/10/P